Concise Handbook of
EXPERIMENTAL METHODS for the BEHAVIORAL and BIOLOGICAL SCIENCES

Concise Handbook of
EXPERIMENTAL METHODS for the BEHAVIORAL and BIOLOGICAL SCIENCES

Jay E. Gould

CRC PRESS

Boca Raton   London   New York   Washington, D.C.

Library of Congress Cataloging-in-Publication Data

Gould, Jay E.
Concise handbook of experimental methods for the behavioral and
biological sciences / Jay E. Gould.
      p. ; cm.
     Includes bibliographical references and index.
     ISBN 0-8493-1104-7 (alk. paper)
      1. Life sciences--Research--Methodology--Handbooks, manuals, etc. 2. Social
sciences--Research--Methodology--Handbooks, manuals, etc. 3.
Psychology--Research--Methodology--Handbooks, manuals, etc. I. Title: Experimental
principles and methods for the bahavioral and biological sciences. II. Gould, Jay E.
Principles and methods of experimental research. III. Title.
     [DNLM: 1. Research--methods--Handbooks. 2. Behavioral
Sciences--methods--Handbooks. 3. Biological Sciences--methods--Handbooks. WM 34
G697c 2001]
  QH315 . G68 2001
  570'.7'2—dc21

2001037817
CIP

This book contains information obtained from authentic and highly regarded sources. Reprinted material is quoted with permission, and sources are indicated. A wide variety of references are listed. Reasonable efforts have been made to publish reliable data and information, but the author and the publisher cannot assume responsibility for the validity of all materials or for the consequences of their use.

Visit the CRC Press Web site at www.crcpress.com

No claim to original U.S. Government works
International Standard Book Number 0-8493-1104-7
Library of Congress Card Number 2001037817
Printed in the United States of America   2 3 4 5 6 7 8 9 0
Printed on acid-free paper

The original version of this *Concise Handbook* was printed in 1991 under the title of *Principles and Methods of Experimentation: A Handbook for the Behavioral Sciences*. An extensive revision was printed in 1992 under the modified title of *Principles and Methods of Experimental Research: A Handbook for the Behavioral Sciences*. The current edition of the Handbook is the result of significant modifications to enhance its clarity, completeness, and utility.

Dedication

Dedicated to my students, whose endeavors to achieve proficiency

in experimental methodology inspired the development of this handbook

Preface

This Handbook has been designed to assist students as well as professionals in the behavioral and biological sciences to understand and apply the principles and methods of experimental research. The Handbook is comprehensive but is written in a concise, outline format. This allows a considerable amount of material to be presented in an explicit, compact form that facilitates the perception and understanding of the hierarchical and parallel relationships among concepts. As a result, learning, retrieving, and applying the information should be greatly enhanced.

The Handbook is organized into three parts. *Part One* covers the philosophy of science, forms of scientific research, steps of the scientific method, variables in research designs, and the initial and final phases of research; *Part Two* covers research ethics and experimental control; and *Part Three* covers experimental design, sampling and generalization, and finally hypothesis testing and statistical significance. There is repetition with variation and cross-referencing throughout in order to assist the reader in understanding the material and also to illuminate the interrelatedness of the various principles and methods of experimental research.

The major *illustrations* are grouped at the back of the Handbook with other useful information as *Appendices*. Placing these materials together allows the reader to readily locate and use them from anywhere within the manual. These can be used for quick reference and study, along with the *Review Summary* and *Review Questions* at the end of each chapter. Feedback about mastery of the experimental principles and methods is provided by having the answers to the "Review Questions" contained within the correspondingly numbered items of the "Review Summary." (Naturally, a Review Summary should not be read immediately before trying to answer the associated Review Questions.)

To further enhance the Handbook's usefulness, *italics* are employed to emphasize key words and phrases in the text. Moreover, alternative terms are frequently provided, and these are separated by *slash/diagonal lines* for compactness. To reduce distraction, most outline entries are not preceded by traditional numerals or letters, except when the items represent a list. *Instructive examples* are provided throughout the Handbook but are kept succinct so that readers are not burdened with unnecessary details. (*Collections of scientific readings* have been published that can be used for more elaborate illustrations of research.) Coverage of *statistics* occurs at appropriate points in the Handbook, but is kept brief and primarily conceptual because there are already excellent handbooks on this subject.

References and Resources

This Handbook builds on the work and writings of many others. The following are the major sources consulted during its preparation. Superscript numbers are placed in the text to indicate where these are specifically referenced. Additional references are placed with the pertinent material within the body of the book.

1. American Psychological Association. (1982). *Ethical principles in the conduct of research with human participants*. Washington, D.C.: Author.

2. American Psychological Association. (1992). Ethical principles of psychologists and code of conduct. *American Psychologist*, 47, 1597-1611.

3. American Psychological Association. (1993). *Guidelines for ethical conduct in the care and use of animals* [Brochure]. Washington, D.C.: Author.

4. American Psychological Association (1994). *Publication manual of the American Psychological Association* (4th ed.). Washington, D.C.: Author.

5. Aronson, E., & Carlsmith, J. M. (1968). Experimentation in social psychology. In G. Lindzey & E. Aronson (Eds.), *The handbook of social psychology* (2nd ed.). Reading, MA: Addison-Wesley.

6. Berscheid, E., Baron, R. S., Dermer, M., & Libman, M. (1973). Anticipating informed consent: An empirical approach. *American Psychologist*, 28, 913-925.

7. Bracht, G. H., & Glass, G. V. (1968). The external validity of experiments. *American Educational Research Journal*, 5, 437-474.

8. Campbell, D. T., & Stanley, J. C. (1963). *Experimental and quasi-experimental designs for research*. Chicago: Rand McNally.

9. Christensen, L. B. (1997). *Experimental methodology* (7th ed.). Boston: Allyn and Bacon.

10. Corso, J. F. (1967). *The experimental psychology of sensory behavior*. New York: Holt, Rinehart, & Winston.

11. Cozby, P. C. (1997). *Methods in behavioral research* (6th ed.). Mountain View, CA: Mayfield.

12. Crutchfield, J. P., Farmer, J. D., Packard, N. H., & Shaw, R. S. (1986, December). Chaos. *Scientific American*, pp. 46-57.

13. D'Amato, M. R. (1970). *Experimental psychology: Methodology, psychophysics, and learning*. New York: McGraw-Hill.

14. Ellingstad, V., & Heimstra, N. W. (1974). *Methods in the study of human behavior*. Monterey, CA: Brooks/Cole.

15. Elmes, D. G., Kantowitz, B. H., & Roediger III, H. L. (1999). *Research methods in psychology* (6th ed.). Pacific Grove, CA: Brooks/Cole.

16. Helmstadter, G. C. (1970). *Research concepts in human behavior*. New York: Appleton-Century-Crofts.

17. Kelman, H. C. (1967). Human use of human subjects. *Psychological Bulletin*, 67, 1-11.

18. Kelman, H. C. (1972). The rights of the subject in social research: An analysis in terms of relative power and legitimacy. *American Psychologist*, 27, 989-1016.

19. Kerlinger, F. N. (1986). *Foundations of behavioral research* (3rd ed.). New York: Holt, Rinehart, & Winston.

20. Kling, J. W., & Riggs, L. A. (1971). *Woodworth & Schlosberg's experimental psychology* (3rd ed.). New York: Holt, Rinehart, & Winston.

21. Lindquist, E. F. (1953). *Design and analysis of experiments in psychology and education.* Boston: Houghton Mifflin.

22. Linton, M., & Gallo, Jr. P.S. (1975). *The practical statistician: Simplified handbook of statistics.* Monterey, CA: Brooks/Cole.

23. Lubin, A. (1961). The interpretation of significant interaction. *Educational and Psychological Measurement,* 21, 807-817.

24. Martin, P., & Bateson, P. (1993). *Measuring behavior: An introductory guide* (2nd ed.). New York: Cambridge University.

25. Matheson D. W., Bruce, R. L., & Beauchamp, K. L. (1978). *Experimental psychology: Research design and analysis* (3rd ed.). New York: Holt, Rinehart, & Winston.

26. McGuigan, F. J. (1997). *Experimental psychology* (7th ed.). Englewood Cliffs, NJ: Prentice Hall.

27. Miller, N. E. (1985). The value of behavioral research on animals. *American Psychologist,* 40, 423-440.

28. National Academy of Sciences and Institute of Medicine, Committee on the Use of Animals in Research. (1991). *Science, medicine, and animals.* Washington, D.C.: National Academy Press.

29. Rubin, Z. (1973). Designing honest experiments. *American Psychologist,* 28, 445-448.

30. Solso, R. L., Johnson, H. H., & Beal, M. K. (1998). *Experimental Psychology: A Case Approach* (6th ed.). New York: Addison-Wesley.

31. Tinbergen, N. (1963). On aims and methods of ethology. *Zeitschrift für Tierpsychologie,* 20, 410-433.

32. Yaremko, R. M., Harari, H., Harrison, R. C., & Lynn, E. (1982). *Reference handbook of research and statistical methods in psychology: For students and professionals.* New York: Harper & Row.

33. Zinser, O. (1984). *Basic principles of experimental psychology.* New York: McGraw-Hill.

Acknowledgments

I would like to express my appreciation to Dr. Sam Mathews and Dr. Bill Mikulas for reviewing portions of this Handbook and making constructive suggestions. I also want to thank Karen Barth of Information Technology Services at The University of West Florida for her assistance in the production of several figures in the appendices. My loving wife Lynn worked with me as well in the production of some appendices as well as text figures, and she has provided continual encouragement, support, and advice throughout this project. Finally, I would like to give thanks to all those students who have been kind enough to make recommendations that led to enhancement of this Handbook.

Contents

Part Two   Research Ethics and Experimental Control

Part Three   Design and Analysis of Experiments

Part One

Science: The Philosophy and Components

1

Philosophy of Science — The Scientific Approach to Knowledge

CONTENTS

Philosophy and its Relationship to the Sciences

A *common philosophy* is shared by all of the sciences, such as psychology, biology, chemistry, and physics. Therefore, this Handbook begins by briefly describing *philosophy*, the *philosophy of science*, and what this philosophy *contributes to the sciences*. Later in this chapter are discussions of the various *approaches to knowledge*; the relationship of non-scientific approaches to the *scientific method and its ideal*; and then delineation of science's *goals, assumptions, and requirements*.

1. **Philosophy Briefly Described**

   — Philosophy means *"love of wisdom."*

      • The origin of the term is the Greek word *philosophos*, meaning *"lover of wisdom."*

         ∞ *Scientists*, too, are ardent seekers of knowledge/wisdom — what is different, as we shall see, is the *approach* taken.

      • As a discipline, philosophy is said to have *originated* in the West more than 2,500 years ago, and it is Western Civilization's oldest known form of *systematic, rational, and critical inquiry*.

   — Philosophy has been defined as the *"study of the ultimate reality, causes, and principles underlying being and thinking."* [Chernow, B. A. & Vallasi, G. A., eds. (1993). *The Columbia encyclopedia*, 5th ed., (p. 2136). Chicago, IL: Columbia Univ. Press.]

   — Traditionally, this discipline has been *divided into several branches*:

- *Metaphysics* inquires into the nature and ultimate significance of the universe.
- *Logic* is concerned with the laws of valid reasoning.
- *Epistemology* investigates the nature of knowledge and the process of knowing.
- *Ethics* deals with problems of right conduct.
- *Aesthetics* attempts to determine the nature of beauty and the criteria of artistic judgment.

— The branches of philosophy are *relevant to science*.

- *Epistemology* is an aspect of philosophy that is very important to science, since it is the *branch of philosophy* that involves the *theory of knowledge* — its sources (i.e., foundations and presuppositions), nature, criteria, limits, and validity (certainty).

  ∞ In simpler terms, epistemology asks: *what can be known, how can it be known, and how certain can we be*?

  - *Episteme* is the Greek word for *knowledge*, and *logy* signifies a *branch of science or learning*, i.e., *the theory or study of something* (as in psychology, biology, etc.)

  ∞ *In science*, epistemology deals with the *basis/justification* and the *objectivity* of scientific knowledge.

  - The *epistemology of science* is clearly one of the important elements of the philosophy of science.

- *Logic* and *ethics* are also aspects of philosophy that are clearly important to science.

  ∞ The scientific method is a *fundamental logic of inquiry* (explained later in this chapter and elsewhere in the Handbook).

  ∞ In addition to being a logical process, scientific research should always be conducted in an *ethical manner* (see Chapter 6).

- *Metaphysics of science* is yet another component of the philosophy of science, but one that is not directly pertinent to the subject matter of this Handbook; i.e., it relates to the *results* rather than the principles and methods of scientific research.

  ∞ The metaphysics of science is concerned with the *philosophically puzzling aspects* of the *supposed reality uncovered by science*, such as those of modern quantum mechanics. [See Papineau, D. (1995). Science, problems of the philosophy of. In T. Honderich (ed.), *The Oxford Companion to Philosophy* (pp. 809-812). Oxford: Oxford University Press.]

- *Aesthetics*, i.e., beauty and artistic judgment, are also of interest to science (especially the behavioral sciences such as psychology), but this too is not directly relevant to the issues addressed in this Handbook.

  ∞ It should be noted, however, that many see a certain *beauty* in both the methods and findings of *science*.

— Broadly speaking, philosophy can be defined as *the search for truth,* i.e., *the pursuit of reality, knowledge, or wisdom.*

  • In this general sense, philosophy is similar to *science,* as well as to *theology.*

2. Philosophy Compared with Science and Theology

— *Philosophy* differs from *science* in that the scientists base their theories on *empirical evidence* (objective observations), whereas philosophy also covers areas of inquiry where objective experience is not available — using *logical reasoning* in place of *genuine observation.*

  • Example: Philosophy, as well as theology, is more applicable than science when it comes to trying to answer the classic question, *"What is the true meaning of life?"*

— *Philosophy* is said to be distinguished from *theology,* on the other hand, in that it ignores *dogma,* and deals with *speculation* rather than *faith.*

  • Example: Some theologians might say that the *Bible* explains the creation of the universe and the many life forms on Earth, and that's all there is to it, whereas philosophers, as well as scientists, would speculate about other mechanisms.

3. Science's Relationship to Philosophy

— In ancient times, *science* as such did not exist, and *philosophy* covered the entire field of study involving natural phenomena.

— Later, however, as more *facts* became known and *tentative certainties* emerged, the sciences broke away from *metaphysical speculation* to pursue their different aims.

  • Because of the historical relationship, many of the greatest *philosophers* were also *scientists* and vice versa, and philosophy has both influenced and been influenced by nearly all the sciences — including the behavioral and biological disciplines.

  • *Philosophy* still considers the *methods* of science (as opposed to the *contents* of science) to be its province, and hence the two disciplines continue to be associated.

  • *Philosophy of science* is the term designating those aspects of *philosophy* that today remain directly related to the *sciences.*

4. Philosophy of Science — Description and Functions[10]

There are *three interrelated general functions* and thus *descriptions* of the philosophy of science: It furnishes *guidance for scientists, the frame of reference of science, and the general method and attitude of scientists.*

a. Guidance for scientists

  • Philosophy of science *defines and systematizes* the specific basic *concepts/ propositions* that all the sciences are built upon, so that students, as well as professionals, are provided with *instruction.*

∞ Note, however, that to become a *skilled researcher* it is also necessary to *practice*; i.e., *experience* is very important!

- *Fundamental features* of science that are *specified and elucidated* by the philosophy are as follows (see later in chapter for details):
  ∞ *Goals/objectives* that are shared by all the sciences,
  ∞ *Assumptions* about the nature of the real world that are held by the sciences,
  ∞ *Principles and methods* of science, including the *requirements* for observations and thus data to be considered scientific,
  ∞ *Adequacy* of the procedures used by scientists with respect to maximizing scientific progress.
    - Philosophy of science developed, in fact, for the purpose of *examining and evaluating* the *methods* of science, the *forms* of scientific evidence, and the *ways* that the sciences best progress toward their shared *goals*.

b. Frame of reference of science

  - Philosophy of science involves a *system/manner of thinking and operating* within which intellectual curiosity is likely to lead to scientific discovery and confirmation.

  - *Divergent definitions* exist, however, for science and its method.
    ∞ Different scientists have different *strategies*.
    ∞ Moreover, individual scientists are *not always consistent* in the strategies they use.
    ∞ Thus there is *no one prescribed routine for scientific research*.
      - Example: The *inductive method* versus the *deductive method* (discussed in Chapter 3).
    ∞ Nonetheless, it is important to note that the *various strategies* of science are not contradictory — rather they are *complementary* in terms of their *strengths/advantages*.

  - *General principles* of science prevail despite divergent definitions.
    ∞ Indeed, the *critical and distinguishing feature of all science* is its general method, or approach; science is defined by its *shared procedures and associated modes of thinking*.
      - These provide a *common foundation* and thus the *frame of reference*, for the activities of all scientists in all the different scientific disciplines.
    ∞ The wide variety of specialized *tools* and the different *techniques* and *strategies* used in the various fields of science represent just *diverse means* of implementing the *general scientific method*.
      - Examples of different tools and techniques are the telescopes and spectral analyses used in *astronomy*; the microscopes and staining procedures used in *biology*; and the Skinner boxes, surveys, and brain-recording techniques used in *psychology*.

   c. General method and attitude of scientists

- Philosophy of science, in its essence, consists of the *general scientific approach to knowledge,* i.e., the *means* to the *ends* (goals) of science.

  - ∞ Fundamentally then, the *philosophy* of science is simply the *methodology* of science, which includes a particular *attitude* (described later).

  - ∞ But the *philosophy* of science does *not* involve the *data* or *body of knowledge* of science nor its *applications;* i.e., it's *not the ends* themselves, *just the means* to the ends.

  - ∞ Hence science is a *unique kind of activity* — the scientific method and attitude — as well as the *results* of that activity.

- IN SUMMARY: The *central aspect* of the philosophy of science is the *scientific method,* which provides the *frame of reference* for science and thus much of the *guidance for scientists.*

5. Scientific Method Broadly Characterized

— The scientific method is a *fundamental logic of inquiry* that represents a *unique process* for solving problems and generating a body of knowledge.

- Basically, it consists of those *principles, procedures, and modes of thinking* that historically have been *most productive* in the acquisition of *systematic and comprehensive knowledge* (these are listed below under "Approaches to Knowledge, Science," and they are elaborated throughout this Handbook).

  - ∞ Since these *carefully honed components* of the scientific method have been most successful in advancing knowledge in the *past,* it is expected that their continued application will also most effectively yield new information in the *future.*

- Note: Although the *rudiments* of the scientific approach to knowledge (i.e., its method) have been present throughout history, the *era of true, modern science* began with the *Renaissance* and especially with the writings of Sir Francis Bacon (1561-1621). *Today* science has expanded to the point that there are now *more* scientists alive than in *all* of previous history combined.

Approaches to Knowledge[9,16]

The scientific method, of course, is not the only approach to knowledge. There are at least *six basic approaches: superstition, authority, intuition, rationalism, empiricism,* and *science.* These are presented roughly in order of increasing strength and usefulness. Note, however, that none of these approaches is infallible.

1. Superstition

— Knowledge acquired through *mere exposure* to the beliefs of others is called *superstition.*

— Superstitions are *irrational beliefs* that result from ignorance, fear of the unknown, trust in magic or chance, or possession of a false conception of causation. [A Merriam-Webster. (1993). *Webster's New Collegiate Dictionary* (10th ed.). Springfield, MA: G. & C. Merriam.]

- *Chance associations* among events are frequently misinterpreted by individuals as representing *causal relationships*.

  ∞ Example: A *black cat* walks across someone's path and the next day something terrible happens to that person; thus begins a superstition about black cats and bad luck.

- Superstitions are often *held tenaciously,* despite much evidence to the contrary.

  ∞ Persistence of superstitions is due in part to *repeated exposure*, which leads to familiarity and habit.

  ∞ General Examples: Beliefs in all sorts of *magic charms and spells*, such as a rabbit's foot bringing good luck (obviously not for the rabbit); as well as beliefs in *common sayings*, such as "You can't teach an old dog new tricks."

2. Authority

— Knowledge acquired from a *highly respected source* is called *authoritative*.

- Although useful when the authority has particular *expertise*, it must be noted that such information or ideas can be *incorrect*, and that this is more likely when individuals are *authoritarian* — dictating *unquestioning acceptance* of the "facts" as stated.

  ∞ Example: "Our great leader, Mao Tse-tung, says that it is so, therefore it *must be accepted as true.*"

- Information should be very *suspect* when we are denied the right to *doubt, question, or test* its accuracy.

  ∞ Example: *The Catholic Church convicted Galileo as a heretic* in 1633 for promoting the Copernican theory that the Earth is not the center of the universe, which was a contradiction of the Church's teachings and therefore a challenge to its *authority.* For this Galileo spent the last nine years of his life under house arrest. It was not until 1992 (over 300 years later) that a papal commission acknowledged that the Church was wrong in condemning Galileo.

3. Intuition

— Knowledge acquired in the *absence of conscious reasoning/inferring* is called *intuitive*.

- *Subconscious* processes are likely involved in intuition, but meaningful outcomes can result.

  ∞ Example: A *hunch or sudden insight* into the solution of a problem could involve the unconscious use of information and cognitive skills *developed from previous experiences*.

4. Rationalism

— Knowledge acquired through *reasoning* represents the process called *rationalism.*

- Different well-meaning and intelligent individuals, however, can arrive at *disparate conclusions* as a result of different reasoning.

- *Thought experiments* are those that in fact involve rationalism, rather than the actual gathering of *new empirical data.*

  ∞ Example: Using *"logic,"* the famous Greek philosopher *Aristotle* reasoned that *heavier objects* should fall at a faster rate than *lighter objects* — which, as it turns out, is incorrect (especially if air resistance is controlled for), even though this idea is based on such *common sense* notions that still to this day many people believe it to be true.

    - *Face validity* is what such beliefs are said to have (covered later in Chapter 4, Types of Validity and their Assessment).

5. Empiricism

— Knowledge acquired through *experience* represents the process called empiricism.

- Example: *Galileo* doubted the validity of Aristotle's logic about heavier objects falling faster than lighter ones, and so he decided to actually make some *observations* (taking advantage of the Leaning Tower of Pisa). What he found through this *empirical approach* was that in fact heavier and lighter objects fall at the *same rate.*

- It should be noted, however, that our *observations/experiences* just like the preceding approaches to knowledge, do *not* necessarily reflect *reality,* they too are subject to *individual differences, inconsistency, and inaccuracy,* or in other words, *variation and error.*

  ∞ We might be *tired or careless,* our attention and perception can be *influenced by past events and desires,* we have *sensory limitations,* we experience *illusions* and even *delusions or hallucinations,* and our *memory can be flawed.*

  ∞ Example: Consider the many *visual illusions* that are shown in various books and how we are fooled by *magic acts.*

6. Science

— Knowledge acquired through a *specific, methodical logic of inquiry* (described below) represents the process called *science.*

- The *scientific method* is considered to be the *most advanced and sophisticated approach,* and it has been the *most successful.*

  ∞ In comparison to the other approaches, consider the *enormous contributions* of science to both our understanding of the principles and mechanisms of the universe and to the betterment of our lives.

  ∞ Of course, science *cannot answer all of our questions,* and thus there are times when we are limited to other approaches, such as *rationalism* which is involved in *speculation* about *metaphysical phenomena.*

- Example: Does *God* exist, and if so, what is the nature of God?
- All the *preceding routes* to knowledge are *incorporated* within science — especially empiricism and rationalism — and thus the scientific approach *builds* on the other approaches:
  - ∞ *Superstition* can play a role in the form of *sources of ideas* for scientific research.
  - ∞ *Authority* often plays a role in the form of *influences* that more experienced and renowned scientists have on the thoughts and activities of other investigators.
  - ∞ *Intuition* plays a role in the form of *hunches* about fruitful avenues of research, the *formulation* of hypotheses, and the occurrence of *sudden insights* about the meaning of results.
  - ∞ *Rationalism* plays an important role in the form of the *logical development* of hypotheses and theories, as well as the *organization and interpretation* of data.
  - ∞ *Empiricism* plays an important role in the form of *systematic, objective/ unbiased observations* carried out during the collection of data.
- *Additional* distinctive and crucial elements are involved in the process of science (these are discussed in more detail later):
  - ∞ *Steps of the scientific method*, which represent the general, methodical, sequential procedures of science;
  - ∞ *Control and manipulation of variables*, which minimizes the masking and confounding of studied relationships/effects;
  - ∞ *Operational definitions*, which are used to produce clarity, objectivity, concreteness, and repeatability;
  - ∞ *Replication of procedures and objective observations* by potentially *different individuals*, which minimizes bias and permits the verification of results;
  - ∞ *A questioning, critical, and yet open-minded attitude*, stressing *empiricism and rational impartiality* in all of science's activities, which leads to new discoveries, the self-correction of deficiencies and errors, and more accurate and complete understanding.
— Significant advantages of the scientific approach

There are *three substantial benefits* of using the scientific method.

1) *Data are more accurate* when obtained through science.
   - ∞ This is because science's observations must be not only *objective, i.e., empirical*, but also *repeatable and public* (discussed later in this chapter under "Requirements of Scientific Observation").
     - Thus scientific observation goes beyond mere empiricism, and because the observations must be repeatable and capable of being observed by others (i.e., public), they are *less subject to the perceptions, biases, prejudices, opinions, emotions, and fraud* of individuals, in comparison to the other five approaches to knowledge.

2) *Correction of erroneous beliefs* is handled more powerfully through science.

∞ Not only must there be *good, objective supporting data* for a hypothesis or theory, but an additional test of validity is its *consistency* with the *totality of the current body of scientific knowledge.*

• Example: The *claims of astrologers* that they can analyze a person's character and predict a person's life course based on the position of celestial bodies is *not supported* by either *contemporary* scientific research or the *vast accumulation* of scientific knowledge about the universe. [See, e.g., Carlson, S. (1985). A double-blind test of astrology. *Nature*, 318, 419-425.]

∞ Furthermore, it should be noted that although scientists might *individually* be swayed by some *prevailing point of view* to look for certain experimental results rather than others or to develop some broad theory that they then seek to find support for, the *scientific community as a whole* strives to judge the work of its members by evaluating the *objectivity and rigor* with which that work has been conducted.

• In this manner, the *scientific method prevails*, and *errors* are likely to be *detected and corrected.*

3) *Superiority of one belief* (hypothesis or theory) over other beliefs can be established with *greater confidence* through science.

∞ This is because *clearly defined comparisons* are made under *systematically controlled conditions,* and conclusions about the accuracy of hypotheses and theories are *rationally* based on *empirical* data designed to be both *reliable* and *valid.*

• Example: The *empirical study* made by Galileo on whether heavier objects fall faster than lighter ones was *not*, in fact, a true scientific experiment. It *lacked controls for undesirable, extraneous variables,* such as the influence of air resistance (imagine the effect on a feather) and the possibility of *experimenter bias* (which might have led to the dropping of one object before another). Hence Galileo's study did not meet the *rigorous standards* incorporated within the scientific method.

— IN SUMMARY: Science can be concisely defined as the *methodical and controlled, rational and empirical search for truth or reality.* (Science is further defined below under "Goals of Science.")

• *The Scientific Ideal,* simply put, is *objective knowledge* based on *evidence* and *reason* (i.e., empiricism and rationalism, as opposed to just superstition, authority, or intuition.)

• *The Scientific Attitude* (as noted earlier) is one that is *questioning, critical, and yet open-minded,* stressing *empiricism* and *rational impartiality* in all of science's activities.

∞ This attitude and the scientific method lead to, among other things, *new and better approaches* to old problems, the *discovery of errors and deficiencies*, and *more accurate and complete understanding* (the first of the four general goals of science, which are discussed next).

Goals of Science[10,25]

There are *four general goals of science: understanding, prediction, control, and systematization*. Moreover, the goal of understanding has two levels, *description and explanation*. These goals are common to all the sciences. Furthermore, they are also similar to the broad goals that we all have in our *everyday lives*. For example, when beginning to work for a new company, we would want to understand more about its operation, predict events before they occur, exert some control, and systematize/organize the knowledge gained. Now consider a hypothetical example in science; imagine that all of a sudden a number of adolescents in some town started to act very strangely; specifically, they would break out in hysterical laughter whenever an adult told them how to behave. The scientists brought in to investigate this phenomenon would first need to thoroughly describe the behavior and the circumstances under which it occurred. Then they would try to explain why this behavior was occurring, predict when it would occur, and try to control it. Last, they would systematize the knowledge gained.

1. **Understanding**
   - The *first goal* of all approaches to knowledge (as indicated above) is *understanding*, with its two levels: *description*, which is a lower level (but nonetheless essential); and *explanation*, which is the higher level.
     - This distinction between two levels of understanding is a very important one that relates to the *two major forms of scientific research: descriptive versus explanatory/experimental* (covered in Chapter 2).
   - *Scientific understanding* is typically defined as the *tentative acceptance* of an *explanation* for the occurrence of some phenomenon.
     - Acceptance is *tentative* because it is recognized that further research might uncover equally good *alternative explanations*, and eventually a *better explanation*, i.e., one that is *more complete and/or accurate*.
       - ∞ Science, in fact, involves the search to find *all the necessary and sufficient conditions* for phenomena, as well as the *precise roles* that these factors play.
         - Example: Science *continues to search* for a more complete and accurate understanding of *violent behavior*; including all the specific roles played by our *nature* (the genes that we inherit) and our *nurture* (environmental experiences), plus the *interaction* of these influences.
       - ∞ Science is thus a *continuous and cumulative process*, but it should be noted that in addition, from time to time there are major *paradigm shifts* in our thinking and understanding: e.g., consider the *scientific*

revolutions attributed to Copernicus, Newton, Darwin, Einstein, Freud, and others.

- *Better research strategies and theories* are periodically developed, *new facts* are added, and *old beliefs*, which always should be *provisional*, are modified when the data gathered indicate that they are incomplete or inaccurate.

∞ Science is therefore also a *self-corrective system*.

- *Errors* are commonly discovered and then rectified.

  Example: It was once thought that *autism* in children is caused by having *cold and uncaring parents*, rather than being due to *biological determinants*, as subsequent research has shown to be the case.

∞ Science does *not* accept the idea of *absolute certainty*. It is understood in science that *nothing is ever truly proven* — only supported/strengthened/confirmed (or not) by the data obtained through research.

∞ Science therefore encourages a *doubting/skeptical attitude* (as noted earlier, along with the benefits).

∞ IN SUMMARY: Science is a *tentative, cumulative, self- correcting, consensus system*, in which the scientific community continually strives to be *objective and open to change*, rather than authoritarian.

- It is expected that *all the evidence*, both old and new, will be considered and assessed in an *impartial manner*, such that each piece of additional data is granted significance according to *methodological considerations*, i.e., the quality of the research that led to the findings, while also taking into account *consistency* with the totality of the current body of scientific knowledge.

— Explanation

- The *higher* of the two forms of understanding is *explanation*, as opposed to just description.

- Explanation involves *specifying the antecedent conditions* necessary to produce some phenomenon — or in other words, it involves finding the *causes or mechanisms*.

  ∞ Example: What are the antecedent conditions, i.e., the *causes*, of schizophrenia? Is there a genetic predisposition, a virus, environmental stress, or some interaction of these factors?

- Explanation permits *prediction* (but is not a requirement) and also usually *control* — both of which are additional goals of science.

  ∞ Hence explanation is a *tool* as well as a goal of science.

- *Many levels* of explanation are possible.

  ∞ Example: *Learning* can be explained either at the experiential, neuroanatomical, electrophysiological, or biochemical level (or some combination of these), as well as at the multicellular level or the single-cell level.

- ∞ Reductionism
  - The point of view which holds that complex phenomena are to be understood and explained by *analyzing* them into ever *simpler* and ultimately strictly elementary, components is called *reductionism.*
  - This principle also holds that the subject matter of one science can be *presented* in terms of other *more basic sciences.*

    Example: Reductionism attempts to explain all *psychological processes* (such as emotion) as *biological processes* and then to explain all these biological processes by the same *physical laws* that chemists and physicists use for *inanimate matter.*

  - Reductionism, however, is *not* always advantageous, i.e., efficient; it has been described sometimes as *trying to read a newspaper using a microscope.* Depending on the *specific application*, different levels of explanation, possibly involving different disciplines, might be most useful.

    Example: Understanding *learning and memory* at the neuro-chemical level is no doubt very important to finding a cure for Alzheimer's disease, but it wouldn't be particularly important for house training a pet.

  - Another *weakness of reductionism* is that it does not take into account the fact that the whole is often *different* (greater or less) than the sum of the parts.

    Example: With regard to the *mind–brain problem* (i.e., the nature of the relationship between the two), one of the most accepted viewpoints is the *emergent property position*, which holds that the mind emerges from neural tissue as a *new property* when a nervous system reaches a certain *complexity and organization.*

— Description
- A *lower form of understanding* than explanation — but nevertheless an *indispensable* one — is *description.*
- It consists of *observations* of the characteristics of phenomena, including their values, with little extrapolation from them.
  - ∞ Description specifies the *way* things are, without explaining *why* the phenomena exist or have their particular properties. Thus it does *not* involve the specification of *causes.*

    Examples: Describing the *symptoms* but not causes of schizophrenia or attention-deficit disorder.

- Description typically *precedes* explanation and for good reasons that should be clear now that explanation has been discussed.
- a) It is *easier to describe* a phenomenon than to explain it.

b) A *thorough description* is usually *necessary* for explanation. How could you explain what has *not yet* been described?

- *Many levels* of description are possible, as is true for explanation.
    - ∞ Example: *Clinical depression* can be described at the behavioral level, the neuroanatomical level, or the biochemical level.

- Types of description

    There are *three forms/types* (as opposed to levels) of description: *classifying, ordering, and correlating*.

    a) Classifying

    - To *classify* is to *organize* a multitude of phenomena into a smaller, more manageable number of *units/categories* based on observed *similarities* in one or more properties.

        Example: Classifying all *animal species* into several *phyla* based on similarities and differences in their characteristics of anatomy, physiology, and behavior.

    - Classifying is the *most fundamental* form of description.

    - Scale of measurement

        Nominal: This kind of scale reflects only the property of *difference* and not the property of magnitude — it literally means "*naming*." (For further discussion of this and other measurement scales listed below, see Chapter 4 under Scales of Measurement.)

        Example: Specifying that some organism is *animal versus plant* (in contrast to measuring how big or heavy one animal is compared with another).

    b) Ordering

    - To *order* is to *systematically arrange* phenomena on a *continuum* based upon a *shared, measurable dimension*.

        Example: A scale of the *intensity* with which various wild animals are *feared* by an individual or group.

    - Scales of measurement

        Ordinal: This kind of scale is dependent on the property of *magnitude*, but *rank orders* indicate only *greater or lesser than*, not the size of a difference.

        Example: Rank ordering individuals from most to least (or first to last) in *judged attractiveness*.

        Interval: This kind of scale is dependent on the property of having *equal-sized differences in the scale magnitude* of numbers reflecting *actual* equal-sized differences in the *attribute magnitude* that is being measured.

        Example: Comparing individuals on intelligence using *IQ test scores* (an interval scale at best), in which if the IQ of individual

A = 120, B = 100, and C = 80, then in fact the *actual difference* in intelligence between A and B really is the same as that between B and C, i.e., 20 units in each case.

Ratio: This kind of scale is dependent on the additional property of having a *true zero point* (i.e., zero on the scale truly represents zero amount of the measured attribute), so that the size of *scale ratios* reflects the actual size of the *attribute ratios.*

Example: Measuring the *duration* of events, or the *time* it takes individuals to complete a task, where if it takes individual A 10 min to complete a task and it takes individual B 5 min to complete the same task, then it is accurate to say that it takes A twice as long as B (10/5 = 2), because zero on the *time scale* truly represents zero time.

c) Correlating

- To *correlate* is to specify the *form* and the *degree of relationship* or *strength of association* between phenomena.

    Correlations can be either *positive/direct* or *negative/inverse relationships*, as well as *linear or nonlinear* (i.e., either a straight or a curved line, respectively, when the relationship is graphed).

    Examples: 1) A correlation could be computed between the amount of *studying* done by students for a course in school and the *grade* they received. A *positive/direct relationship* would be expected; i.e., *higher* amounts of studying would likely be associated with *higher* grades.

    2) In contrast, a *negative/inverse relationship* would be expected for correlations between the amount of *marijuana* used by students and their *school grades*; i.e., *higher* amounts of drug use would likely be associated with *lower* grades.

- Scales of measurement

    Correlation coefficient and *coefficient of determination* are described later in Chapter 2 under "Correlational Study" and also in Chapter 11 under "Statistics."

d) IN SUMMARY: There are *three forms of description. Classifying* phenomena into categories involves the nominal (naming) scale of measurement. *Ordering* phenomena along some measurable dimension/continuum involves either the ordinal, interval, or ratio scale of measurement. These four scales for classifying and ordering are listed in a sequence of increasing properties of measurement and thus increasing amounts of information provided. *Correlating* to determine the degree of relationship/association between phenomena is the third form of description, and involves the correlation coefficient and coefficient of determination scales.

- Description permits *prediction* — another goal of science.

∞ Hence description is also a *tool* as well as a goal of science.

- Examples: 1) From previously described *correlations* between the degree of cigarette smoking, exercise, and quality of physical health, it is possible to predict, with some accuracy, the physical health of individuals from knowledge of either or both of the other two variables.

  2) Mendeleyev classified and ordered the known chemical elements into his now famous *Periodic Table of Elements,* which most students learn about in high school. From the location of gaps in the table when he first assembled it, Mendeleyev was able to *predict the properties* of new elements that he fully expected to be discovered later.

- Note that understanding at the level of *explanation* is required for *control* (see later) but not for *prediction*.

2. Prediction

— The *anticipation* of an event prior to its occurrence is *prediction*.

— Formally stated: Prediction is the process whereby based on a *previously described relationship* between two or more events — *antecedents and consequents* — their probable *future relationship*, in a new but somewhat similar situation, is specified.

- Examples: 1) Results from studies with *animals* are used to predict the effects of treatment conditions for *humans,* such as the efficacy of different medical drugs or the outcome of intermittent versus continuous reinforcement on behavior.

  2) Results from carefully controlled research conducted in *laboratories* are used to predict behavior in the "*real world*."

— *Hypotheses and theories,* based on our *understanding* (a goal), are *tested* in science by using *predictions* that are derived from them.

- Hence prediction is also a *tool* as well as a goal of science.

- If a prediction is *supported* by the data of scientific studies, then the associated hypothesis or theory and hence our understanding are *confirmed*.

3. Control

— *Control* is a process in which the *antecedent conditions* influencing/determining a phenomenon are *manipulated* in order to *cause* a desired outcome, specifically, the phenomenon of interest, or *consequent event*.

- The ability to *predict* some phenomenon, when based on *understanding* at the level of *explanation* (i.e., knowing the antecedent conditions), usually leads to the ability to *control* the phenomenon (a major exception being astronomical events).

— *Behavior of organisms* is something that scientists often strive to find ways to control, but the control is typically *indirect* rather than *direct*.

- Scientists investigate and attempt to *regulate/adjust* the variables that *result* in behaviors of interest — i.e., they manipulate what are thought to be *antecedent conditions* that *lead to* the behavior; they do not usually directly manipulate the behavior itself.
    - ∞ Examples: 1) In behavior modification therapy, *incentives* might be provided to *indirectly* increase certain desired behaviors. 2) Alternatively, *information* might be provided to produce a change in *attitude*, and thereby *indirectly* bring about a change in some *behavior*, such as smoking.
    - ∞ Many individuals *do not understand* this indirect mechanism of control, and thus science, e.g., psychology, sometimes has a *threatening image* in the public's eye.
    - ∞ Scientists, however, recognize that they have an *obligation* to ensure that the ability to control behaviors — as well as to describe, explain, and predict them — is *used responsibly*.
- Scientific testing of hypotheses and theories through *research* is commonly done under *controlled conditions*.
    - Hence control is also a *tool* as well as a goal of science.
    - Furthermore, the ability to *control* is dependent on the ability to *predict*, and both are dependent on *understanding* at some level — therefore control, in addition to prediction, is a *tool* that serves as an *important check* on the quality of our understanding.

4. Systematized Knowledge

- Formulation of a *systematic body of knowledge* is a very important goal of science, which is accomplished by *organizing* a number of discovered facts (along with the methods used to obtain them) into more general *models, theories, principles, and laws* (this is discussed further in Chapter 3 under "Theory").
    - Examples: 1) In psychophysics, *Fechner's Logarithmic Law* and *Stevens' Power Law* each mathematically relate psychological perceptions (such as brightness and pitch) to physical stimulus values (such as intensity and frequency) based on a body of research covering a variety of sensory modalities.

      2) *Darwin's Theory of Evolution by Natural Selection* was the result of integrating years of painstaking research in biology, geology, and other fields.
- *Organizing a body of knowledge* into theories, laws, etc. enhances our *understanding*, as well as our ability to accurately *predict* and exert *control*.
    - Indeed, as a result of systematization, knowledge is more readily *recalled, communicated, applied* and *expanded upon*.
    - Hence systematized/organized knowledge is also a *tool* as well as a goal of science (as was noted for all the other goals).
- IN SUMMARY: The *goals of science* — understanding (description and explanation), prediction, control, and systematized knowledge — are *interrelated*, and in all cases they serve as *means* to ends (i.e., tools), as well as *ends* in themselves.

Assumptions of Science[9,10,25]

There are *four assumptions* that form the *foundation* upon which all of science seeks to achieve the goals that have just been discussed. These assumptions are: 1) *the reality of space, time, matter and energy*; 2) *order*; 3) *determinism*; and 4) *discoverability*, which incorporates the *principle of parsimony*. Just as with regard to the goals of science, the assumptions of science are *interrelated* and are also part of our *everyday lives*, even when we are not aware of them.

1. Reality of Space, Time, Matter, and Energy

— Scientists assume that *objects* (and events) exist *externally* to us; i.e., that they have an *independent existence* and are not figments of our imagination; that they have *real properties* (in various amounts); and that they occupy *real space* (usually thought of as tridimensional, although it is now believed that there are many more dimensions) and dwell in *real time* (an additional dimension with the properties of simultaneity and irreversible succession).

- This *independent reality* assumption is considered *fundamental* to all the other assumptions of science.

- However, it should be noted that this is only a *supposition* and *cannot be proven*; moreover, events might be only *probabilistic* in occurrence rather than *absolute*.

  ∞ Example: 1) Our knowledge of the position and momentum of subatomic particles at any given time is probabilistic (this is the *Heisenberg Uncertainty Principle* of quantum mechanics; whether similar *indeterminacy* also applies to phenomena in areas such as psychology is unknown — see later under "Statistical/Probabilistic Determinism").

- In addition, there are the problems that our *perception and memory* of external reality are not necessarily *reliable or valid*, i.e., consistent, accurate, and complete (discussed earlier under "Empiricism," an approach to knowledge).

  ∞ Examples: 1) *Perceptual illusions* of a variety of forms are experienced, such as the *moon illusion* in which the moon appears to be much larger when it is seen near the horizon than when it is seen overhead, even though the retinal-image sizes are the same.

  2) *Sounds, colors, and smells* do not really exist outside our *minds* — there are only physical pressure variations in air or water, electromagnetic radiations of different wavelengths, and various chemicals in the environment.

- These issues are of *great concern* to philosophers and scientists, as well as others.

2. Order (uniformity)

— Scientists believe that the reality of the universe is *lawful*, not haphazard; that there is *regularity, consistency, stability, and repeatability* of the phenomena and relationships of nature; that phenomena are *systematically related*, that events follow one another in an *orderly sequence*, and that this can be *replicated*.

- *Without order,* science — and even existence — would be impossible.
- *Science,* in part, is a search for the *rules of order* in the universe.

3. Determinism (causality)

— *Determinism* is probably the most interesting and for some the most controversial of all the assumptions, because it often appears to conflict with the notion of *free will* (which is discussed below).

— Scientists believe that the occurrence of an event is *determined by* (dependent on) *prior/antecedent events,* i.e., scientists believe that phenomena have *causes.*

- Example: What are the antecedent events that are the determinants/causes of your *reading this Handbook*?

— This notion of determinism is implicit in the *assumption of order.*

- How could there be any order *without* determinism/causality?

— Both determinism and order are *prerequisites* for the *scientific goals* of understanding, prediction, control, and even systematized knowledge (with its models, theories, principles, and laws).

- Thus there can be no *science* without *acceptance* of these assumptions.
  - ∞ *Science,* in fact, is a search for the *rules of order and causality* in the universe.

— Scientists typically assume that *all behavior is determined.*

- Because a phenomenon could have *many* or only *obscure* antecedent conditions, we *might not currently know* every or even any of the determinants of a particular behavior, but that doesn't preclude the *eventual discovery* of all the causal factors.
  - ∞ Example: *Schizophrenia* is one of many mental disorders for which the *causes* have remained obscure for centuries. However, taking advantage of significant technological advances, we are now gradually learning more about the biological as well as environmental *determinants.*

— Statistical/probabilistic determinism

- Rather than *strict* determinism, in some cases there might be only *probabilistic* determinism, but that is still determinism.
- *Probabilistic determinism* means that under specifiable conditions a certain result will occur, but only in an *explicitly stated percentage of cases.*
  - ∞ Furthermore, this percent value is only *converged upon,* in a *random manner,* over the *long run.*
    - Example: The expectation that when a coin is flipped it will come up heads 50% of the time is likely to be *more closely approximated* the more times the coin is flipped.
  - ∞ This is *statistical lawfulness;* i.e., it is a *probability law.*

- In such cases, even though there are determinants, *no prediction can be made with certainty* about the outcome of *individual events*.

    Example: How *a coin* will fall on any given *flip* is uncertain.

∞ Note, however, that even probabilistic events are in fact most likely *totally determined*.

- Events may *appear* to be just *probabilistically determined* because it is difficult to specify *all* of the determinants with *sufficient precision* to make accurate predictions regarding individual events.

    Example: If a *feather is dropped* from a great height, you probably could not predict exactly where it would land. Nevertheless, you would not claim that the feather has *free will*. Instead you would recognize that there are *determinants* of where the feather will fall, even though the determinants are difficult to specify precisely enough to make an accurate prediction and even though the feather lands at different apparently random points each time.

— Free will

- *Free will* is a concept that usually is *contrasted* with *determinism*, but in fact the two notions are not necessarily incompatible.

- Definitions of free will:

∞ Purposeful activity that is *neither* determined nor random.

∞ Philosophical and religious doctrine that attributes the cause of behavior to volition and independent decisions of the person, rather than to *external* determinants (note the implication here of *internal* determinants). [Wolman, B. B. (Ed.). (1973). *Dictionary of Behavioral Sciences*. New York: Van Nostrand Reinhold.]

∞ Ability to choose between alternatives so that the choice and action are, to an extent, creatively *determined* by the conscious subject (note that here free will is *explicitly* stated as being *determined*). [A Merriam-Webster. (1979). *Webster's New Collegiate Dictionary*. Springfield, MA: G. & C. Merriam.]

- Interpretation of free will

∞ Many individuals believe that humans, unlike other organisms, have a *soul* and thus *free will*; and therefore manipulations can have an effect only by *choice* of the person involved.

- This, of course, would make the scientific investigation of human behavior *substantially more difficult*.

∞ But note that although some causes of behavior are *within the organism* — e.g., the structure and function of the nervous, endocrine, and immune systems — these are still *determinants*.

- Moreover, the *organismic determinants* are themselves actually determined by *external factors*, i.e., the *genetic* contributions of parents and the *experiences* of the individual in his/her physical and social environments.

Thus, if there is free will, it is itself determined!

— Predeterminism

- *Predeterminism* should not be *confused* with determinism.

- Definition

  ∞ *Predeterminism* is determination *beforehand*, such that events *cannot be modified* and will occur no matter what happens; i.e., they are *foreordained or predestined.*

  ∞ *Determinism*, in contrast, allows for *flexibility*, if for no other reason than that, for all practical purposes, some determinacy may be just *probabilistic* (see "Chaos Theory" — next).

— Chaos theory[12]

The following discussion of *chaos theory* is supplemental. It is presented because this relatively new concept is very important and because it is relevant to the *assumptions of order and determinism.*

- Description

  ∞ Chaos theory is about events that *appear to be random* but which nevertheless are governed by *strict rules.*

- Principles of chaos theory

  ∞ *Simple deterministic systems can generate random-like behavior.*

    - Example: Computer-generated *random-numbers tables* are deterministic but appear random.

  ∞ *"Randomness"* thus can paradoxically be *deterministic.*

  ∞ Hence there is *order in apparent randomness* and vice versa.

    - *"Randomness"* generated in this manner, i.e., by *deterministic systems*, is referred to as *chaos.*

      Example: The variation in time that occurs between the *drips of a leaky faucet* appears random but is actually deterministic and thus chaos.

    - Such random-like behavior is *fundamental*, in that gathering additional data will not eliminate it.

      Example: Advances in technology are used to try to precisely predict *weather changes*, which nevertheless continue to appear probabilistic.

  ∞ In principle, the future is completely determined by the past, but in actual practice, there is *great sensitivity to initial conditions* for chaotic systems.

    - More technically speaking, chaotic systems are *nonlinear*, i.e., the effect/result is *not proportional* to the cause.

- *Small uncertainties*, due to errors or noise, are *amplified exponentially* (i.e., raised to some power) over time and space (as in the graph of a *curve* that sweeps upward).

- Therefore, even though behavior might be *predictable* in the *short term*, it is *unpredictable* in the *long term*.

   Example: When a boulder is repeatedly rolled off a mountain top from the "same" point, in the "same" direction, and with the "same" force, where it lands will appear random due to the variation/errors in measurement at the start that are amplified on the way down the mountain side.

   Note: Quantum mechanics theory suggests that the *initial measurements* we make are always *uncertain*, to some degree.

   ∞ IN SUMMARY: *Chaos theory implies* that there are new *fundamental limits* on our ability to make *predictions*. On the other hand, it also implies that many "random" phenomena are *more predictable* than previously thought. Hence this theory is used to *increase* our ability to predict.

- Nature and chaos theory

   ∞ Nature might employ chaos for *constructive purposes*.

   ∞ Through the amplification of small fluctuations (errors or noise), chaos can provide natural systems with access to *novelty*, such as controlling the escape behavior of prey, producing genetic variation, and generating creativity.

   - *Creative thoughts* in some cases might be *decisions* or what individuals consider to be the exercise of *free will*.

   - Thus *chaos provides* a mechanism allowing for *free will* within a world that is governed by *deterministic laws*. [Preceding material paraphrased from: Cruchfield, J. P., Farmer, J. D., Packard, N. H., & Shaw, R. S. (1986, December). Chaos. *Scientific American*, 46-57.]

4. **Discoverability and the Principle of Parsimony**

   — Scientists assume that the universe is *rational* and hence that *answers can and will be found* to scientific questions, i.e., that *relationships and causes/determinants* of events are *discoverable*.

   - Thus, it is believed that the *rules of order and causality* in the universe *can be uncovered*.

   - If this were *not* thought to be true, why would anyone choose to be a scientist?

   — It is also assumed that the *Parsimony Principle* is usually *correct*.

   - This principle is to seek and prefer the *most general (broadest) and simplest explanations* that account for sets of phenomena (elaborated in Chapter 3 under "The Hypothesis," and Chapter 12).

   — IN SUMMARY AND CONCLUSION: It should now be apparent that the *four assumptions of science are interrelated*. The *independent reality* assumption is fundamental to all the other assumptions, and the *determinism* assumption is implicit in the assumptions of both *order* and *discoverability*. Moreover, all of these assumptions are prerequisites for achieving the *four goals* of science.

Science can be defined as a search for the *rules of order and causality* in the universe. It should be noted, however, that there might be *limits* on the ability of science to *discover* all the relationships and causes and therefore *boundaries* on our capacity to *understand, predict, control,* and develop a *systematic body knowledge*. On the other hand, it is also entirely possible that if there are limits, they are merely *temporary.*

Requirements of Scientific Observation[25]

Fundamental to finding answers to questions is *observation*. But to be considered *scientific*, there are *three standards* of observation to be met: the observations must be *empirical*, *repeatable*, and *public*.

1. **Empirical**

   — By *empirical* it is meant that the observations must consist of the *objective* experiences of events.

   • In other words, to the maximum extent possible, measurements must be *uninfluenced by our expectations, biases, and desires*.

   ∞ Phenomena that are observed must be *real*, not figments of the *imagination* and not *selective* so as to support some *preconception*.

   ∞ Note: An *attitude of rational impartiality* should characterize *all* the activities of science (as noted earlier).

   • Observations must therefore be *distinguished* from a scientist's *inferences* about them.

   ∞ Example: When making recordings of electrical activity in the brain, we must distinguish this *empirical data* from any *interpretations*, such as regarding ongoing mental processes.

   ∞ It should be pointed out that *in a research paper* a scientist's *expectations/biases*, i.e. hypotheses and theories to be tested, are placed in the introduction section, whereas *empirical observations* are placed in the results section and the *inferences* are placed in the discussion section — hence the biases, observations, and inferences are *separated out*.

2. **Repeatable**

   — By *repeatable* it is meant that observations must be *capable* of being *replicated*.

   • This is necessary for *verification* and determination of *reliability*.

     ∞ To permit replication of observations, it is essential to have *accuracy, completeness, and clarity* in the *methods section* of research logs and reports.

- When results *cannot be duplicated,* it indicates that the initial findings might have been due in part to *chance* or, more specifically, to some *uncontrolled factor,* resulting in *error.*

- However, failure to reproduce results could also be due to some *critical variation/modification* of the original research materials or procedures or in the selected samples of research participants.

- Yet another possibility for failure to replicate findings could be a *real change* occurring over time in the phenomenon under study.

3. Public

— By *public* it is meant that observations must occur in a *manner, place, and time* such that it is also possible for *others* to make them.

- This is necessary for *independent verification* and hence *validation* of results.

     ∞ The potential for others to make the same observations is an important *remedy* for the problems of *faulty or dishonest research.*

- Note that observations do *not* have to be made in a *public place* in order for others to also make them, but the research must be *communicated* to other scientists through oral or written reports.

— Introspection

- *Subjective observations* of what is occurring in *one's own mind* is referred to as *introspection.*

     ∞ It involves the systematic *self-observation* of mental processes/ experiences.

- Introspection is *not public,* and hence it is *not considered scientific observation.*

     ∞ Other individuals *cannot directly share* the experiences.

     ∞ Instead, others can only *observe reports* of the experiences and then *draw inferences* about the actual mental events.

     ∞ The *problem* with introspection is that the reports might not be true reflections of the experiences — they might be *false, inaccurate, or incomplete*; and the terms used to describe the experiences might be *imprecise or ambiguous.*

         - Example: When individuals say they *love* someone, are they necessarily being truthful, and if so, what might they really be experiencing — is it *admiration, devotion, affection, lust,* or some combination of these feelings?

         - For these reasons, as well as *individual differences,* trying to have the *"same experiences"* as another person in order to verifying them would not be satisfactory.

- Note, however, that *reports without inferences* can be taken as *legitimate scientific data*, unlike the supposed actual experiences.
  - ∞ Example: *Studies* of *sensation and perception* would suffer from the problems of introspection unless scientists defined the terms so that they deal with *observable responses* rather than *assumed sensory experiences*.
    - *Absolute threshold*, e.g., is defined as the minimal stimulus value, along some dimension, that 50% of the time the research participant/subject *reports* perceiving.
    - Also, the reports are kept *simple* so they are more likely to *accurately* reflect experience, e.g., "yes" versus "no" or "greater" versus "lesser" responses to the stimuli.
  - ∞ Through these and other means, such as designing clever and well-controlled cognition studies and recording the activity of the brain, science can *objectively* study *evidence* about events of the mind — even though there would still be some uncertainty about the *actual experiences* themselves.

Review Summary

1. *Philosophy* can be defined as the *search for truth*, i.e., the pursuit of reality, knowledge, or wisdom. Originally the *sciences* were just parts of philosophy, but as *facts* became available and *tentative certainties* emerged, the sciences broke away from *metaphysical speculation* to pursue their different aims. However, philosophy still considers the *methods of science*, as opposed to the contents of science, to be its province. Hence there is a *philosophy of science*, a very important component of which is the *epistemology of science*, which deals with the *justification*, or basis, and the *objectivity* of scientific knowledge.

2. The *philosophy of science* is essentially the scientific approach to knowledge — it is a unique method and attitude. The *scientific method* can be broadly defined as a *fundamental logic of inquiry*, consisting of the *procedures and modes of thinking* that historically have been most productive in the acquisition of *systematic and comprehensive knowledge*.

3. There are six different *approaches to knowledge*:
   a. *Superstition* — knowledge acquired through *mere exposure* to beliefs
   b. *Authority* — knowledge acquired from a *highly respected source*
   c. *Intuition* — knowledge acquired in the *absence of conscious reasoning*
   d. *Rationalism* — knowledge acquired through *reasoning*
   e. *Empiricism* — knowledge acquired through *experience*
   f. *Science* — knowledge acquired through a *specific, methodical logic of inquiry*

4. *Science* incorporates the other approaches to knowledge, but it also includes *additional, distinctive and crucial elements*: steps of the scientific method; control and manipulation of variables; operational definitions; replication of procedures and objective observations; and a questioning, critical, and yet open-minded attitude that stresses rational impartiality.

5. Three *significant advantages* of the scientific approach are: (a) *Data* of science are *more accurate* (because observations are objective/empirical, repeatable, and public — and thus less subject to biases, prejudices, and opinions); (b) science provides a more powerful method for *correcting erroneous beliefs* (because the data must be consistent with the totality of scientific knowledge, and the research is judged by its objectivity and rigor); and (c) the *superiority of one belief* over others can be established with *greater confidence* (because clearly defined comparisons are made under systematically controlled conditions, with conclusions rationally based on empirical data that are designed to be both valid and reliable).

6. There are four general *goals of science*.

   a. *Understanding*, which has two levels

      i. *Explanation: Tentative acceptance/specification* of the *antecedent conditions* necessary to produce some event/phenomenon — i.e., specification of its *causes*.

      ii. *Description: Observations* specifying the way things are without explanation as to why. A *lower form* of understanding in that there is no specification of the causes of phenomena.

   b. *Prediction*: On the basis of a *previously described relationship* between two or more events, *antecedents and consequents*, their probable *future relationship* in a new but similar situation is specified.

   c. *Control*: Process in which the *antecedent conditions* influencing/determining a phenomenon are *manipulated* or *altered* to produce/cause a desired outcome, the *consequent event*.

   d. *Systematized knowledge:* Formulation of a systematic body of knowledge by *organizing* discovered facts (and the methods used to obtain them) into more general *models, theories, principles, laws*, etc.

7. Science is a *self-corrective system* in that it encourages a questioning, critical attitude, which leads to the discovery of errors, deficiencies, and new and better approaches to old problems.

8. Scientists do not control behavior directly but rather *indirectly*. They do this by attempting to control the variables that result in the behavior of interest; i.e., they manipulate what they believe to be the *antecedent conditions* that *cause* the desired behavior, not the behavior itself.

9. There are four interrelated *assumptions of science* that form the foundation upon which all of science seeks to achieve the goals previously noted.

   a. *Reality of Space, Time, and Matter*: Scientists assume that objects and events exist externally to us, i.e., that they have an *independent existence* and are not figments of our imagination.

b. *Order (uniformity)*: Scientists believe that the reality of the universe is *lawful*, not haphazard; e.g., that there is *regularity, consistency, stability, and repeatability.*

c. *Determinism (causality)*: Scientists believe that the occurrence of an event is *determined by* (dependent on) *prior/antecedent events.* This assumption is implicit in the assumption of order. It should be noted that the concept of *free will* is not necessarily incompatible with the assumption of determinism; i.e., free will can be thought of as consisting of *internal determinants* of behavior (e.g., nervous and endocrine system activity).

d. *Discoverability and the Principle of Parsimony*: Scientists assume that *answers can and will be found* to scientific questions — that the rules of order in the universe will be found. It is also assumed that seeking the *fewest and simplest explanations* for phenomena (parsimony) will usually be correct.

10. There are three *requirements of scientific observation.*

a. *Empirical*: Observations must consist of the *objective* experiences of events. In other words, to the maximum extent possible they must be *uninfluenced by expectations, biases, and desires.* Moreover, the observations must be distinguished from the scientists *inferences* about them.

b. *Repeatable*: Observations must be capable of being *replicated*, which is necessary for *verification* and determination of *reliability.*

c. *Public*: Observations must occur in a manner, place, and time so that it is also possible for *others* to make them, which is necessary for *independent verification* of results.

11. *Introspection*, the subjective *self-observation* of what is occurring in one's own mind, is *not public*, and thus it is not scientific observation.

Review Questions

Note: To provide feedback about mastery of the Handbook material, answers to the review questions for this and all following chapters are provided by the correspondingly numbered items in the "Review Summary." More extensive answers can be found in each chapter's text. For obvious reasons the review questions should not be answered immediately after reading the associated review summary.

1. Briefly define philosophy, and explain its relationship to the sciences.
2. Give a broad definition for the *scientific method.*
3. List and briefly define six different *approaches to knowledge.*
4. How does *science differ* from the other approaches?
5. What are the two *significant advantages* of the scientific approach?
6. List and describe the four general *goals of science* (including any subcomponents).
7. How is science a *self-corrective system*?

8. How do scientists *control* behavior?

9. List and describe the four *assumptions of science,* and discuss any *interdependence* among the assumptions.

10. List and describe the three *requirements of scientific observation.*

11. Why is *introspection* not scientific?

2

Forms of Scientific Observation and Research

CONTENTS

Naturalistic Observation Used in Descriptive Research

There are two basic forms of scientific observation: *naturalistic observation*, which is best used to *describe* phenomena and their relationships, and *experimental observation*, which is the best means for *explaining* phenomena and their relationships. Hence, the corresponding forms of scientific research are called descriptive research and explanatory/experimental research. These relate to the two aspects of the *scientific goal of understanding* discussed in Chapter 1, Description and Explanation. We will discuss experimental observation after first covering naturalistic observation and its various techniques.

1. **Characteristics of Naturalistic Observation**
   — Observation of events as they occur *naturally*, i.e., independent of the observer's behavior, is what is meant by *naturalistic observation*.
   - Examples: Observing and recording the *migrations of birds* or the *behaviors of schizophrenics* as they occur naturally, *without any intervention* by researchers or clinicians.
   - *No attempt* is made to produce, influence, or control events by *direct manipulation of variables/factors*.
     ∞ This is just the opposite of *experimental* observation.
     ∞ Note: *Variables* are *characteristics* of organisms, objects, environments, or events that have values which can *vary/change* across individuals, items, conditions, or time (for more specifics, see Chapter 4 under Variables in General).
   — *Descriptive research* uses this *naturalistic* form of observation.

2. **Functions of Descriptive Research**

The achievement of *several goals of science* is advanced by descriptive research.

a. *Description* of particular events or phenomena in *accurate detail* is the primary purpose of descriptive research, and this is an element of the scientific goal of *understanding* (see Chapter 1).

- *Variables* (features/properties) would be identified, and their *values* (e.g., magnitudes) determined.

  - ∞ Example: Delineating the specific properties of *Alzheimer's Disease* — a condition involving progressive, irreversible loss of memory, deterioration of intellectual functions, disorientation, apathy, speech and locomotor disturbances, etc.

- *Relationships* (correlations/associations) that exist *between* phenomena or events also are often identified and characterized.

  - ∞ Example: Ascertaining the possible association of *clinical depression* with the experience of *specific stressors* in life.

b. *Prediction*, another goal of science, is made possible when *associations* are described.

- Example: Predicting the expected level of *lawlessness* in some area from the *determined relationship* between *poverty and crime*.

c. *Systematized knowledge*, yet another goal, can be developed partly through description, but would be enhanced by explanation.

- However, achieving the goals of *explanation*, and thus *control*, would be difficult without the purposeful and direct manipulation of isolated variables under controlled conditions — which is what takes place in *experiments*, not descriptive research.

- It is thus *experimental research* (discussed later in this chapter) that is usually used to determine *causality* (and thus to achieve *explanation*), which in turn can typically lead to *control*.

d. *Hypotheses* about *cause-and-effect relationships*, which later might be tested experimentally, can, however, be developed through careful descriptive research — thus advancing the goal of *explanation*.

3. **Reasons for Needing to Conduct Descriptive Research**

There are *three primary rationales* for descriptive research.

a. *No alternative exists* to naturalistic observation in many cases.

- It is simply *impossible* to *influence* certain phenomena by *direct manipulation of variables*, and thus to study them experimentally.

  - ∞ Example: Investigations of the effects of *sibling birth order* on personality characteristics, such as competitiveness.

b. *Unethical behavior* would be involved if certain factors were *experimentally manipulated* in order to determine their effects.

- Such variables must therefore be studied as they occur *naturally*.

∞ Example: It would be unethical to purposely and directly manipulate the *withdrawal* of *parental affection* to determine its effects on children.

∞ Note, however, that it might be considered ethical to experimentally study such phenomena in *nonhuman species*.

c. It's *advantageous* in the *early stages* of investigation usually to do descriptive research *prior to experimental manipulation*.

- Strengths of naturalistic observation preceding experimentation

  ∞ *Description* of behaviors can be made *first* as they occur *naturally.*

    - This can lead to *reduced errors* in experimental research.

      Example: Because it has been found through *naturalistic observation* that cockroaches *innately* prefer dark environments, it would have been foolish to *experimentally investigate* whether cockroaches can *learn* by studying whether they can be *trained* to run from a light compartment into a dark compartment to avoid an *electric shock.*

  ∞ *Phenomena that require an explanation*, through experimental observation, can be *identified* by naturalistic observation.

    - Example: The finding that individuals frequently behave differently when they are in *groups*, e.g., mob behavior.

  ∞ *Prediction of behavior* can be made from data obtained about associations/correlations between variables (as noted earlier).

  ∞ *Causality inferences* (to be followed up on experimentally) can also be made by observing and measuring associations between phenomena and their antecedents (noted earlier).

    - Example: There is a positive correlation between the incidence of *schizophrenia* and being born during the *winter* — this is possibly due to the autumn increase in *viral infection*, which would occur during the second trimester of gestation, which is very important for brain development.

4. **Weakness of Naturalistic Observation**

— There are usually many *simultaneously occurring and varying antecedents* to behaviors and other phenomena, and these antecedents are *confounded*, i.e., inextricably mixed together.

- Therefore, by itself naturalistic observation is a *very inefficient and imperfect* way to determine *causality* — and thus to achieve the scientific goal of *understanding* at the level of *explanation*, which might lead to the goal of *control* (as stated earlier, these goals are best accomplished through experimental observation).

  ∞ Example: Can descriptive research identify the true causes of *alcoholism, suicide,* or *success in life*? How could naturalistic observation, which lacks the direct manipulation and full control of variables that takes place in *experiments*, discover both the *necessary and sufficient conditions* for these phenomena?

- But note that *some antecedents* might be *implausible causes*, thus permitting some weeding out of *alternative explanations*.
  - ∞ Example: Older children have *larger bones* than younger children, but this is unlikely to be the reason for their *greater use of drugs* (alcohol, nicotine, etc.).

Techniques of Naturalistic Observation Used in Descriptive Research

There are *eight different techniques* covered here, and each has its own particular advantages and applications.

1. **Naturalistic/Natural-Environment/Field Study**
    — In field studies, organisms are observed in the *wild*, i.e., in their natural environment, *without* attempting to manipulate any variables.
      - Note: This is different from a *field experiment*, which would involve the purposeful and direct manipulation of certain variables, and the control of other variables, in a natural setting or environment.
    — *Descriptive* field research can lead to entirely *new insights* about the behavior of a species.
      - Example: The discovery of *tool manufacturing* among chimpanzees.

2. **Contrived or Artificial Natural-Environment Study**
    — In artificial natural-environment studies, organisms are brought into a *constructed facsimile* of their home environment, e.g., large aquariums and modern naturalistic zoos, where research is done.
    — Strengths and advantages over natural-environment studies
      - *Closer and less obstructed observation* is possible, which permits more reliable, accurate, and complete data to be gathered.
      - *More comfortable conditions of observation* are usually possible.
      - *More cost effective research* is possible in many instances.
    — Deficiencies
      - Size of the facsimile is typically *smaller* than the actual home environment, thus *limiting freedom*, which could affect behavior.
      - *Modification or omission* of other important characteristics of the home environment is also likely in the facsimile.
        - ∞ Therefore, *generalizing* the findings to the home environment must be done with *caution*, and should be *confirmed* through *natural-environment studies* (generalization is discussed extensively in Chapter 11).

3. **Case Study/Case History/Clinical Method**
    — In case studies, *individuals or social units* (e.g., families) are the focus of study.

— Typically, *tests are run* and the *past is probed* by interviews, questionnaires, and the searching of records.

— This is usually done to try and help *mentally or physically ill persons* by *gathering data* from them, and those who know them, in order to determine the cause(s) of their problem.

— Strengths

  • *Rare phenomena* can be described.

  • *Counterinstances* of widely accepted principles might be found.

  • *Hypotheses about causality* can be generated.

— Deficiencies

  • *Memory* is often the basis for the data gathered, and thus the information might not be complete or even accurate.

  • *Interpretations* that differ in important ways are often possible, since the data are frequently rich and complex.

  • *Generalizability* of findings to other individuals would be questionable, since the data would likely not be representative.

4. Archival/Secondary Record Study

— In archival studies, *previously gathered and saved data* are analyzed by researchers who are *not* the ones who originally compiled the records.

  • *Sources of data* include birth, school, and census records; police and hospital records; and computer files, letters, and photos.

    ∞ Example: Investigating whether there is a relationship between the *month in which people are born* and the *incidence of schizophrenia* later in life, which would support a viral theory of schizophrenia.

— Although very useful conclusions can be derived from analyses of such data, which might not be otherwise available (e.g., records showing the incidence of mental disorders or crimes), there are the twin pitfalls of *selective storage* and *selective survival* of records.

  • Thus the data might be *incomplete and nonrepresentative*, and hence the analyses and conclusions could be *distorted and misleading*.

5. Survey Study

— In surveys, an *oral or written questionnaire* serves as a probe to determine selected features of *populations* at certain points in space and time.

— *Samples* are used in an attempt to answer *questions* of how many or how much; as well as who, what, when, where, why, and how.

  • Example: A survey of *drug abuse* and likely *contributing factors* might be conducted among *distinct segments of society* to *describe* the relative degree of abuse of *different drugs* by groups having *diverse life histories* and living under *assorted conditions* — thereby developing hypotheses to *explain* the abuse.

— Deficiencies and difficulties

- *Sampling error* can be large if the sample is *very small* and/or *not representative* of the population (see Chapter 11 for elaboration).

  ∞ Example: Surveying only parents who made the effort to attend a *PTA meeting*, in order to determine whether a school bond issue would likely be approved by the voters as a whole.

- *Truthfulness* of the survey answers is questionable, especially when the truth is in conflict with *socially acceptable answers*.

  ∞ Example: Consider a survey about the incidence of *washing hands after going to the bathroom.*

- *Time, effort, and expense* of collecting, coding, and analyzing the data can be substantial.

- *Construction of the questionnaire* (open-ended versus closed- ended questions, and the wording and ordering of the questions) as well as the *method of data collection* (face-to-face, telephone, or mail) often strongly influence the results of the survey.

- *Constraints on the practical length* of a survey questionnaire exist: if too long, the respondents might lose interest and thus become less cooperative (face-to-face interviews can usually be longer than surveys by mail; those by telephone should be short).

6. **Longitudinal and Cross-Sectional Studies**

   — *Developmental studies* use the related longitudinal and cross- sectional techniques.

   - These represent different approaches for determining changes that take place over *time* in association with *age*.

   - Typically they are employed in *descriptive research*, but they also can be used for *experimental research*.

   — Longitudinal technique

   - In the longitudinal technique, a group of individuals is measured *repeatedly over time* at selected intervals, in order to determine changes in specific characteristics (e.g., cognitive skills) that occur as they *mature/age*.

     ∞ In addition, different *types of individuals* can be compared over time.

   — Cross-sectional technique

   - In the cross-sectional technique, individuals in *different age groups* are measured at the *same point in time*, and the differences found across the samples are *assumed* to reflect changes that would occur if just *one group* of individuals were observed *longitudinally* (over time) as they age.

     ∞ In addition, different *types of individuals* can be compared across ages.

   — Deficiencies

   - *Results* generated by the two techniques *aren't always the same.*

   - Both techniques contain *serious confoundings* — which can explain why the results of the two techniques might be different.

∞ *Cross-sectional studies* are quicker, but they are less direct and they involve *assumptions of group equivalence* that might be wrong.

- Example: The sample groups could differ not only in age, but also in *generational experiences*, such as the educational opportunities that were available when they grew up.

∞ *Longitudinal studies* similarly confound changes in *age* with changes that occur over time in the *world*.

- Example: *Attitudes toward sex and race* have been changing, and will likely continue to change.

7. Correlational Study

— In correlational studies, observations are made on a number of individuals to determine the *values* of one variable (e.g., amount of alcohol consumption) *associated* with the values of another variable (e.g., frequency of violent behavior) to *describe* the degree and form of *relationship between these variables*.

- Correlations (*r*) can be *linear or nonlinear* (a straight versus curved line when graphed), and either *positive or negative*, with a *correlation coefficient* that can range from +1 through 0 to –1. (Correlations are also discussed in Chapter 1 under Description, and in Chapter 11, under Statistics.)

∞ A coefficient of +1 is a perfect *positive/direct* relationship, where higher scores on one variable are associated with higher scores on the other variable; –1 is a perfect *negative/inverse* relationship, where higher scores on one variable are associated with lower scores on the other; and 0 indicates that there is no relationship at all, just *randomness*.

∞ Example: A *negative correlation* has been found between the *grade-point averages* of college students and the *amount of tobacco* they smoke — those smoking more have lower grades.

∞ Note: It is *very unlikely* that correlations will be a *perfect –1 or +1* in any given instance, since the variation (variance) of any particular factor/variable would very rarely be entirely accounted for by the variation of only *one* other factor — instead, phenomena typically have *multiple determinants*.

- Correlational studies can be conducted either in the *laboratory* or in the *field*.

— Strengths

- *More formal and methodical means of description* are provided by correlational studies than by the other naturalistic techniques used to determine relationships; hence the *data and systematized knowledge* are *more precise and complete*.

- *More accurate predictions* are thus possible from the descriptive information provided about the relationship between variables.

- *More credible hypotheses about causality,* which might lead to *explanation and control*, can therefore be *inferred* from correlational data; these hypotheses can later be tested using the more powerful approach of *experimental observation*.

- IN SUMMARY: The *potential contributions* that can be made by *all* of the techniques for naturalistic observation are *particularly strong* for the correlational study technique.
— Weakness/deficiency
  - *Understanding* at the level of *explanation*, and thus *control*, typically is *not achieved* by this technique; i.e., *causality* usually is *not established* (due to the problems described below) even by high correlations that indicate strong relationships/associations.
    a. Directionality problem
      - Even assuming there were a causal relationship, the correlation would *not indicate* which variable was the *cause* and which was the *effect (A→ B or A ← B)*, or whether the causality was *bidirectional* (A ↔ B), in which case there would be the unanswered question of *preponderant causality.*

        Example of great social and political importance: A positive correlation has been found between the *number of handguns* in a community and the *number of homicides* involving handguns, but are more handguns, in a community responsible for the increase in homicides, or do more homicides lead to more people having handguns for protection? Could not, in fact, the causality be *bidirectional*? Moreover, as discussed next, could not some *third factor*, such as socioeconomic class, be responsible in part for both?

      - *Logic* and *timing* can sometimes help determine the direction of a *possible* causal relationship.

        Example: Given a positive correlation between *smoking and lung cancer*, it is highly unlikely that lung cancer could be the cause of smoking — especially because smoking invariably *precedes* the incidence of lung cancer, and *causes must precede effects.*

    b. Third variable problem
      - There might be *no causal relationship* between the two variables: Instead they might both be under the *influence of some third factor* (A ← C → B), whose variation causes their variation, and hence is the cause of their association/relationship. (See also Ex Post Facto Study, next.)

        Example: The positive correlation between *watching violent television shows* and displaying *aggressive behavior* could be due to a third variable, such as genetic brain disfunction or child abuse, that in turn increases the probability of both personal and vicarious (watching it on television) aggression.

    c. Coincidence problem
      - Phenomena that *covary* (i.e., vary together) might do so only by *chance*, or, in other words, just by coincidence.

- This is a problem also for *experiments*, but is less likely due to greater control of variables in experimental research.

    Example: Could not the fact that historically the *stock market* goes up when the *hemlines of skirts* go up — and vice versa — just be a coincidence? Even if there were some causality, wouldn't there still be the problems of directionality and third variables?

d. *Post hoc* error
    - "After this, therefore because of this" (*post hoc ergo propter hoc*) is an *assumption* that can be fallacious.
    - It represents the connected dangers of inferring *causality* and *directionality* using the *logic of timing*.
    - This error is a function of the *third variable* and *coincidence problems* associated with correlational data.

        Example: *Thunder follows lightning*, and the intensity of thunder is positively correlated with the intensity of the preceding lightning. Nevertheless, thunder is not caused by lightning — they are both caused by a *third variable*, i.e., electrical discharges in clouds.

8. Ex Post Facto Study

— An ex post facto study is little more than a *correlational study* that *resembles an experiment*, and thus it is sometimes *confused* with experimentation.

- Usually just a *few values* (two or three) of one of the variables are investigated, as is the case in *experiments,* rather than many values being studied, as is done in most *correlational research.*

- However, the variable of interest is not directly manipulated, as it would be in an experiment, but rather its values are *chosen after the fact (ex post facto)* by what is called *measured-selection manipulation*, which is also the case for correlational studies.

    ∞ Specifically, two or more groups of participants are *selected* such that *when measured they already differ* on a variable of interest, and then some other characteristic or behavior is observed in order to determine if there is an associated difference on that variable.

    - In effect, the research participants *assign themselves to the different conditions*, rather than the experimenter assigning them.

        Example: A study found that *managers earn less money* if they happen to have *wives who also work.*

    - This *lack of control* leads to *problems* regarding *causality* (see below under Deficiency).

— Note: The ex post facto technique is often *combined* in studies with the use of *experimental manipulation* for one or more other variables, and such combinations can add to the confusion over descriptive versus experimental research (see later under *Semi-Experiments*).

— Deficiency

- *Causality is usually not established* by this research method

    a. Directionality problem (as for correlational studies)

    - Example: In a news brief titled "Too Much Television is Hazardous to Your Health," *Vitality* magazine in July 1994 reported that "A study of 4,280 young people by Kaiser Permanente found a link between *watching television* and *obesity, hostility, depression and poor health habits.* The study found that those who watched television 4 or more hours a day were twice as likely as those who watched for 1 hour or less per day to be sedentary, to be cigarette smokers, and to score high on hostility tests. They also were 71% more likely to be obese and 54% more likely to be depressed. The issue is whether watching TV for several hours a day is *the cause* of these problems, or *the effect.*

    - Note that in ex post facto studies, as is true for correlational studies, *logic and timing* might help to determine the direction of a *possible causal relationship,* but we must watch out for possible *post hoc errors.*

        Example: If there were a relationship between one's *genetic sex* and the *need for affiliation* (being with others), then any causality between the two could only be in one direction, given the *arrow of time.*

    b. Third variable problem (as for correlational studies)

    - There might be *no causal relationship* between the variables of interest: Participants with different values of one characteristic *could also differ in other ways,* and one or more of these might be an *extraneous variable* that is the *true cause* for observed differences in the behavior(s) studied (as noted earlier, this is *confounding*).

    - The reason for this problem is that, in naturalistic observation, the investigator does not *directly manipulate* variables, and does not assign the different values to the research participants in such a way as to ensure that *in all other ways* the participants are likely to be *equivalent.*

        Example: In a news brief titled "Secondhand Smoke Affects IQs," *Vitality* magazine in July 1994 reported that "*Children whose mothers smoked during pregnancy have significantly lower IQ scores than children of nonsmokers,* according to researchers at Cornell University and the University of Rochester in New York. Tobacco smoke contains 4,000 chemical components that may damage the nervous system of a developing fetus. Smoking may also reduce the flow of oxygen and nutrients to the fetus."

        But in such an ex post facto study it would be necessary to take into account, as the investigators did, a *wide range of interrelated third variables* — such as parents' age, diet, drug use, education, and IQ. Interestingly, when they did so, they still found an effect of four IQ points, which is similar to the effects of moderate levels of lead exposure on children.

   c. Coincidence problem (as for correlational studies)
- Any apparent association between variables might be just a *chance correspondence*.

Experimental Observation Used in Explanatory/Experimental Research

Experimental observation is a much more effective research approach than naturalistic observation used in descriptive research for determining *causality*, and thus for achieving the scientific goals of *understanding* at the level of *explanation*, as well as *control* and *systematized knowledge*.

1. **Characteristics of Experimental Observation**
   - *Manipulation of variables* (either in the external or internal environment) in a *systematic, purposeful,* and *direct fashion* is the *cardinal feature* of experiments, i.e., it distinguishes them from naturalistic observation used in descriptive research.
     - Because variables are under the *direct influence* of the researcher, critical events occur at a *specified time and place*.
       - ∞ Thus the investigator is *fully prepared* for precise observation and measurement of the *effects* of the manipulated variables.
       - ∞ Therefore, researchers are also assured of the opportunity to *test repeatability* for *verification* and for determining the *degree of reliability* of the results.
   - *Control of variables* in the external and internal environment is another critical feature of experiments.
     - While certain key factors are controlled by systematically varying/manipulating them in a purposeful and direct manner (noted above), *extraneous factors* — and this is crucial — are controlled by ensuring that they are *made equivalent*, or at least *minimized*, so that they do not *confound* (get confused with) or *mask* (hide) the *measured effects* of the manipulated variables.
       - ∞ *Control techniques* for these *unwanted, extraneous factors* include: eliminating, holding constant, balancing, matching, yoking, taking repeated measures, randomizing, pretesting, using control groups or additional treatment groups, and conservative designs (see Chapters 7 and 8).
   - *Explanatory/experimental research* uses this form of observation.

2. Functions of Explanatory/Experimental Research

— Determination of *causality* — the scientific goal of *understanding* at the level of *explanation*, which might lead to the goal of *control* — is the *primary function* of experimental research (in addition to achieving the goals of *description, prediction, and systematized knowledge*).

- Through the *purposeful and direct manipulation* of one or more factors/ variables, along with the *control* of extraneous factors, it is possible to ascertain with a high degree of confidence just which events *cause certain effects*, and to determine the precise *forms of these relationships*.

    ∞ This discovery of *cause-and-effect relations* is the *essence and primary advantage* of experimental observation, versus the naturalistic observation used in descriptive research.

    ∞ Note, however, that experimentation is rarely if ever perfectly executed, and thus *mistakes* can occur when claiming a *causal relationship*.

    - *Minimal probability* of this error equals the *significance/confidence level* used in statistical analysis of the data, which equals the *likelihood of declaring an effect* when the obtained differences actually are due only to chance.

    - Other forms of *error* are also possible (covered later in Chapters 7, 8, and 12).

3. Major Variables of an Experiment [See Appendix 1]

There are *three major types of variables* in any experiment; *independent variables* are manipulated to determine effects on *dependent variables* that are measured, while *extraneous variables* are controlled by making them equivalent or minimal (Chapter 4 discusses variables in detail).

a. Dependent variable

- This is the *response of interest*, either an overt behavior or a physiological event, that the researcher *measures, describes*, and tries to *explain*.

    ∞ It is the *output* of the organism — an *activity* of the *participant*.

- Its value is *dependent* (hence the name of this variable) on the *conditions present* — including hopefully the manipulated independent variable(s) of the study (described below).

    ∞ Variation of the dependent variable is the potential *effect* in a *cause-and-effect relationship*.

b. Independent variable

- This is the *stimulus/antecedent condition* that the experimenter purposefully and directly *manipulates or varies* in order to determine its effect (if any) on the dependent variable, which would be a consequent event or result.

    ∞ It is the *input* to the organism — an *activity* of the *investigator*.

- Its value is *independent*, at least relatively so (hence the name of this variable), from the behavior that is measured — the dependent variable(s).

∞ Variation of the independent variable is a potential *cause* in a *cause-and-effect relationship.*

• IN SUMMARY: The *goal* of an experiment is to demonstrate that *independent* manipulated stimulus conditions (*potential causes*) lead to *dependent* measured responses (*actual effects*) (see Chapter 4 for further discussion).

∞ Example: Experiments have been conducted to determine the effects of *aversive stimulation,* such as electric shock (an independent variable), on the frequency of some *undesired behavior,* such as smoking cigarettes (a dependent variable).

c. Extraneous variable

• This is any *additional factor* that is not an independent variable, but which nevertheless might affect the dependent variable; e.g., differences among individuals in genetics or past experiences.

∞ It is an *extra* potential influence in a study.

• It's an *extraneous* (hence its name), *unwanted variable.*

∞ An extraneous variable must be *controlled* (minimized or made equivalent), thus it is also called a *control variable* Chapters 7 and 8, for more information).

• Usually there are *several* extraneous factors, but note that certainly *not everything,* in addition to the independent variable(s) of a study, could affect the dependent variable(s) of interest.

• Note also that the *extraneous variables of one study* may be the *independent variables of other studies,* which is how it would be determined that they can affect certain dependent variables, and thus need to be controlled.

Example: The gender, appearance, and behavior of investigators running studies could be manipulated as *independent variables,* and, if found to affect dependent variable measures, then they would be controlled in other studies as *extraneous variables.*

4. Basic Components of an Experiment with Respect to Its Variables

IN SUMMARY: There are *three basic elements* to every experiment:

a. Purposeful, direct manipulation of *independent variables;*

b. Control of as many as possible *extraneous variables;*

c. Measurement of *dependent variables.*

• Putting it all together: If the aspect of behavior measured, the *dependent variable,* changes significantly for different values of the purposefully and directly manipulated *independent variable,* then it is asserted with confidence that a *causal relationship* exists between them — assuming, of course, that there was proper and adequate control of the *extraneous variables.*

∞ But, as was indicated earlier and will be reiterated later, there is always the *possibility of error* — the conclusions can be wrong.

5. Levels of Manipulation: Experimental versus Naturalistic

The following description of the *three levels of manipulation* that are possible is both a *summary and elaboration* of information already given.

a. Purposeful, direct manipulation

This level of manipulation involves the following:

- *Purposeful determination* by the investigator of what *values* of an *independent variable* will be studied;
  - ∞ This selection of values may be *random*, but it's *intentional*;
- *Direct creation/production* of the chosen values, i.e., *conditions*;
- *Assignment of participants or subjects* by the *investigator* to receive the different values and conditions of the independent variable;
  - ∞ This is frequently done in a *random* manner, but, nevertheless, it's still *intentional*.

- Application
  - ∞ *Explanatory research* involving *experimental observation* employs this level of manipulation in the *control* of *independent variables* (as already discussed) and in the *assignment of participants* to the different independent variable conditions.
    - Example: *Purposeful and direct production* of different schedules of reinforcement administered to different *randomly assigned groups* of research participants.

- Consequences
  - ∞ *Causality can be determined* because variation in the *dependent variable(s)* could most logically be attributed to manipulation of the *independent variable(s)* when their control by purposeful, direct manipulation is combined with the *control* of other, undesired, *extraneous variables* (including participant, experimenter, and environmental), which might also have effects in a study.

b. Purposeful, measured-selection manipulation

This is a *lower level of manipulation*, and involves the following:

- *Purposeful determination* by the investigator of what *values* of a variable will be studied (as for the preceding level);
  - ∞ This selecting of values may be *random*, but it's *intentional*;
- *Selection of measured values* from among *already available existing ones*, by *choosing individuals* who *already possess* the desired values of the *participant variable* (rather than manipulation through direct creation);
- *Self-determination by participants*, who in effect *assign themselves* to values/ conditions based on their own characteristics or experiences.

- Application

- ∞ *Descriptive research* involving *naturalistic observation* (e.g., correlational and ex post facto studies) often employs this level of manipulation for *participant-characteristic variables*.
    - Example: Studies of differences in language and math abilities for *males versus females*, or for those who have attended *public versus private* schools.

- Consequences
    - ∞ Causality usually *cannot* be determined for the following reasons:
        - *Independent variables* aren't controlled by direct manipulation. Thus the *direction of effect* usually cannot be determined, even if there is a causal relation, given *timing uncertainties* (i.e., not knowing which are the *antecedent versus consequent events*).
        - *Participants* with their extraneous variables are not as controlled as with the preceding level of manipulation, although matching participants on *some* characteristics as a method of control is possible.
        - *Extraneous variables* of other sorts (e.g., those in the environment) also are usually not as well controlled in research that uses measured-selection manipulation of independent variables.
        - Hence, extraneous variables (participant as well as other factors) might *systematically vary* along with the values of an independent variable, and hence their effects on a dependent variable would be *inseparable* and thus *confused* with those of the independent variable. *Confounding* is what this is called in research.

            Example: When studying the relation *of gender* to such things as aggressiveness or promiscuity, the different *socialization experiences* of males versus females would ordinarily not be controlled, and hence these — rather than the *biological distinctions* of the sexes — could be the cause of any differences found in their behaviors.

c. No manipulation

This *lowest of levels* involves the following:

- *No purposeful determination* of what *values* of a variable will be studied;
- *No manipulation* of the values, *either directly or by selection;*
- *Self-determination* of values *by participants* who just happen to exhibit or experience them.

- Application
    - ∞ *Descriptive research* involving *naturalistic observation* of the *purest form* (e.g., natural-environment and artificial natural- environment studies, as well as many correlational studies) employs this level of manipulation.
        - *Observational research* is the name commonly given to this descriptive approach involving *no manipulation*.

- • Example: Carefully observing and recording the various behaviors of a group of chimpanzees as the events *naturally* take place in the chimps' home environment.
- • Consequences
  - ∞ Causality is *least likely* to be determined, since there is the *least control* in such research having *no manipulation* of variables.

6. Semi-Experiments

— Different levels, or forms, of manipulation can be *combined* in a single study when more than one independent variable is included.

- • *Experimental independent variables* would have their values *purposely and directly manipulated.*
- • *Correlational independent variables,* in contrast, would have their values *simply selected* from among those already available naturally, e.g., *participant variables* such as their sex, age, or IQ.
  - ∞ Note: Correlational independent variables (e.g., education level) are often really just *"extraneous variables"* that are being *controlled* by measured-selection manipulation, balancing, and analysis of their effects in order to remove their variance from the statistical error term — thereby increasing the *power/sensitivity* of the design and analysis to find an effect of the experimental independent variable(s) (discussed in Chapter 8, under Treatment Groups and Systematizing Extraneous Variables).
    - • This is in *contrast* to extraneous variables that are controlled by other techniques, e.g., by elimination, by being held constant, or by randomization.

— An important *reason*, in addition to control of extraneous variables, for designs to employ *correlational* as well as experimental methods is to determine how *different types of individuals* respond to the same purposely and directly manipulated (i.e., experimental) independent variable(s).

- • Example: For *males versus females,* how might the *need for affiliation* be influenced by a *threatening/stressful situation?*

— *Quasi-experiment* and *mixed design* are terms that have been used to refer to these kinds of studies, but they should not be used since these terms have other, more common, usages (covered in Part Three).

— *Experiment* is the term that in fact is most often used, so long as at least *one* independent variable is *purposely and directly manipulated,* regardless how the other independent variables are manipulated.

- • Unfortunately, this terminology can increase the likelihood of *misinterpreting* the data with respect to evidence of causality.
  - ∞ Only for *experimental independent variables,* not for correlational independent variables, can there be *confidence* regarding the determination of *cause-and-effect relationships.*

— *Semi-experiment* is a term that, if used, would help avoid confusion.

7. **Criticism of the Experimental Method**

   — Experimentation is sometimes *criticized* because when events (variables/factors) are brought into the *laboratory* for controlled, experimental study, it is felt that their nature is *changed*.

   - Events do not naturally occur in *isolation* in a *sterile environment*.
     - ∞ Therefore, what is studied is *unnatural*, artificial, and not relevant.
       - This problem relates to the *interactions* that may occur among independent and extraneous variables, and hence the question of *generalizability* of results to different situations (see Chapters 10 and 11).

   — *Artificiality and non-relevance* criticism such as this isn't entirely fair.

   - When attempting to determine *causality*, it is necessary *first* to learn what the effect of one or more variables is on some other variable *without* interference from *uncontrolled factors*; thus the *weakness* of laboratory experiments flows from what needs to be done to determine causality, which in fact is the technique's *strength*.
   - Once all of the relevant independent variables have been studied in *isolation* and in *specific controlled combinations*, and their influences on some dependent variable and on each other are determined, then there can be an *understanding* of the "natural event" that would not be possible without *experimental control*.

   — *Generalizability* of laboratory findings to the "real world" of multiple interacting influences is, however, a *legitimate and crucial concern*.

   - *Differences* between laboratory settings and natural settings must be *minimized* in order to maximize the generalizability of results.
   - *Increased interest* exists today in conducting experiments *in the field*, i.e., the "real world," in order to make them more relevant.
   - *Field experiments* have, in fact, confirmed the *generalizability* of laboratory research — hence *blunting criticism* of experimentation.
   - Note that field experiments should involve not only *natural settings*, but also *natural treatments* and *natural behaviors*, i.e., all variables should be defined (operationally) in *real-world* terms.

8. **Interrelationship of Experimental and Naturalistic Observation**

   — Experimental and naturalistic approaches are *complementary* in terms of their *relative strengths and weaknesses*.

   - Predictions that are generalizations from *well-controlled laboratory experiments* can be tested in *less-controlled naturalistic studies*, which, however, involve more normal/typical conditions — and the reverse sequence can be done as well.
     - ∞ Example: For a number of years *naturalistic observations* in the field have been integrated with field and laboratory *experiments* to unravel the mystery of *how bees communicate* to each other at the hive both the direction and distance, as well as other features, of a food source that has

been found. What has been learned through all this work is that the bees actually communicate through a dance that has special characteristics which signal the information.

Review Summary

1. *Naturalistic observation* is used in descriptive research.

   It is the observation of events as they occur *naturally,* i.e., independent of the observer's behavior. No attempt is made to produce, influence, or control events by direct manipulation of variables, which is in contrast to experimental observation.

2. *Descriptive research,* using naturalistic observation, attempts to *identify* the variables present and their values, and sometimes to determine the *relationships* (correlations/associations) that exist between variables.

3. *Reasons* for conducting descriptive research are as follows:

   a. *No alternative* might exist to naturalistic observation, since it might be impossible to influence the phenomena of interest by direct manipulation of variables, and thus to study them experimentally.

   b. *Unethical behavior* might be involved if certain factors were experimentally manipulated in order to determine their effects.

   c. *It's advantageous* in the *early stages* of investigation usually to conduct descriptive research *prior to experimental manipulation,* since it permits: 1) reduction of errors, 2) identification of phenomena that require explanation, 3) prediction of behavior, and 4) development of inferences about causality.

4. The *weakness* of naturalistic observation is that there are usually many *simultaneously occurring and varying antecedents* to behaviors and other phenomena, and these antecedents are *confounded,* i.e., inextricably mixed together, so that it is difficult if not impossible to determine *causality.*

5. There are eight or more different *techniques* of naturalistic observation used in descriptive research: 1) natural-environment study, 2) contrived natural-environment study, 3) case study, 4) secondary record study, 5) survey study, 6) longitudinal and cross-sectional studies, 7) correlational study, and 8) ex post facto study. These techniques have different advantages and applications.

   The latter two techniques are sometimes confused with *experimentation,* but the variable of interest is not directly manipulated, as it would be in an experiment, but rather its values are chosen *after the fact* by what is called *measured-selection manipulation:*

   a. *Correlational studies* consist of observations that determine the *values* of one variable that are *associated* with those of another variable in order to *describe* the degree and form of the *relationship between them.* However, *causality* (explanation at the level of understanding) is not usually established due to the *directionality problem* and the *third variable problem*, as well as the *coincidence problem*. This is also true for the following type of descriptive research.

    b. *Ex post facto studies* are little more than *correlational studies* that resemble experiments, which is because usually just a *few values* of one of the variables are investigated, as is done in experiments, rather than many values, as in most correlational research.

6. *Post hoc error* is the assumption, when wrong, that "after this, therefore because of this." It represents the connected dangers of inferring *causality* and *directionality* using the *logic of timing*. This error is a function of the *third variable* and *coincidence problems* associated with correlational data.

7. *Experimental observation* is used in explanatory/experimental research. It is a much more effective research approach for determining *causality* (cause-and-effect relationships), and thus for achieving *control* over behavior, than are the techniques of naturalistic observation used in descriptive research.

8. *Explanatory/experimental research* is characterized by:

    a. *Purposeful and direct manipulation* of antecedent conditions, called *independent variables*, in order to determine their effects;

    b. *Control* of other, unwanted factors, called *extraneous variables*, by keeping them *equivalent* or *minimizing* them;

    c. *Measurement* of responses, called *dependent variables*, in order to determine the effects, if any, of independent variable manipulations.

9. There are three *levels of manipulation* used in research:

    a. *Purposeful, direct manipulation* of independent variables and assignment of participants to conditions, as done in experimentation;

    b. *Purposeful, measured-selection manipulation* of variables with self- assignment of participants to conditions, as done in descriptive research using naturalistic observation techniques, e.g., correlational and ex post facto studies;

    c. *No manipulation* of variables, as done in descriptive research involving naturalistic observation of the purest form, e.g., natural-environment and artificial natural-environment studies.

10. A *semi-experiment* is what one might call a study that combines the purposeful, direct manipulation of one or more *experimental* independent variables with the purposeful, measured-selection manipulation of one or more *correlational* independent variables (such as gender or age).

    This can be used to investigate how *different types of participants* respond to the same experimentally manipulated independent variable, as well as to *control* for extraneous participant factors that are not eliminated or held constant.

11. A common *criticism of the experimental method* is that when events (variables/factors) are brought into the *laboratory* for controlled, experimental study, their nature is *changed*. Events do not naturally occur in *isolation* in a *sterile environment*. Therefore, it is said that what is studied is *unnatural* and thus *irrelevant*.

Although *generalizability* of laboratory findings to the "real world" is a legitimate and crucial concern, the criticism is not entirely fair. This is because the weakness stems from the *strength* of the experimental approach, which involves the *control of extraneous variables* so that the effects of purposeful, direct independent variable manipulations can be determined with confidence. It should also be noted that experiments can be done in the *field*, i.e., the "real world," to make them more relevant. Moreover, the results of well-controlled laboratory experiments can be confirmed by less-controlled naturalistic studies done in the "real world," and the reverse is also possible.

Review Questions

1. Define *naturalistic observation*.

2. State the *functions* of *descriptive research* using naturalistic observation.

3. What are three *reasons* for conducting descriptive research (these should be different than the functions noted above; be sure to explain your answers)?

4. What is the *weakness* of naturalistic observation?

5. List eight different techniques of naturalistic observation used in descriptive research, and describe the two that are most similar to, and sometimes confused with, experimentation. Also, explain how the variable of interest is *manipulated* in the latter two types of descriptive research. Finally, give the reasons that these two typically do not establish *causality*, i.e., describe the *problems*.

6. Describe and explain *post hoc error*.

7. What are the two advantages/strengths of *experimental observation* compared to naturalistic observation?

8. List and describe the three *basic components of an experiment* with respect to its variables.

9. List and describe the three *levels of manipulation* used in research, and indicate which are involved in experimental versus naturalistic observation.

10. Describe a *semi-experiment*, distinguish between its types of independent variables, and explain why such research is done.

11. State the common *criticism of the experimental method*, and discuss why it is not entirely fair.

3

Steps of the Scientific Method

CONTENTS

Introduction

This chapter begins with a brief coverage of all the *primary and secondary steps of the scientific method*, and then examines Primary Steps 1 and 2 in detail. The other primary steps are elaborated later: Primary Step 3 in Chapters 4, 9, and 10; Steps 4, 5, and 6 in Chapters 11 and 12; and Step 7 in Chapter 5. With regard to the additional steps: Secondary Step 1 is elaborated in Chapter 11, whereas Secondary Steps 2 and 3 were discussed earlier in Chapter 1. The following are some important *general points*:

1. There is a good deal of *overlap and interrelatedness* among the steps.
2. Scientists, therefore, often *move back and forth* between these steps.
3. The steps might be *listed somewhat differently* by different scientists.
   — Such variations, however, just represent *complementary approaches* to the general scientific method.
     - Example*: Inductive method versus deductive method* are two important complementary strategies used in science (covered later in this chapter under Hypothesis Formulation).

Primary Steps

There are *seven major steps* involved in the scientific method of research: 1) formulate a problem, 2) formulate a hypothesis, 3) design a study, 4) collect and organize the data, 5) summarize and statistically analyze the data, 6) evaluate the results and draw conclusions regarding the hypothesis, and then 7) communicate the findings.

1. **Formulate a Problem**

   — The *starting point* of any scientific inquiry is a *research problem/question* about the nature/description of, or the explanation for, some phenomenon.

     • Example: What is the *cause* of clinical depression?

2. **Formulate a Hypothesis**

   — Next, a *hypothesis* is put forth as the possible *answer/solution* to the problem's question.

     • Typically, this is a statement about the *conjectured relationship* among the factors that will be investigated (the independent and dependent variables).

       ∞ Example: A *deficiency* in the serotonin neurotransmitter mechanisms of the brain is the cause of clinical depression.

3. **Design a Study**

   — The *design* is a *plan* for gathering data to *test the hypothesis*.

     • Scientists must determine whether the hypothesis is *probably true or probably false.*

       ∞ In other words, does the hypothesis appear to *solve the problem* — does it *answer the research question?*

     • *Experimental designs* are the most powerful designs for research (discussed earlier in Chapter 2).

4. **Collect and Organize the Data**

   — *Empirical observations/data* are the primary criteria for testing hypotheses (and theories); however, *thought experiments* — involving *rationalism* as an approach to knowledge — are also sometimes used.

     • Example: *Einstein* used thought experiments in working out and evaluating (i.e., "testing") his *Theories of Relativity.*

5. **Summarize and Statistically Analyze the Data**

   — *Descriptive statistics* are used to *summarize* the data.

     • Examples: *Measures of central tendency* — mean, median, and mode; *measures of variability* — standard deviation and variance; and *measures of relationship/association* — correlation coefficient (covered in Chapter 11, under "Statistics Versus Parameters").

   — *Inferential statistics* are used to *analyze* the data.

- Examples: chi-square test, *t*-test, and analysis of variance.
    - ∞ Such *tests of significance* are used to determine the *reliability*, not the magnitude, of the findings; specifically, to get a measure of the *probability* that the *association(s)* found between variables are only due to *chance* (tests of reliability versus magnitude are covered in Chapter 12, under "Statistical Significance Versus Practical Significance").

6. **Evaluate the Results and Draw Conclusions Regarding the Hypothesis**
    — This step involves the processes of *interpretation* and *inference*.
    - Data are *compared* to the hypothesis, which is a *prediction* of the results, to determine whether the hypothesis is or is not *supported* by the empirical evidence — and hence whether the hypothesis is *probably true or probably false*.
        - ∞ In experiments this can lead to *explanation*, i.e., specification of *causal relationships* among variables/phenomena.

7. **Communicate the Findings**
    — Normally this last step involves the writing and publication of a *research report*.
    — Instead or additionally, communication can involve an *oral or poster presentation* at a convention/conference.
    — Communication increases the *utilization and impact* of the results.
    — It also allows *independent* evaluation and verification of findings.
    - This possibility is a *requirement* of scientific observation, i.e., the observations must be *public*.
    — Although communication represents the *final major step* of the scientific method, it also might represent a *beginning*.
    - Research reports serve as an important source of *new research problems and hypotheses* in the *ongoing process* of science.

Secondary Steps

In addition to the seven primary steps of the scientific method, there are *three closely related ancillary steps*: 1) generalization, 2) prediction, and 3) replication.

1. **Generalization**
    — This secondary step is *closely allied* with the primary step of *drawing conclusions*.
    — Scientists want to *extend* a study's findings from the *specific* set of *conditions* and *participants* used, i.e., they want to make as *general* a statement as possible about the *implications* of their results.
    - However, they must be careful not to generalize *too broadly* since this could lead to *error*.

∞ Example: Generalizing research results obtained with a *sample* of U.S. college students to the *population* of all humans on the planet could be a mistake, since the sample might not be *representative* of the diverse world population (see Chapter 11 for an extended discussion).

2. Prediction

— Predicting is *closely related* to the secondary step of *generalization*.

   • Generalizations, in fact, are actually a form of *prediction*.

— Scientific predictions often refer to the *proposed results* of a test of some prior *hypothesis/theory*, or a modification of it, under *new conditions* and/or involving *different types of participants/subjects*.

   • Example: Proposing that the research results found for *rats* under *laboratory conditions* would also be found for the human species in our *complex environments*.

—   This brings the scientist back near the beginning of the scientific method.

   • Note: A significant piece of research, through generalization and creative thinking, will not only *generate new questions/problems and predictions/hypotheses*, but it will provide *theoretical implications* relevant for explaining *other phenomena* occurring in analogous situations.

3. Replication

— Repeating a study involves a special form of *prediction*.

   • In this case, a study is *repeated* as it was *originally* conducted, in order to see if the results can be *reproduced*, and thus *confirmed/verified* — which would be the logical *prediction*.

— *Generalization* is typically involved to some extent since, due to practical limitations, *replication is rarely complete or precise*.

— *Partial replication*, rather than "complete" replication, is said to occur when something *other than* just the participants, experimenter, or location is *changed*, and/or when anything is *added or deleted*.

Problem Formulation (Step One of the Scientific Method)[26]

Developing a meaningful research problem is the first step in conducting scientific research (*meaningfulness* is discussed under items 5 and 6 below).

1. Best Way to State a Research Problem

— Usually it is *clearest* when a problem is put in the form of a *question*.

   • Generally, research questions ask *how or why* something occurs, and specifically they ask *what the relationship* is between two or more variables — usually *independent and dependent variables*.

- However, in descriptive research the question might simply ask *what are the properties* (i.e., characteristics and their values) of some phenomenon.

2. **Ways to Find a Research Problem**

    There are *three common approaches* to coming up with research problems to study:

    a. Be alert to *personal experiences* and *newspaper or magazine reports*.

    - *Everyday events* at work, school, home, athletic competitions, etc., can present practical issues that need solutions — so be *inquisitive*, develop a *questioning attitude*.

        ∞ Example: What are the *motivations* for some unusual behavior(s) you have observed among fraternity members?

    b. Read *journals and books*

    - *Previous research and theories* raise many new questions that need answers.

        ∞ Example: Given the *Theory of Evolution by Natural Selection,* how can *altruistic/helping behavior* — which might lead to the *death* of the helpful individual — be explained?

    c. Attend *conferences*, take *courses*, and have *discussions*

    - *Keeping current* in your knowledge and *tapping the minds of others* can lead to interesting ideas for research.

3. **Ways in Which Research Problems Become Evident**

    There are *three types of situations* that make us aware of research problems that need to be addressed: a) contradictory results of studies, b) a gap in knowledge, or c) a fact that needs explaining.

    a. Contradictory results of studies

    - Two or more studies attempting to answer the *same question* might obtain *different results* and thus *different answers*.

    - This naturally raises questions as to *why* this has happened.

        ∞ Example: Investigations over a number of years have found opposite results for the effects of *heightened sexual arousal* on the level of *subsequent aggression* (sometimes finding that it increases it, and in other cases that it decreases it).

    - Possible reasons for contradictions

        1) *Extraneous variables* might have been *inadequately controlled* in one or more experiments, thereby causing different results.

            - Probably the *most common cause* (see Chapters 7 and 8 for coverage of extraneous variable types and their experimental control).

            - As a result, one or more of the studies might have been *poorly conducted*, and thus the results would be questionable, or the *value* of one or more extraneous variables simply might have *varied* between the studies.

            - *Additional research* is usually needed to determine the cause of contradictions.

Example: In a study it was found that "Right handers *moved their eyes* leftward when solving spatial problems and rightward for verbal problems *when the questioner sat behind them*. But, *when facing the questioner*, the same participants moved their eyes predominantly in only one direction, either right or left, regardless of problem type. The *results indicate* that the cerebral hemispheres, though specialized for problem type, are also preferentially activated within an individual."

What is important for us, however, is that this study cleared up the inconsistent results appearing in the literature regarding direction of eye movement in response to problem type. The discrepant results were apparently due to an *extraneous variable* in the procedures used, specifically, *experimenter location*, which varied between the two studies. [Gur, E. R. (1975). Conjugate lateral eye movements as an index of hemispheric activation. *Journal of Personality and Social Psychology*, Vol. 31 (4), 751-757]

2) *Independent variable* values might have been different.

- This could lead to different results in two or more studies if the *relationship* between the independent and dependent variables is *nonlinear*, rather than being linear. It occurs when the value of the dependent variable *does not change by a constant amount* as the independent variable is changed by a constant amount — thus a graph of the relationship would be *curvilinear*, rather than a straight line.

  This is particularly problematic when the relationship is *non-monotonic*, i.e., when the *direction* of the independent variable effect *changes* for different portions of its range/continuum — in which case, as the independent variable increased in value the dependent variable value would first increase and then decrease in value, or vice versa (see figure in Chapter 7 under "Maximizing the Primary Variance").

3) *Dependent variables* might have been different.

- When this occurs, the studies could be *measuring* different things, at least to some degree, or the studies could differ in their *sensitivity or range of possible values*.

  Example: Measuring the *speed* versus the *accuracy* of dependent variable responses is really very different.

4) *Participant populations* sampled might have been different.

- Hence, they could differ in their *responses* to conditions.

  Example: Studying *college students* versus the *general population*.

5) *Designs* might have been different.

- Some designs are *more powerful or sensitive* than others. (Examples are covered in Chapters 9 and 10.)

6) *Statistical analyses* might have been different.
- Here too, some are *more powerful/sensitive* than others. (Examples are covered in Chapter 12.)

7) *Type I or Type II Errors* might have occurred.
- These errors are covered in Chapter 12 under Potential Errors During Statistical Decision Making.

b. Gap in knowledge
- Being aware of *what we do not know,* as a result of what we do know, is another way that research problems become evident.
  - ∞ It is often said that: "The more we *learn,* the more we realize we *don't know.*"
  - ∞ Example: Knowing that *marijuana* has a debilitating effect on *short-term memory* raises questions about what other effects this drug might have on *cognition.*

c. Fact that needs explaining
- Being *aware* of a fact but not understanding *why it is so* is yet another way in which research problems become evident.
  - ∞ A fact needs explaining when it *does not fit in with* — i.e., cannot be *related to* — existing knowledge.
  - ∞ It *demands* explanation, and thus *collection of further data.*
- Problems of this kind might lead to *major discoveries.*
  - ∞ Example: Knowing that *opiates,* such as morphine, relieve pain and are very addicting led to a search for their mechanisms and the eventual mapping of the distribution of opiate receptors in the brain, and subsequently to the even more important discovery that the brain produces its own *endogenous opiates* that regulate pain, mood, and learning.

4. **Basic Types of Problems — Classes of Questions**[24,31]

There are four fundamental types of research problems. Distinctions made here between *proximate versus ultimate questions* help us to *avoid confusion* about the mechanisms/causes of behaviors and mental processes (cognition, emotion, and motivation) and help to *organize and guide research.* However, the proximate and ultimate mechanisms are *interrelated and complement one another* — they are not competing explanations, rather they are *different levels of explanation.*

a. Proximate causal mechanisms (two "how" questions)
1) Immediate causation (stimuli and mechanisms)
  - ∞ What are the *external environmental factors* (e.g., discriminative, sign, or releaser stimuli) and the *internal psychological, physiological, and anatomical factors* (e.g., goals, endocrine, and neural events) that are involved in the short-term regulation (i.e., elicitation, control, and execution) of the behavior or mental process?

    2) Development (ontogeny)

        ∞ What are the *hereditary/genetic/innate factors* and the *environmental/experiential/learning factors* that are involved in assembling the behavior or mental process over longer time spans in the lives of individuals, and what are the relative roles? (Note: This relates to the classic Nature-versus-Nurture Problem.)

  b. Ultimate or distal causal mechanisms (two "why" questions)

    1) Adaptive significance (function)

        ∞ What is the current role of the behavior or mental process in *promoting survival and reproduction*, i.e., what are its advantages for gene replication — How is it adaptive?

    2) Evolutionary history (phylogeny)

        ∞ What factors *influenced the evolution* — and what was the evolution — of the behavior or mental process during the species' history over geological time?

Example: What might the answers be to the four different types of questions above for the phenomenon of *human language*?

5. Restriction on Problems

— Meaningfulness

- To be meaningful, the question put forth by the problem *must be capable of being answered* with the tools *available* to the *scientist*.

    ∞ Thus a *meaningful* problem is a *solvable* problem.

- Present meaningfulness

    ∞ Research problems with present meaningfulness can be solved with the tools *presently available* to the scientist.

        - Only this type of problem/question can be considered for *current* research.

- Potential meaningfulness

    ∞ Research problems with only potential meaningfulness are *not presently meaningful*, but it is considered *possible* that *in the future* the tools necessary to solve them will become available to the scientist (see below).

        - This type of problem/question should, therefore, be placed in a *wait-and-see category*.

6. Meaningless or Presently Meaningless Types of Problems

There are *four reasons* for problems *not* being presently meaningful: 1) there is no empirical reference, 2) it is presently impossible to obtain relevant data, 3) the problem is unstructured and vaguely stated, or 4) the terms are inadequately defined.

  a. No empirical reference

- For such problems it is impossible to obtain *relevant objective data* in order to solve them, hence they are *meaningless*.

- ∞ These problems do not deal with properties *observable* by the ordinary senses of human beings, even using instruments.
    - Example: "What is the *mind of God* like?"
- ∞ Hence such problems are *not* subject to scientific investigation.
- These problems *only can be solved* by divine vision, revealed truth, or similar mystical power; by intuition; or by reasoning, i.e., *rationalism*.
    - ∞ Such problems, therefore, are described as being not scientific, but rather *metaphysical, mystical, theological, or superempirical*.

b. Presently impossible to obtain relevant data

- Although not presently solvable, such problems *could have an empirical reference*, and thus be at least *potentially meaningful* for the future, after advances in research capabilities.
    - ∞ Example: "What is the biochemistry of life forms *outside our solar system*, if there are any, and what rules govern the behavior of these life forms?" Although we *cannot presently* gather the necessary data, advances in space travel or communication *might* allow this in *the future*.

c. Unstructured and vaguely stated

- In such problems, it is *not clear* what the *relevant variables/ events* are and thus what the *appropriate observations* would be.
    - ∞ Example: "How does the *universe* operate?"
    - ∞ Such questions are *too general*, and thus it is *uncertain* what is meant by them — making the problems *meaningless*.
    - ∞ Note, however, that it might be possible to *reformulate* the question so that it is *more precisely stated*, and thus *more specific*, and hence *answerable*.
        - Example: "How do stars form and how does *life* evolve?"
    - ∞ *Experimental problems* should be *clearly stated* in terms of specific, observable, *independent and dependent variables*.
        - Example: "What are the effects of a *stimulus-enriched environment* on the *structure and functioning of neurons* in the brain?"

d. Terms inadequately defined

- Poorly defined terms or variables are a source of *vagueness*, hence this cause for problems being meaninglessness is really a *subset* of the preceding reason.
    - ∞ Once again, it is unclear exactly what the necessary *relevant observations* would be.
- *Clear specification* is needed of the meaning of the problem's important terms if productive research is to be conducted.
- Major cause for the *ambiguity* of terms is that our everyday language is full of *words with multiple definitions*.
    - ∞ Examples: Without precise definitions, how can we study the possible causes of *love,* or the effects of *pornography?*

∞ To make concepts clear and unambiguous, science uses *operational definitions*, which give terms/variables an *empirical basis* (discussed later near the end of this chapter).

7. **Other Considerations Regarding Research Problems**

   a. Interest

   • Is the problem *intriguing enough* to merit consideration?

   ∞ Problems are said to be *uninteresting* if the answer to the question put for by the problem is *obvious*.

   • Caution: The obvious is not always the *actual outcome* of carefully controlled research — *science has its surprises*.

   Example: In studies on the increase in the number of neurons that occurs during the development of the nervous system, who would ever have expected that *before birth half the neurons in the brain actually die?*

   b. Importance

   • Is solving the problem *worth the time and effort*?

   ∞ Caution: Experiments on what *appears to be a trivial problem* could produce results that are *very important*.

   • Example: Studying the *behavior of bears and coyotes* has led to non-harmful ways of controlling the undesirable behaviors of harming humans or killing livestock. One approach developed is to eliminate open garbage dumps, and another is to produce conditioned taste aversion. The psychologist John Garcia has done much useful *basic and applied research* in this area.

   c. Expense

   • Problems might be quite important and interesting but *too costly*.

   ∞ Example: Are there *other intelligent life forms* in the universe, and just what are their *intellectual abilities?*

   d. Basic/pure research Versus applied research

   • Are either basic or applied research problems *more crucial*?

   ∞ Basic research

   • This category of research investigates problems relating to *fundamental questions* about the principles and mechanisms of various phenomena.

   Example: What are the anatomical and biochemical mechanisms in the brain of *learning and memory?*

   ∞ Applied research

   • This category of research investigates problems whose solutions are expected to have *immediate application*.

   Example: What events in the brain are responsible for the memory loss occurring in *Alzheimer's disease?*

- Research on both types of problems is essential.
  - ∞ *Basic research* establishes the foundation for applied research.
  - ∞ *Applied research*, in turn, can improve the quality of life of humans and other species, and also raises new questions for basic research.
  - ∞ It would be *very shortsighted* for society to concentrate on applied research at the expense of basic research, since applied research is *dependent* on basic research.

e. Chance of contributing new knowledge

- Some problems are more likely to *advance science* than others.
  - ∞ All things being equal, the scientist should *focus* on those problems that are most likely to yield *productive outcomes*.
    - Example: When a problem develops from *studies yielding contradictory results*, to conduct *just another experiment* wouldn't seem worthwhile if the only outcome would be to *score another point* for one side in the controversy.

      To be of *real value*, the new study would have to use an *innovative, fresh approach*, one that would be likely *to explain* the contradictory results, and thus *more thoroughly* answer the research question.

      In other words, it is important to know what, exactly, were the *critical differences* in the earlier studies that *account for the contradictory findings*.

      More specifically, were different studies investigating the effects of different ranges of an *independent variable*, or measuring somewhat different *dependent variables*, or were *extraneous conditions* different?

Hypothesis Formulation (Step Two of the Scientific Method)[26]

Having developed a research problem that is expected to be solvable, and thus meaningful, the next step is to propose a *meaningful hypothesis*.

1. Definitions of Hypotheses and a Restriction

a. Hypothesis

- Broadly speaking, a hypothesis is a *potential answer* to the question put forth by a problem.
  - ∞ It is a *tentative solution* to the problem posed by a scientist.

b. Meaningful hypothesis

- A *meaningful* hypothesis is a *testable* hypothesis.

∞ It is a *testable proposition* that might be a problem's solution.

- As was noted for problems, *scientific hypotheses* also should be *restricted* to those that are meaningful.

- It must be possible, in other words, to determine a *degree of probability* (there are few absolutes) as to whether a hypothesis is true or false — and thereby to *reject or fail to reject* the hypothesis.

- Note: A *problem* is considered meaningful (solvable) if there is a meaningful (testable) *hypothesis* that is relevant to it — hence this is a very useful way of *checking* on whether a *problem* is in fact *meaningful*.

c. Meaningful experimental hypothesis

- This type of hypothesis is an *experimentally testable* statement of a *potential relationship* between *independent and dependent variables* that might be the/ a solution to a research problem.

∞ It must be testable *empirically* through the *purposeful, direct manipulation* of antecedent conditions, and the *measurement* of consequent conditions, under *controlled* conditions.

∞ Note that the definition of a meaningful *experimental* hypothesis incorporates the preceding definition of a *meaningful* (i.e., testable) hypothesis, which in turn incorporates the basic, general definition of a *hypothesis*.

2. Best Ways to Write Hypotheses

a. Logical form of the general implication

- This manner of writing a hypothesis is what is commonly called a *conditional statement*: "If ____, then ____."

∞ Generally put: *If* certain *antecedent conditions* occur, i.e., some specified *manipulation* of an *independent variable(s)*; *then* certain *consequent conditions* will occur, i.e., some specified *measured value(s)* of the *dependent variable(s)*.

∞ It is important to note that this represents a *prediction*.

- To be *testable*, hypotheses must make *predictions* about *observable events*, not about unobservables.

∞ Conditional statements are usually *clearer* and *more concise* than the typical, less structured ways in which hypotheses are *commonly stated* by students, as well as professionals.

- If not initially phrased as a conditional statement, it is nonetheless *possible and advantageous to rephrase* such hypotheses in the "If ____, then ____" form.

Example: Paying someone for work they usually do voluntarily will result in a reduction of the quality of the work that they do." This can be restated more clearly and concisely as: "*If* someone is paid for work they usually do voluntarily, *then* the quality of their work will be reduced."

Note that by using this *logical format* it becomes very clear what will be *done* and what is *expected*, i.e., what will be *manipulated* and what will *be measured*.

b. Quantitative form of a functional relationship

- Stating a hypothesis as a *mathematical function, or equation*, is very desirable, but often it is not yet possible to do so in the behavioral sciences, or sometimes even the biological sciences.

  ∞ The general form of the *mathematical function (f)* would be

  dependent variable = *f*(independent variable or variables)

  - Example: Psychological perceptual responses (ψ), e.g. perceived loudness, are a *specific function* of physical stimulus values (l), e.g. sound intensity — such as was proposed in *Fechner's Log Law*, i.e., ψ = k(log l) + C (Note: k and C stand for constants that are computed).

  ∞ Note: The quantitative method of writing a hypothesis is also a *conditional statement*, but it is a *more precise* "If ____, then ____" expression.

  - Rather than just stating, e.g., that *"if* the independent variable increases, *then* the dependent variable will increase;" it is stated that *"if* the independent variable is increased by X amount, *then* the dependent variable will correspondingly be increased by an amount equal to Y."

3. Principle of Parsimony

— According to this principle: If *different hypotheses* are presented to answer the question posed by a given problem, then the one that is *least wasteful of resources* — i.e., *most parsimonious* — is preferred.

- This is a *basic rule and assumption of scientific thinking*, which is usually right, about the organization/operation of the universe.

— Two components

- Occam's Razor

  ∞ *General explanations* are to be preferred over those that are appropriate to a more limited range of phenomena, according to this component of parsimony.

  - Thus if two hypotheses (or theories) are equally complex, the one that can *explain more results* is to be preferred.

  - In other words, explanatory principles should not be *needlessly multiplied*.

    Hence it should be *assumed* that *one set of principles* explains a wide range of phenomena under a variety of conditions, until *empirical evidence* indicates that this assumption is incorrect.

This axiom is appealed to when *generalizing* research *results* and their *explanations* across different types of individuals and species, different independent and extraneous variable conditions, and different dependent variable measures.

- Morgan's Canon
  - ∞ *Simplest explanation* (hypothesis or theory), i.e., the one involving the *fewest explanatory concepts and assumptions,* is to be preferred when it accounts for the same amount of data just as well as do more complicated explanations.
    - Thus, this component is to never appeal to a *higher or more complex process* for explanation when a *lower or simpler one* (physically, psychologically, developmentally, evolutionarily, etc.) will do the job equally well.
  - ∞ Example of both Occam's Razor and Morgan's Canon: If mice are discovered in a barrel of flour when none were there before, we shouldn't jump to the explanation/theory of *spontaneous generation* of life from inanimate matter. *A more general and simpler explanation or hypothesis* is that there was a *small opening* in the barrel through which the mice were able to crawl.
- IN SUMMARY: *Seek the smallest possible number of simplest principles that can successfully explain the greatest number of phenomena studied.*
  - ∞ Note, however, that this *assumption* that *"less is more"* can sometimes lead to *error* — i.e., while usually correct, the principle of parsimony is *not always correct.*
    - As Albert Einstein (1879–1955) once said: "Everything should be made as simple as possible, but not simpler."

4. Theory

— Definition

- *Theories* are general *propositions* used to *explain* a phenomenon or, more commonly, a set of phenomena (empirical relationships).
  - ∞ Example: *Signal-Detection Theory* states that "Sensitivity to a stimulus depends not only upon the stimulus intensity but also upon the experience, expectation, and motivation of the person being tested."[*The Encyclopedic Dictionary of Psychology* (3rd ed.). (1986). Guilford, CT: Dushkin.]
- Hypothesis (idea, view, or notion) and theory are both terms referring to tentative explanations, and thus they are closely related and sometimes used interchangeably — nevertheless, technically these terms do have different meanings.
  - ∞ *Theories* are explanations of how *larger pieces of the puzzle* go together, as opposed to smaller pieces, i.e., they are usually explanations of *groups of phenomena* — thus theories are *more general* than hypotheses.

∞ *Theories* are also more likely to have *considerable supporting evidence* — thus theories are *less tentative*.

- *Principle*
  - ∞ This label/designation is typically used after a theory becomes more firmly established by the accumulation of *additional supporting evidence*.
    - Example: *The Phylogenetic Principle* that, in general, ontogeny (development of the individual) recapitulates phylogeny (evolutionary development of the species).

- *Law*
  - ∞ This label is commonly used after enough evidence has been obtained so that there is essentially *no doubt* about the veracity of some specified empirical relationship(s).
    - Example: *Law of Effect,* by E. L. Thorndike (1874–1949) and followed up by B. F. Skinner (1904–1990), that a response which is followed by a *pleasant consequence* is more likely to be repeated, and a response followed by an *unpleasant consequence* is less likely to be repeated.

— Functions

Theories have *two general purposes*. They are used to *organize established facts*, and they are used to *guide future research*. Hence theories look both to the *past* and to the *future*.

1) Organize established facts
   - ∞ Theories are an important aid for arriving at a *systematic body of knowledge*, and thus advance the other goals of science as well.
     - They *clarify knowledge* by providing a basis for the *organization of data* resulting from *many studies*.

       Example: It was theorized that there are separate systems for *short-term versus long-term memory*, based on such things as differences in storage capacity and rates of forgetting, as well as the effects of blows to the head, electro-convulsive shocks, and damage to brain areas.

     - Theories also *facilitate retrieval of knowledge* in a manner analogous to remembering a *rule* rather than all the *individual facts*.

       Example: *Drive-Reduction Theory*, which holds that motivated behavior arises from drives or needs, and that responses which satisfy these drives/needs tend to be reinforced, and thus are more likely to reoccur.

2) Guide future research
   - ∞ Theories also generate *testable predictions* by way of *specific hypotheses* derived from the theories, which can advance the scientific goals of *explanation and control* of phenomena.

- To be testable, a theory must make predictions about *observable events* (as also noted earlier for hypotheses).

- *Scientifically meaningful theories* are *testable theories*.

   Example: *Darwin's Theory of Evolution*, where the environment can be varied and effects observed on the survival and reproductive success of individuals as a result of changes in anatomical, physiological, or behavioral characteristics of successive generations.

— Meaningless theories

There are *two reasons* that theories might be meaningless, i.e., untestable: they might be *vague* or *too general*.

1) Theories are scientifically meaningless when stated in such *vague terms* that it is *unclear* what operations should be performed in order to *test* them — which is done by gathering *empirical evidence* that either supports or refutes the theory (consider the following: *immorality* is caused by *godlessness*).

   ∞ This point also applies to *hypotheses*, and is similar to the vagueness concern noted earlier for research *problems*.

2) Theories are also considered scientifically meaningless when they are *so general* that they can explain *any possible outcome*, and thus are *not refutable*, which means that they are *not testable*.

   ∞ Some theories are not refutable because their explanations are after the fact (*post hoc*), i.e., *no predictions are made*.

      - Example: *Freud's Personality Theory* involves the id, ego, and superego that vie for control. But the decision as to which is dominant at any time tends to be *after the fact*, thus the theory does *not* generate testable predictions.

      - It should be acknowledged, nevertheless, that a *scientifically meaningless theory* still can be *therapeutically meaningful*, i.e., useful. From the practitioners point of view, a theory is only as meaningful as the effective repertoire of *techniques* that it generates.

— Evaluation criteria

The *quality* of a theory is based on how *completely and accurately* it fulfills the *two* functions of theories noted earlier.

1) How well does the theory account for *past* research findings, i.e., how well does it *organize established facts*?

   ∞ It should be possible to generate *old results* from a theory.

   ∞ The *more* results that can be *explained*, and the *simpler* (i.e., fewer explanatory concepts) and more *precisely* this can be done, then the more *parsimonious* and *better* the theory.

2) How well does the theory predict *new* research findings, i.e., how well does it *guide future research*?

∞ It should be possible to generate *novel results* from a theory.

∞ The *more* results that can be *predicted*, and the *simpler* and more *precisely*, then the *better* the theory.

   • Hence, *exact mathematical statements* are superior to more general verbal statements.

∞ *Strong tests of theories*, in contrast to weak tests, pit one theory against another.

   • This is done by having them *generate different predictions* that are then compared against the data to determine which theory *best accounts* for the findings (see example later in this chapter under Deductive Method).

   • This *comparative approach* is a *superior* means for evaluating the *relative quality* of theories or hypotheses.

— Additional considerations

There are *three other concerns.*

1) Mutual exclusiveness

   ∞ An important principle to always keep in mind is that different theories/hypotheses about the same phenomena are *not necessarily mutually exclusive*, i.e., if one is *correct*, it is not necessarily the case that the other is *incorrect*.

      • When there is *good support for more than one theory*, the different theories might each explain *different aspects* of the same phenomena, or *different subsets* of the data — thus, both could be correct, at least in part.

      • Example: *The Trichromatic Theory of Color Vision* (developed in the 19th century by Young and Helmholtz) and the *Opponent-Process Theory* (developed about the same time by Hering) were later both found to be correct, but for *different levels* in the visual system, i.e., peripheral retinal photoreceptors versus more central levels of the nervous system, respectively.

2) Mutual exhaustiveness

   ∞ Another important consideration — which is related and complementary to mutual exclusiveness — is whether or not proposed theories are *mutually exhaustive*, i.e., whether there might be *additional or better explanations*.

      • Example: The *Place Theory of Pitch Perception* and the *Frequency (or Telephone) Theory* (developed in the 19th century by Helmholtz and Rutherford, respectively) were both found to be correct, but for coding *different ranges* of frequencies, i.e., relatively high frequencies versus low frequencies, respectively. However, later on, the *Volley Theory of Pitch Perception* was developed (in the 20th century

by Wever and Bray). This theory *complements* the earlier two theories by *best accounting* for coding of the "middle" range of auditory frequencies.

3) Data versus theory

∞ Finally, a very important point is that *research data should always dom-
inate theory* — not vice versa.

• If research results (empirical observations) do not support a theory,
then the theory probably needs to be modified or even discarded
— i.e., *the facts can't be changed to accommodate a theory (or hypothesis)*.

• However, a fact of life is that a *bad theory* usually is *not eliminated*
just because it is poor, but rather only because a *better theory* has
been formulated that can take its place.

Example: Darwin's Theory of Evolution by *Natural Selection* was
needed to replace Lamarck's Theory of Evolution by *Inheritance
of Acquired Characteristics* (the necks of giraffes were said to have
become longer with each generation because the young inher-
ited the stretched necks of their parents who were reaching
higher for leaves to eat).

5. Inductive versus Deductive Methods[33]

— General features

• The inductive and deductive methods represent *two variations* on the
general scientific method.

• They are associated with the two different *functions of theories*:

∞ *Induction* is used for *organizing/systematizing established facts*.

∞ *Deduction* is used for *guiding future research* through the generation of
predictions.

• The two methods actually represent *complementary strategies* for research,
with the *weaknesses* of one being the *strengths* of the other (as discussed
later).

• Moreover, although these different approaches proceed in *opposite directions*, they are *strongly interrelated*.

∞ *Observations* (experimental or naturalistic) lead through *induction* to
new *general principles or theories*, which in turn suggest through *deduction* what *further observations* should be made — thus coming full circle
(this is elaborated later):

$$\text{observations} \rightarrow \textit{induction} \downarrow$$

$$\uparrow \textit{deduction} \quad \leftarrow \quad \text{theories}$$

∞ Investigators therefore do not necessarily use one approach to the
exclusion of the other, although they might *prefer* one.

— Inductive method

- This approach is likely to be used when studying topics that have received little or no previous attention, i.e., *relatively unexplored areas* — thus little data would have been gathered, and hence good general principles or theories would not yet have been formulated.

- Induction moves from the *specific to the general*.

  ∞ It begins with relatively *specific empirical observations or reflections*, which are then followed by the formulation of a *problem* (a question) and perhaps also a *hypothesis*.

  ∞ Next it moves through the collection, organization, analysis, and summarization of a substantial *database*, obtained under varying conditions, onward to the eventual identification of *generalities* — the formulation of *general principles or theories*.

    - Example: Development of the theory that there are at least *two general forms of memory* — *procedural* (for how to do things) versus *declarative* (for facts and events) — based on observations of many amnesiac patients, which were followed up on by several additional studies.

      To elaborate: The observation was made that people suffering from amnesia typically remember *how* to do things, such as the use of language and driving a car, but they do not remember *who* they and others are, or *what* their past experiences were. This led to the question/problem of the nature of memory, which led in turn to the gathering, analysis, and summarization of additional data on amnesias that occurred under a variety of circumstances.

      Ultimately, this research led to the development of a *theory* that there are at least *two forms of long-term memory*, and that they must be different in their anatomical and/or physiological bases. The two forms were labeled *procedural memory* (implicit*)* versus *declarative memory* (explicit), with the latter having two components, *semantic* (for facts) and *episodic* (for events), and being much more susceptible to disruption. [Squire, L. P. (1987). *Memory and Brain.* New York: Oxford University.]

- Induction can lead to use of the deductive method as a means for further research to enhance knowledge in the subject area.

— Deductive method

- This approach is likely to be used after a *substantial database* has been developed, and thus one or more *general principles or theories* would have already been formulated.

- Deduction moves from the *general to the specific*.

  ∞ It begins with a generalization — *a general principle or theory* — from which a specific *hypothesis* is logically *deduced* as a potential solution to a *problem* (an answer to a question).

∞ Next it moves through the collection, organization, analysis, and summarization of *observations* (data) in a *test* of the hypothesis and its prediction(s), in order to empirically verify the theory.

• Example of this and another point: There are three major theories for the *acquisition of language,* and they can be evaluated through *strong tests*, i.e., by comparing gathered data against differential predictions of the theories (as noted earlier, this is a very powerful approach to evaluating competing theories).

Traditional learning theory proposes that the principles of reinforcement and conditioning are the explanations for language development; *cognitive learning theory* proposes that rule learning is involved; and the *biological theory* proposes that there is an evolutionarily built-in capacity from birth for language acquisition and use. These different theories are *testable* by measuring, respectively, the effects of reinforcement and conditioning on language acquisition, the occurrence of rule learning, and the linguistic capacities present at birth. (The previously discussed considerations of *mutual exclusiveness* and *mutual exhaustiveness* should be remembered.)

• Note: Use of the deductive method is likely to lead to the use of a form of *induction* when making judgments about the *implications* of data/results for the theory — thus the two methods are *reciprocal.*

∞ When results *confirm the hypothesis* of a study, it is typically concluded that the *general principle or theory* from which the hypothesis was logically derived has been *strengthened.*

∞ When the results *do not confirm* the hypothesis of a study, it is typically concluded that the *validity* of the general principle or theory from which the hypothesis was deduced has been *weakened* — note that in either case the conclusions represent *induction* from the specific results to the general theory.

— Strengths and weaknesses of the two methods

• Investigators do not always agree on which approach would be best in a given *research area* at a given *stage of data acquisition.*

∞ It might be argued by some, e.g., that the *database* is not yet adequate to warrant the *development of theories,* and that a switch from the *inductive* to the *deductive method* might divert investigators away from the discovery of empirical relationships that are *not implied* by the prevailing theories that had been developed.

• In other words, *too early a switch* from the inductive method could *bias* the subsequent application of the deductive method.

• Inductive method strengths

∞ Allows development of knowledge in subject areas that are *unexplored or poorly explored*, and thus when good general principles or theories have not yet been formulated.

∞ Investigators are less apt to be bound by some *conceptual system or theory* when gathering and interpreting data.

- *Expectations* about the nature of results, in other words, are less likely to occur.

- *Preparedness* to recognize and accommodate findings as potentially significant is thereby greater, in contrast to the likelihood of rejecting as unimportant certain results just because they don't *conform* to expectations.

- *Opportunities* for new discoveries are thus maximized.

- Inductive method weakness

∞ Tends to leave knowledge *unsystematized*, at least for a time, while sufficient data are collected, organized, analyzed and summarized for the formulation of one or more general principles or theories.

- Therefore, the scientific literature provides *less direction* to investigators in generating new research studies.

- Deductive method strength

∞ *Theories* direct future research by suggesting problems or hypotheses that need to be investigated.

- Deductive method weaknesses

∞ Theories can *limit the problems* scientists choose to investigate (as noted earlier).

- Therefore, potentially *more important problems* might not be investigated.

∞ Theories also create *expectations* about the nature of the results.

- Hence potentially important findings might not be recognized or accepted (correctly interpreted) simply because they are *not suggested* by the theory and its derived hypotheses.

- Note from the preceding how the *weaknesses* of the deductive method are the *strengths* of the inductive method, and vice versa — in other words, the two approaches are *complementary*.

Review Summary

1. There are seven *primary steps* of the scientific method:

   a. Formulating a problem

    b. Formulating a hypothesis

    c. Designing a study

    d. Collecting and organizing the data

    e. Summarizing and statistically analyzing the data

    f. Evaluating the results and drawing conclusions about the hypothesis

    g. Communicating the findings of the study

2. There are three *secondary steps* of the scientific method:

    a. Generalization

    b. Prediction

    c. Replication

3. There are three ways in which *research problems* become evident:

    a. Contradictory results are found in two or more studies attempting to answer the *same* question.

    b. Gap in knowledge becomes apparent as a result of what has been learned.

    c. Fact needs explaining since it does not fit in with existing knowledge.

4. There are four *basic types of problems* — classes of questions regarding behaviors or mental activity.

   Two of these involve *proximate* causal mechanism: 1) What are the *immediate causes* in terms of external environmental factors and internal psychological, physiological, and anatomical factors, and 2) what are the *developmental factors* in terms of heredity and experience (Nature-Nurture Problem)? The other two types of problems involve *ultimate or distal* causal mechanisms: 1) What is the *adaptive significance* (function) in terms of promoting survival and reproduction, and 2) what was the *evolutionary history* for the species?

5. To be *meaningful* a research problem must be *capable of being answered* with the *tools available to the scientist.*

   Four reasons for research problems being *meaningless or presently meaningless are*: 1) there might be *no empirical reference*, in that it is impossible to obtain relevant objective data since the problem does not deal with observable properties; 2) it might be *presently impossible* to obtain relevant data, although the problem might be *potentially meaningful*; 3) the problem might be *unstructured and vaguely stated*, so that it is not clear what the relevant variables/events are (often correctable by being more specific); or 4) the terms might be *inadequately defined* — a specific source of vagueness (correctable by using operational definitions).

6. *Basic research* investigates problems relating to *fundamental questions* about the principles and mechanisms of phenomena.

   Applied research, in contrast, investigates problems whose solutions are expected to have *immediate application*. Research on both types of problems is essential, however, since basic research establishes the foundation for applied research, which in turn can improve our lives.

7. A *meaningful experimental hypothesis* is an *experimentally testable* statement of a *potential relationship* between independent and dependent variables that *might be the (or a) solution* to a research problem.

    Thus it is more than a *hypothesis in general,* which is simply a *potential answer* to the question put forth by a problem.

8. The *best ways to write a hypothesis* would be as a *conditional statement* ("If ___ then ___") or *quantitatively* as a functional relationship.

9. *The principle of parsimony* is a basic rule and assumption of scientific thinking that states that the hypothesis that is *least wasteful of resources* is to be preferred.

    There are two components of this principle: *Occam's Razor,* that *general explanations* (those that can explain more results) are to be preferred; and *Morgan's Canon,* that the *simplest explanation,* or the one involving the *fewest assumptions,* is to be preferred when it accounts equally well for the same amount of data. Thus, one should seek the *smallest* possible number of *simplest* principles that can successfully explain the *greatest* number of phenomena studied.

10. *Theories* are general propositions or sets of propositions used to *explain* phenomena.

    They differ from hypotheses in that they usually have *more supporting evidence* and are *more general.* The two major *functions* of theories are to *organize established facts,* and thus to arrive at a *systematic body of knowledge* — and hence *description,* and to *guide future research* by generating testable *predictions* — which can lead to *explanation.* As for hypotheses, theories can be scientifically meaningless, or they can be *meaningful,* i.e., scientifically *testable.* Theories are *evaluated* on the basis of how completely and accurately they fulfil their two functions.

11. Different theories/hypotheses about the same phenomena are not necessarily *mutually exclusive.* In other words, they could both be correct — each explaining *different aspects* of the same phenomena, or *different subsets* of the data. A complementary consideration is that a proposed theory (or set of theories) might not be *mutually exhaustive.* That is, there might be *additional or better explanations.*

12. The *inductive method* and the *deductive method* are two *variations* on the general scientific method. They are associated with the two different *functions of theories:* induction is used for *organizing/systematizing* established facts, and deduction is used for *guiding* future research through predictions. They are thus *complementary strategies.*

    The *inductive method* moves from the specific (empirical observations or reflections) to the general (principles or theories) and is more likely to be used when studying topics that have received *little previous attention,* and thus before good general principles or theories have been formulated. The *deductive method* moves from the general (principle or theory) to the specific (empirical observations) and is more likely to be used after a *substantial database* has been developed, and thus when one or more general principles or theories have already been formulated.

13. The inductive and deductive approaches are *strongly interrelated.*

Although they proceed in opposite directions, one often leads to the other. *Induction* can lead to use of the deductive method as a means for further research. On the other hand, *deduction* is likely to lead to the use of induction when making judgments about the implications of results for a theory. Moreover, the *weaknesses* of one are the *strengths* of the other. With the *inductive method*, investigators are less apt to be bound by some *conceptual system or theory* when gathering and interpreting data, but knowledge is left *unsystematized* while sufficient data are collected to formulate general principles/theories. With the *deductive method*, theories are used to *direct* future research, but they can *limit* the problems chosen to be investigated.

Review Questions

1. What are the seven *primary steps* of the scientific method?

2. What are the three *secondary steps* of the scientific method?

3. List and explain the three ways in which *research problems* become evident.

4. List and describe the four *basic types of problems* — classes of questions — regarding behaviors or mental activity.

5. List and explain four reasons for *research problems* being *meaningless or presently meaningless*.

6. Distinguish between *basic/pure research* versus *applied research*, and explain the relationship between them.

7. Define a *meaningful experimental hypothesis* and contrast it to a hypothesis in general (i.e., state the distinguishing characteristics).

8. What are the two best ways to *write hypotheses* (give an example of each)?

9. Describe the *principle of parsimony* and its two components.

10. Describe the two *major functions of a theory*.

11. Explain the difference between *mutually exclusive hypotheses* and *mutually exhaustive hypotheses*.

12. Describe both the *inductive method* and the *deductive method*, and indicate when each is most likely to be used.

13. State one *strength* for the inductive method and one for the deductive method, and discuss the *relationship* between these two approaches.

4

Variables in Research Designs

CONTENTS

Design Construction (Step Three of the Scientific Method)

Having formulated a meaningful research problem and hypothesis, the next step is to design a study, i.e., to develop a *basic plan* for gathering data to test the hypothesis or hypotheses. This chapter focuses on the variables of experimental designs, particularly the *independent and dependent variables. Intervening variables*, which are hypothetical constructs, are also covered. Extensive discussions of *extraneous variables* and their control in experiments occurs in Chapters 7 and 8; and detailed coverage of designs for experiments, with *all their components*, is presented in Chapters 9 and 10. The remaining steps of the scientific method are also addressed in later chapters, as listed at the beginning of Chapter 3.

1. **Elements of Experimental Research Designs**

    There are *three primary elements* of research designs, and they are directly related to the *three basic components of an experiment* (covered earlier in Chapter 2).

    a. Selection of *independent variables* (antecedent conditions whose effects are to be determined) and specification of their *levels*, the method of selecting the levels (i.e., systematic versus random), and the means for their purposeful, direct manipulation.

    b. Choice of *dependent variables* (overt behavioral responses or physiological events of interest to the investigator) and specification of the methods for their *measurement*.

   c. Determination of *extraneous variables* (extra, unwanted variables that might affect the dependent variable) and specification of the techniques for their *control.*

   d. *Additional elements* are choice of *participant/subject population(s)*, and specification of the sampling technique, sample size, and method of assignment to conditions, as well as *statistical analyses.*

2. **Relationship of Design Construction to Earlier Steps**

   — As mentioned in Chapter 3, there is a good deal of *overlap and interrelation* among the steps of the scientific method.

     • For example: *General* specification of the independent and dependent variables actually *precedes the design stage*, because they are part of the formulation of the *research problem and hypothesis* (steps one and two of the scientific method).

Variables

Before discussing the specific principle variables of experiments, it is important to first consider the *definition* and various *forms* of variables: *qualitative versus quantitative, and discrete versus continuous.*

1. **Definition of Variables**

   — *Variables* are *characteristics* of organisms, objects, environments, or events that have values which can *vary/change* across individuals, items, conditions, or time (noted earlier near beginning of Chapter 2).

2. **Qualitative/Categorical versus Quantitative Variables**

   — *Qualitative/categorical variables* vary in *kind/type.*

     • Examples: Kinds of animal species, types of psychoses, forms of visual stimuli, and states of consciousness.

     • These are measured on a *nominal scale* (see later in this chapter under "Scales of Measurement for Variables").

   — *Quantitative variables* vary in *amount/magnitude.*

     • Examples: Age, reaction time, intensity of a sound, degree of neurosis, and rated tastiness of foods.

     • These can be measured on *ordinal, interval, or ratio scales.*

3. **Discrete versus Continuous Variables**

   — *Discrete variables* are those that come only in *whole units or categories* — which includes all the qualitative variables and some of the quantitative variables.

     • Examples: Presence or absence of an event, and the number of children, marriages, hostile acts, errors or correct responses.

— *Continuous variables* are those whose values lie along a *continuum* and can be represented by both whole and fractional units — which is the case for many quantitative variables but no qualitative variables.

- Examples: intensity of light and amount of sleep deprivation.

- Note that the *measurement* of continuous variables is often actually *discontinuous*.

  ∞ Examples: 1) Although *time* to complete a response (speed/duration) can be measured down to microseconds or even nanoseconds, we normally measure only to the nearest millisecond or hundredths of a second; 2) Although *comprehension* of some subject matter is a continuous variable, we usually measure it in whole numbers representing how many questions are answered correctly on a test.

  ∞ But this does not change the *continuous nature* of the variable — it is just that measurement can occur with various *degrees of precision*, limited only by necessity and the precision of measuring instruments available.

Principal Variables of Experiments[9,26] (See Appendix 1)

Two critical variables for experiments are *independent and dependent. Extraneous variables*, on the other hand, are extra, *unwanted variables* (as noted above and in Chapter 2 and as elaborated in Chapters 7 and 8). All of these variables can be either qualitative or quantitative and discrete or continuous (see above). Figures in Appendix 1 illustrate the *organism within its environment*, receiving *stimuli* and emitting *responses*. As shown in the figures (and noted in Chapter 2), independent variables are the *manipulated stimulus input conditions* to the organism, and dependent variables are the *measured response output effects*.

1. **Independent Variables**

   — Categories of independent variables

   Independent variables fall into *two general groups: environmental/stimulus factors* and *organismic/participant factors*.

   1) Environmental/stimulus factors

      ∞ These can consist of any characteristic of an organism's *external environment* that might influence its behavior:

      - *Instructions* given to the research participants/subjects,

      - *Procedures* used in the experiment,

      - *Tasks* given to the research participants,

      - *Stimuli* presented in the external environment.

      ∞ Usually these environmental/stimulus factors are *manipulated purposefully and directly,* and hence these factors are typically studied in *experiments* (versus descriptive research).

∞ Moreover, these factors represent the *commonest category* of independent variables in the behavioral sciences.

2) Organismic/participant factors

∞ These consist of any *anatomical, physiological, or psychological characteristic* of an organism/participant.

- Examples: Brain damage, hormone levels, emotional and cognitive traits and states, age and gender, etc.

∞ Usually these are *manipulated by measured selection* in studies with *human participants*; hence for humans these factors are typically investigated in correlational or ex post facto *descriptive research*, which involves *naturalistic observation*.

- Alternatively, these factors can be studied as *correlational independent variables* as a component of *semi-experiments* (covered earlier in Chapter 2).

∞ In contrast, studies with *non-human animals* usually involve investigation of these factors by *experimental manipulation*.

- *Participant/subject variable*, it should be noted, is a term that is often used in a *more restricted sense* to mean any characteristic of a participant that can be measured/described, but that *cannot* be experimentally (purposefully and directly) manipulated; e.g., biological sex and age — hence they are just *correlational variables*.

— Means of variation of independent variables

There are *three ways* by which to vary independent variables: through *instructions, events*, and *measured selection*.

1) Manipulation by instructions

∞ Research participants can be given *different information* as a means of independent-variable manipulation.

- Example: In a verbal learning study, one group of participants but not another might be told to use *memory encoding and rehearsal strategies*.

∞ Extraneous variable problems that can occur

- Participants vary in *attentiveness* and thus *perception* of the pertinent information provided in the instructions.

- They might also vary in the *interpretation* of instructions.

∞ Effects of extraneous variables associated with instructions

- *Unintended, undesirable variation* will occur in the outcome of independent-variable manipulation.

- *Error* is thereby introduced into the resulting data.

∞ Control of instruction extraneous variables

- Use *printed instructions* that are kept simple, emphatic, short, and to the point.

- Manipulate *no more than one variable* through instructions in any given study.
- Ask participants for *questions* they have about the instructions.
- Use *practice trials* for demonstration and feedback.

2) Manipulation by events

∞ Several *types* of events can be manipulated as a means of variation of independent variables.

- *Procedures* used in the experiment

  Example: *Massed* (all as once) versus *distributed* (over time) *practice effects* on learning some task.

- *Tasks* given to the research participants

  Example: *Problem-solving individually* versus with a *group* in a study of accuracy and speed of solutions.

- *External* environmental stimuli presented

  Example: Presentation of *words* versus *spatial patterns* to the right or the left half of the visual field on different trials in order to determine specialization of functions in the two hemispheres of the brain.

- *Internal* organismic stimuli introduced

  Example: Varying the level in an organism of some *drug*, such as marijuana, in a study of memory.

∞ Advantages of manipulation by events versus instructions

- *Minimizes* inattention and misinterpretation problems.

  Example: *Giving* participants *practice and feedback* using some memory strategy versus just *instructing* them about a strategy that could be used.

- *Increases* realism, meaningfulness, and impact.

  Example: Manipulating anxiety level in individuals by *showing* other people (actually actors) responding as if they are being *electrically shocked* while participating in a study versus just *telling* individuals that they will be electrically shocked during a study.

- *Increases* representativeness of participant behavior with respect to the "real-world" (versus laboratory situations).

  Example: A study of *helping behavior* where a nonpregnant woman or the same woman made up to appear very pregnant *requests* individuals to give her their seat on a crowded bus versus a study in which an experimenter *just describes* the two situations and asks participants how they *think* they'd respond.

3) Manipulation by measured selection

∞ As a third form of manipulation, research participants can be selected who *already differ* on the independent variable — i.e., some measured *internal state or trait* or some *experience* that they have already had (discussed earlier in Chapter 2).

- This is an *indirect* means of manipulation that *contrasts* with the *purposeful, direct manipulation* by instructions or by events (as described above).

∞ It is used for *correlational and ex post facto studies* in descriptive research or for *correlational independent variables* that can be part of *experimental research* (covered in Chapter 2).

∞ Note that this is the *least desired* means of manipulation of independent variables when the *goal* is determination of causality, i.e., *explanation,* since control of extraneous variables is poor.

- Sometimes, however, this is the *only* possible or ethical means of manipulation, such as for certain variables involving *human* research participants.

  Example: Studies in which *age, IQ, or years of education* is the independent variable or those involving *human brain damage or endocrine system dysfunction.*

— Levels/values of independent variables: the conditions

- *Two or more levels* of an independent variable are *required* in an experiment, because only through *comparison* can it be determined whether there is an *effect* on some dependent variable(s).

∞ Experimental/treatment conditions

- These are levels of an independent variable *other than* either zero or some standard value.

- Treatment conditions are *compared* against each other and/or a control condition(s) to determine the independent *variable effects.*

∞ Control conditions

- These are either a *zero* level or some *standard* value of an independent variable.

  Example: In studying the effects of a noise-independent variable on cognitive performance, a control condition might involve a *typical/standard, non-zero level* because *complete absence* of background noise is *atypical* and thus would be a *treatment* of sorts.

- Control conditions are used to establish a *baseline* against which to compare the effects of one or more experimental treatments and to *control* and *evaluate* extraneous variable effects such as *placebos* (this is elaborated in Chapter 8, under "Control Groups for Secondary Variables").

  Placebo control conditions involve *zero* level of *actual* medical treatment (unbeknown to the recipients), while producing a standard level in the research participants of *expected* therapeutic effects.

- Note: Using control conditions is *not always required*, because experimental/treatment conditions can be compared just with each other.

— Forms of variation of independent variables

There are *three types* of variation: *presence versus absence, amount*, and *type of a variable*.

1) Presence versus absence of a variable

   ∞ *Experimental/treatment* versus *control conditions* are compared in this form of independent-variable manipulation, in order to determine the effect, if any, of the independent variable on some dependent variable(s).

   - Example: A study of the effects on insomnia of *injecting* versus *not injecting* some *sedative drug*.

   ∞ Due to the *nature* of an independent variable and/or *ethical considerations* (see Chapter 6), it is not always feasible or logical to use a *zero value*, as opposed to a *standard value*, for the *control condition*.

   - Example: In a study of the effects on cognition of *environmental or participant body temperature*, you wouldn't want to bring the temperature down to *absolute zero* and thereby kill the research participant — instead you would use some *standard value* as a control condition.

2) Amount of a variable

   ∞ The effects of two or more *treatment conditions* (i.e., non- zero/non-standard values) that differ *quantitatively* are compared in this form of independent-variable manipulation.

   - Example: A comparison of the effects on vigilance of *different dosages* of a new *stimulant drug*.

   ∞ *Control conditions* also can be included.

   - This would combine *two forms of variation:* presence versus absence and amount of the variable.

   - This combination allows determination of *both* the *effect* of a given independent-variable level (by comparing presence versus absence) and also the *different effects* of different amounts of the independent variable.

3) Type of a variable

   ∞ The effects of two or more *treatment conditions* (i.e., non-zero/non-standard values) that differ *qualitatively* are compared in this form of independent-variable manipulation.

   - Example: A comparison of the effects of *different kinds of illicit drugs* on reports of hallucinatory experiences.

∞ *Control conditions* (zero/standard values) can be included.

- This would combine *two forms of variation*: presence versus absence and type of the variable.

- This combination allows determination of *both* the *effect* of a given independent variable level (by comparing presence versus absence) and also the *different effects* of different qualitative types of the independent variable.

∞ *All three forms of variation* might be combined in a study.

- Example: A comparison of the effects on anxiety of different *types and amounts,* including *zero,* of *tranquilizer drugs.*

— Number of independent variables and levels

- Because behavior is *multidetermined* and variables *interact,* important information and clarity are gained by studying the effects of *more than one* independent variable at a time.

- There is no *theoretical or statistical limit* to the number of independent variables that can be varied and investigated in a given study, but there are *practical limits.*

  ∞ With *greater numbers* of independent variables, studies become *more difficult* to design, run, analyze, and interpret.

- Hence, *only include* in a study those independent variables that seem likely to reveal *important relationships* in a *clear fashion,* while staying within your limitations of experience and other resources (e.g., time, energy, and funds for equipment).

  ∞ In other words, *do not unnecessarily include* independent variables in a study just because they *exist* or to make the study appear more *complex* and thus *impressive.*

  ∞ *Elegant studies,* those that are simple and clear, are often the most illuminating.

- *How many levels of variation* of a given independent variable to study is a decision that should involve similar guiding principles (more on this in Chapter 7 under "Maximizing Primary Variance: Several Values of the Independent Variable").

2. Dependent Variables

— Categories of dependent variables

Dependent variables (like independent variables) fall into *two general groups: environmental/overt-behavioral responses* and *organismic/physiological responses.*

1) Environmental/overt-behavioral responses

  ∞ These consist of *externally observable behaviors,* and they represent the *commonest category* of dependent variables studied in the behavioral sciences.

  - Examples: *Number of errors* a rat makes in a maze and the *amount of time* individuals take to respond to stimuli.

2) Organismic/physiological responses

∞ These often consist of *nervous system and endocrine system activities/ events.*

• Examples: *Electrical activity* of the brain (EEG), *estrogen hormone level* in the blood, and *pupil diameter* of the eye.

∞ Note: The *distinction* between overt behavior and physiological responses is sometimes rather *arbitrary.*

• Examples: *Externally visible* pupil dilation and salivation might well be considered *overt* rather than *physiological* behaviors and so might *intelligence,* because it is usually measured by pencil marks or verbal responses on IQ tests.

— Relationships studied between independent and dependent variables

• *Three sources of variables* exist in the behavioral and biological sciences, if the two categories of independent and dependent variables are considered together:

a) *Environmental stimulus* (S)-independent variables,

b) *Environmental response* (R)-dependent variables,

c) *Organismic events* (O) that can be either *stimulus*-independent variables or *response*-dependent variables.

• Note: Organismic independent and dependent variables can be *central nervous system* as well as *peripheral processes,* and each can be in the *sensory* or *motor* system (thus more complex than shown in Appendix 1) — intervening variables are covered later in this chapter.

• *Four categories of cause-and-effect relationships* involving independent variables influencing (→) dependent variables are commonly studied among these three sources of variables:

a) S → R

• Example: reinforcement schedules → rate of bar pressing,

b) S → O

• Example: visual patterns → electrical brain activity,

c) O → R

• Example: brain lesions → speed of maze learning by rats,

d) O → O

• Example: electrical brain stimulation → hormone levels.

∞ R → R is a relationship *not* typically studied in *experiments* but rather in *descriptive studies* (e.g., with the correlational or ex post facto techniques).

• Note: In *correlational studies* all factors are often *dependent variables,* i.e., behaviors or characteristics.

• Example: Relationship between cocaine use and paranoia.

— Types of measures for dependent (and independent) variables

There are *six different forms* of measurement for variables: *magnitude, frequency, latency, speed, accuracy,* and *profile.*

1) Magnitude/amplitude/intensity

   ∞ Examples: *Magnitude* of blood pressure, *amplitude* of the galvanic skin responses, and *intensity* of muscle tension.

2) Frequency/rate/probability/proportion/number

   ∞ Examples: *Frequency* of stuttering, *rate* of heat beats, and *probability/ proportion* of various responses.

3) Latency/reaction time (RT)

   ∞ This is the *time* it takes an organism to *begin/initiate* a response after a stimulus, i.e., *time between* the *onset* of a *stimulus* and the *onset* of a *response.*

     • *Example: Time* taken by humans to *start* pressing a switch/key *after the onset* of some stimulus.

4) Speed/duration

   ∞ This is the *time* it takes an organism to *complete* a response once it has begun, i.e., *time between* the *onset* of a *response* and the *termination* of a *response.*

     • Example: *Time* taken by humans to *complete* a problem.

   ∞ Note: Most *latency/reaction time* measures include not only the time for initiation of a response, but also time to complete a response — but this *speed/duration* of response is felt to be relatively constant in magnitude or small, and thus not of great concern when looking for effect differences.

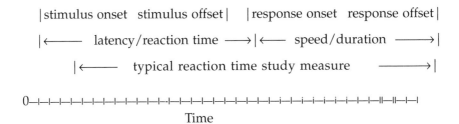

|stimulus onset   stimulus offset| |response onset   response offset|

|←——— latency/reaction time ——→|←—— speed/duration ——→|

|←——— typical reaction time study measure   ——→|

0—|—|—|—|—|—|—|—|—|—|—|—|—|-|—|—|—|—|—|—|-|—|-|—|-|—|-|—|—|—||—||—|—|

Time

     • It is possible to obtain a *purer reaction time assessment* by simply measuring the time taken to move a finger *off* rather than on to, a *switch/key* after the onset of some stimulus — because the *very beginning* of the finger lift could open an electrical contact and thus stop the timer.

5) Accuracy/correctness

   ∞ This is a measure of the number of *correct responses* (or errors), *time on target* (when in pursuit), *distance from target* (when trying to hit/strike something), etc.

     • Example: Number of *correct answers* on an aptitude test.

6) Complex score/profile

∞ This is a more *complicated* assessment than the preceding types of measurements.

• Example: *Profiles* derived from scores on the Minnesota Multiphasic Personality Inventory (MMPI).

— Multiple-dependent-variable measures

• *More than one* dependent variable can be measured per study.

• This might be advantageous since an independent variable is likely to affect a number of *different measures* of some behavior.

∞ Many phenomena or *constructs* that psychologists wish to study are *multidimensional,* and the different dimensions involve different measures that might not be highly correlated (discussion is under *operational definitions* later in chapter).

• Example: *Learning* is a phenomenon that can be measured in several different ways/dimensions, e.g., accuracy, speed of performance, or number of trials to reach some criterion.

∞ If the measures are *highly correlated* (around 95%), then they are measuring nearly the same thing, and thus only *one* (the simplest) would be needed to obtain most of the information.

∞ If, however, the measures are *not highly correlated,* then the dependent-variable measures are either *unreliable* or they are not measuring the same thing; i.e., they are measuring *different aspects* of the *construct* (concept) being studied.

• It would then be possible to obtain statistically significant effects of the independent variable for *some dependent-variable measures* (or combinations) but not for others — hence the value of multiple dependent-variable measures.

• *Multivariate analysis of variance (MANOVA)* is a statistic that analyzes the effects of independent variables on *two or more dependent variables,* without artificially inflating the *significance* of differences when the dependent variables are *correlated.*

∞ Thus it is a *more powerful and valid approach* than separate analyses of variance (ANOVA) for each dependent variable (consult advanced statistics books for more information).

• *Our understanding* of the *complex relationships* that exist between antecedent conditions and consequent behaviors is enhanced by multiple-dependent-variable studies and MANOVAs.

• Regardless of the number of dependent variables investigated in a given study, one should choose the *most sensitive measure(s)* of the effects of the independent variable(s) (discussed later).

∞ *Previous research* can help in making the choice.

— Scales of measurement for variables and their data types[15]

There are *four categories of measurement scales:* ratio, interval, ordinal, and nominal; and *three types of data:* score, rank-order, and frequency. (Note: There was some coverage of measurement scales earlier in Chapter 1 under "Description.")

- The *scale* chosen for measurement depends on the *data type.*
    - ∞ The data type and scale, in turn, determine the *mathematical and statistical operations* that can be performed on the measures and thus the kinds of *conclusions* that can be drawn.

1) Ratio and interval scales for score data
    - ∞ *Score data* involve numerical values that reflect relatively *precise measurement.*
        - Example: Heart rate and blood pressure.
    - ∞ *Ratio scales* of score data possess all four of the properties of measurement: *difference, magnitude, equal intervals,* and *a true zero* (at which no amount of the attribute exits).
        - Thus equal differences and ratios of *assigned scale values* (scores) represent/reflect equal differences and ratios of the *attribute measured.*

            Examples: Comparing differences and ratios of the *time* used to complete some task or of the *number* of words correctly recalled.

            Note that since these scales have a true zero value, the *ratios* among the values are meaningful; i.e., they reflect true ratios along the dimension measured.
    - ∞ *Interval scales* of values possess only the first three properties of ratio scales — *difference, magnitude, and equal intervals* — they do not have a true zero value.
        - Examples: *Fahrenheit temperature scale,* because zero degrees Fahrenheit does not represent zero amount of thermal energy, and perhaps also SAT (Scholastic Aptitude Test) and IQ scores, although these latter two might really be only ordinal scales (discussed next).
    - ∞ It is meaningful to add and subtract (to determine the sum and differences of) *interval or ratio scale* scores, but multiplication and division (to determine products and proportions) are meaningful only for *ratio scale* scores.
        - Hence, only with *ratio scale* scores can you say that something is, e.g., twice as great as something else.
        - With *interval scale* scores you can just say that some *difference or sum* (not ratio) is either the same, greater, or less than another by some amount, e.g., 1/4.

            Example: We can not say that 40°F, is twice as hot as 20°F, because zero degrees is *not a true zero value,* but we can say that the difference between 40 and 20 degrees is twice as great as the difference between 20 and 10 degrees.

∞ *Mean, median,* and *mode* computations of *central tendency* are appropriate for both ratio and interval scale scores (these computations are described in Chapter 11 under statistics and parameters).

2) Ordinal Scales of Rank-Order (Ordered/Ordinal) Data

∞ *Ranks* represent positions along some *ordered* dimension.

∞ *Ordinal scales* of ranks posses only the properties of *difference and magnitude* — not equal intervals or a true zero.

- Examples: Judged *attractiveness* of individuals and a hierarchy of *fears* based on perceived severity.

∞ It is *not meaningful* (mathematically) to multiply, divide, or even to add or subtract ranks, because equal differences between *ranks* do not represent equal differences in the actual *attribute measured.*

- Hence you can only say that something is the *same, greater,* or *less* than something else — you cannot compare the degrees of difference between ranks (a difference of 5 ranks could be less than that of 3 ranks elsewhere).

∞ *Median* and *mode* computations of *central tendency* are appropriate for rank-order values, but the *mean* is not, even though it too is sometimes computed.

- Example: *Grade-point averages* represent *means* rather than the appropriate computation of *medians* for such rank-order data.

3) Nominal scales and frequency data

∞ *Frequencies* represent the number of occurrences of events.

∞ *Nominal scales* are the weakest/least informative, because they have only the property of *difference* — and the differences are just *qualitative,* not quantitative, i.e., in *name* only.

- Examples: *Classes/categories/types of responses,* such as caring versus aggressive or studying versus partying.

∞ Assignment of *numbers* to classes (e.g., 1 for answering yes and 2 for answering no on a survey) would be *arbitrary* because the classes differ only qualitatively — thus it is not meaningful to multiply, divide, add, or subtract them.

- *Quantitative differences* are only in terms of *frequency* of occurrence for the various categories/classes of events.

∞ *Mode* computations are the most common measure of *central tendency* for nominal scales.

- Example: Which class/type of response is *most common.*

- *Median and mean* computations can be done for the *frequencies,* but the mean would not be very useful since it would always simply equal the total number of observations divided by the number of classes/categories of responses measured.

∞ *Frequency data*, it should be pointed out, are not limited to nominal scales (e.g., heart rate is frequency *score data*, which would be measured on a *ratio scale*).

— Participant awareness of the measurement process

• Reactive/obtrusive measurement

∞ This consists of observations that are made when participants are *aware* of and *reacting* to the measurement process itself.

• Hence that which is being measured is *influenced* by the very process of making the measurements.

• *Truism in the social sciences* is that the process of measurement might produce *changes* in that which is being measured.

Example: Observing the behavior of people while in the *same room*, rather than using a *one-way mirror*.

∞ Applications

• Reactive measurement is used when you want to investigate the *specific effects* on performance of having individuals *know* that they are being observed.

• It is also used when it's *necessary* to assess performance with the *full cooperation and support* of the research participants. This can assure *high motivation* and thus measurement of *optimal/best performance*.

Example: Studies of the effects of various warning conditions on the *reaction time* of individuals.

∞ Hawthorne effect (defect)

• This refers to the *improved performance* of individuals that is due just to their *knowledge or belief* that they are being observed and participating in an experiment.

It demonstrates the importance of *social factors, attitude, and expectations* on participant behavior.

Effect is an example of *reactive measurement* and its *pitfalls* (defects) when it is *undesired* and controls are inadequate (relates to "Demand Characteristics" in Chapter 7 under "General Sources of Variance").

• *Basis of the Hawthorne Effect concept* is a series of studies conducted in the late 1920s and early 1930s at the Hawthorne plant of the Western Electric company.

It was determined that the increased productivity produced in the workers was not actually due to any specific aspects of the treatments used by the researchers, because it was found that *every treatment* (changes in working hours, lighting, temperature, methods of pay, rest periods, etc.), whether designed to increase or decrease efficiency, led to *increased productivity.*

Improvements were presumably due simply to the workers' *awareness of being specially treated and observed/studied* — thus illustrating the importance of *social factors, attitude, and expectations.*

∞ Reactive measurement with intervention

- By definition, measurement that is reactive *always involves intervention*, because whatever the individual is aware of and reacting to would be an intervention.

- This represents either *experimental or descriptive research,* depending on whether comparisons are made, the form of independent variable manipulation (purposeful and direct versus measured selection, respectively), and the degree of control over extraneous variables.

- Nonreactive/unobtrusive measurement

  ∞ In contrast to reactive measurement, this consists of observations that are made when participants are *unaware* of and thus *not reacting* to the measurement process.

  - Hence the individual is *not influenced* by the process of measurement, and what is measured is *natural behavior.*

    Example: Observing animal behavior from a *blind.*

  ∞ Application

  - Nonreactive measurement is used when it is considered necessary to assess *uninfluenced, natural behavior.* This can assure the measurement of *normal/typical performance.*

    Example: Observing the *lawfulness* of car drivers.

  ∞ Nonreactive measurement without intervention

  - This involves observing individuals while *remaining undetected* (using a blind or one-way mirror) and *without modifying* environmental (habitat) or organismic variables.

  - It represents *descriptive research* using *naturalistic observation,* e.g., the natural-environment technique.

  ∞ Nonreactive measurement with intervention

  - This involves observing individuals while *remaining undetected* but with *purposeful and direct manipulation* of variables.

  - It represents *explanatory research* involving *experimental observation* of *natural behavior.*

    Example: Observing whether people will take the time and trouble to put an *uncancelled, stamped envelope* in a mail box when they *find it lying in the street,* as a function of some *manipulated variable* such as the expensiveness of the writing stationary.

- Participatory observation
  - ∞ This occurs when a researcher observes and records the behavior of individuals while he/she is *actively participating* with them in the circumstance/context of interest.
    - Such an approach is often used in situations that otherwise *might not be open* to scientific investigation.

      Example: Studying the *social behavior of students in schools* (including their dating, gossiping, aggression, sharing, group studying, and drug use) would be very difficult for a researcher who didn't act like a member of the group and therefore fit in.

    - Note that there is a high risk of introducing *biases* into the observations, i.e., *losing objectivity*, whenever a researcher interacts with those being observed — especially if the researcher *already belongs* to the group studied.

      Example: Researching the behavior of individuals in *singles clubs*, when it happens that the investigator is already a member of such a club.

    - Moreover, the *behavior* of those observed might be *influenced by that of the researcher* — especially if the observed are *aware* of being participants in a study (hence, what we would have is *reactive measurement*).

      Note: Such *intervention* may or may not be planned.

  - ∞ When people are *unaware* that they are being studied, then there is also the *ethical issue of concealment* and the *invasion of privacy* that might occur (see Chapter 6).

  - ∞ It should be pointed out that there are *various degrees* of *participation* possible and even of *concealment* — which impacts on the *ethics* of such research.

    Example: Some of the individuals might be *informed* and asked whether they would *give permission* for concealment.

    - Finally, it should be noted that people often *quickly become used to* the presence of researchers or their video cameras, and then behave naturally (e.g., families when observed in their own homes) — thus *negating the need* for *concealment* and/or *observer participation* in order to conduct effective research.

Accuracy and Consistency of Variables

Accuracy and *consistency* are two very important and interrelated *criteria* that apply not only to the measurement of *dependent variables* but also to the manipulation of *independent variables.* These standards of *accuracy* and *consistency* of variables are formally referred to as *validity* and *reliability,* respectively.

1. Validity of Dependent and Independent Variables
— Definition
  • Validity refers to the *accuracy/correctness/relevancy* of a measure or manipulation.
    ∞ Does, for instance, a *dependent variable* truly *measure* the construct/ concept that it is supposed to be measuring?
      • Example: How would you *correctly* measure *anxiety*?
    ∞ Validity also refers to whether *manipulations* of an *independent variable* actually *produce* the desired values.
      • Example: If *anxiety* were being treated as an independent variable, wouldn't you need to measure whether you had *accurately* produced the levels wanted?
    ∞ Note: *If the variables of a study are not valid, then the study itself is not valid, and thus it is of little, if any, usefulness.*
— Types of validity and their assessment
  Experts in the field of *test construction* have defined *four types/measures* of validity: *criterion-related validity, content validity, construct validity, and face validity.* These concepts are also of value to experimental psychology (note: other forms of validity are discussed in Chapter 9 under "General Principles").
  1) Criterion-Related Validity (Predictive Validity)
    ∞ *Criterion-related validity* asks whether the measure (some test or dependent variable) *compares well* with *external standard measures* that are considered to be *direct (actual) measures* of the characteristic or behavior being investigated?
      • In other words, does some *test or dependent variable* accurately *predict* behavior on the *criterion/true measure?*
        Examples: 1) Does performance on some *aptitude test* accurately predict performance in the *workplace?* 2) Does some measure of learning *nonsense syllables* in a *laboratory setting* accurately predict the ability to learn and apply *new concepts* in the *classroom?*
      • To answer the question, a *correlation coefficient* must be calculated between scores on the *predictor measure* (test or dependent variable) and performance on the *criterion measure* (external standard/direct measure).
        This correlation is referred to as a *validity coefficient.*
      • But establishing an *acceptable criterion* can be a problem.
        Example: What are *true* criteria for *success in life*?

- In any case, criterion-related validity is considered to be *more objective* than content validity (see next).

2) Content validity

   ∞ *Content validity* asks how well does the *content* of the measuring instrument (some test or dependent variable) *sample* the kinds of things about which conclusions are to be drawn?

   - In other words, how *representative* are the test items?

     Example: On *comprehension tests*, such as course exams, does the *content of the questions* relate well to the material that was *actually* taught?

   - *Logical examination* is required to answer this question. *Subjectivity*, however, is a problem when evaluating content validity. *Objectivity* can be increased by having *several judges* who *independently rate* the representativeness of the test items and then keeping only those items where there is *agreement* that they are *highly representative*.

3) Construct validity

   ∞ *Construct validity* asks to what extent do certain *explanatory concepts/constructs*, which are the basis of the theory underlying the measure, *account for performance?*

   - In other words, are the scores on the measure *consistent* with *theoretical expectations?*

   - *Correlations* between the *new* test, or dependent variable measure, and *established* measures of the explanatory constructs can be used to answer this question.

     Example: To validate a new test for the trait of *aggressiveness*, one could correlate the obtained scores on that test with those for various widely accepted measures of hostility, self-control, anxiety, and problem-solving ability — assuming that these were the new test's theorized bases of aggressiveness.

   - Construct validity is actually a *sophisticated form of content validity.*

4) Face validity (common sense or assumption validity)

   ∞ *Face validity* asks whether the test or dependent variable — at face value — *appears* to measure what it claims to measure?

   - In other words, is it a *generally accepted measure?*

     Example: Does *pupil diameter* make any sense to you (does it have face value) as a dependent variable measure of *pleasure or sexual arousal/attraction?*

   - Face validity is *not a rigorous concept*, but yet it *can't be ignored*: its presence in the absence of other standards can be very misleading, while its absence in the presence of other standards can result in disbelief.

Example: *Pupil diameter* can indeed be a *valid measure* of pleasure or sexual arousal/attraction, if extraneous variables are properly controlled.

- Note: The evaluation of *hypotheses and theories* also sometimes involves *face validity* with its inherent dangers.

    Example: The *Copernican Theory of Planetary Motion*, in which it was postulated that Earth circles the sun rather than the other way around, was rejected because it simply seemed absurd — after all, it did not feel as if Earth moved, and furthermore the *Bible* clearly implies that the Earth is the center of the universe and hence stationary.

— Sensitivity and range of the dependent-variable scale

- The *sensitivity and range* of the dependent-variable scale must be sufficient to allow the *effects* of different values of an independent variable to be *distinguished* from one another and thus to provide *valid* dependent-variable measures.

    ∞ *Maximally sensitive* dependent variable measures should be used whenever practical.

    - Generally speaking, *scale data* are preferred over *ordinal data*, and both are preferred over *frequency data*, because of their greater precision of measurement and hence sensitivity.

- Ceiling and floor attenuation/truncation effects

    ∞ *In ceiling and floor truncation* the *effects* of independent variable manipulations are *restrained* in either the *upper (ceiling)* or *lower (floor)* range of a dependent-variable scale.

    - Truncation occurs because the dependent-variable *task* is either *too easy (ceiling effects)* or *too difficult (floor effects)* given the *abilities* of the research participants.

        Example: Scores on a comprehensive statistics exam, for those who have had only one versus two statistics courses, would likely show a *ceiling effect* (many perfect scores) for the two-course group if the test were *too easy for their skill level* but on the other hand a *floor effect* (many zero scores) for the one-course group if the test were *too difficult*.

    - These scale attenuations *distort/skew* dependent variable distributions and hence the measurement of independent-variable effects.

    - More importantly, the restriction of the *range* of obtained scale values reduces *measurement sensitivity* to the independent-variable effects.

2. Reliability of Dependent and Independent Variables[19]

— Definition

- *Reliability* refers to the *consistency/stability/repeatability* of a measure or manipulation — which in turn infers accuracy/precision (elaborated later).

∞ Reliability also relates to the *generalizability* of data from a sample.

∞ *Dependent variable measures* have their reliability influenced by many factors, such as quality of the *apparatus* used, *care* taken when making measurements, and the quality of *instructions* given to participants.

- Example: Using an *old recording apparatus* that functions erratically would induce *variability* in the data and hence *reduce reliability* of the measures.

∞ *Independent-variable manipulations* too must be reliable/consistent in order to obtain reliable *dependent-variable measures*.

- Examples: An *apparatus* used for presenting stimuli must provide consistent stimulus duration, intensity, and clarity; and *instructions* used for manipulating independent variables must also do so consistently — which would not be the case if they were unclear.

— Techniques for assessing reliability

Three methods for determining reliability are: *test-retest technique, alternate-forms technique,* and *split-half technique.*

- *Correlation coefficient statistics* are used to compute reliability using any of the techniques.

  ∞ These indicate the *strength of relationship/association* between two variables or measures (covered in Chapter 2 and in Chapter 11 under "Statistics versus Parameters").

  - A *high positive correlation* between repeated measures on the same variable, taken for each of a number of individuals, indicates that there is a *high* degree of *consistency* and thus *reliability* of measurement.

    Example: When several individuals are measured twice on the same variable, such as performance on a proofreading task, a *high positive correlation* tells us that the individuals' scores on the second measure are *very similar* to those on the first measure (evaluated relative to the group as a whole) and hence the performance measure is judged *reliable.*

  - *Reliability coefficients* are what *correlations* are called when used to evaluate the *consistency of measures.*

1) Test-retest technique

  ∞ Correlating two *successive* measures on the *same test* is called the *test-retest technique.*

  - This is *usually not a desirable technique,* due to the possible effects of *recall* (carryover effects) from the first to second test.

  - Hence if memory is involved in the task being measured, then *artificially high reliability coefficients* could result.

    Example: The test-retest technique would be inappropriate for *problem-solving tasks,* in which the solutions could be remembered, but the technique would be acceptable for determining the reliability of *vigilance/attentiveness measures.*

2) Alternate-forms technique

∞ Correlating two *successive* measures on two *different but equivalent versions (parallel forms)* of the same test is called the *alternate-forms technique.*

  • Example: Determining the reliability of an *aptitude* measure by using different but *comparable/equivalent* sets of questions for two successive measurements.

3) Split-half technique

∞ Correlating two *simultaneously* obtained measures from *equivalent halves* of a *single test* is called the *split-half technique.*

  • Example: Comparing the number of correct answers on the *odd versus even items of one comprehension test.*

• Note that *all the techniques* for measuring reliability involve *replication of measurements.*

• In experimental research, being able to *replicate the results of an entire study* would not only indicate *reliability of the research findings* but would also represent a good test, and hence indication of the *reliability of the dependent-variable measure(s).*

∞ This is because *replication of an independent variable effect* is not likely to be attained if the dependent-variable measure is unreliable.

• Note that *inferential statistical analysis* of the results of a *single study* is in itself a measure or test of the *reliability* of measures (as mentioned earlier).

∞ *An experiment,* in fact, involves *replication* of measurements across the several participants assigned to the various conditions.

3. Components of Measures (See Appendix 2)

Measurement components, reliability, validity, and extraneous variables are further covered in Chapters 7 and 8, and Appendix 5.6.

— *Any measure* can be thought of as consisting of *two components*:

Measure = True Score + Error Score (i.e., measurement error).

• Note: this is only *hypothetical* — it is not actually possible to take a measurement and divide it into these two components.

∞ If it were possible to do so, there would be much less concern about *error,* because then the error component could always be *isolated* from the true score and hence *tossed out.*

• *Experiments* are an important illustration of this principle regarding the *components of any measure.*

∞ When a *dependent variable* is measured, the influences of both an *independent variable* (the true score component) and also one or more *extraneous variables* (the error score component) are added together and thus mixed.

- The *true score component*, it is assumed, taps relatively *stable* characteristics of the individual and thus should not vary substantially unless the testing conditions are changed.

- The *error score component*, on the other hand, is likely to *vary*.

  ∞ This *error score component*, in fact, can itself be *divided into two subcomponents:* Constant Error + Variable Error.

    - Error thus represents any *constant error of measurement* in addition to all the *variable error* due to *chance/random factors,* the latter causing fluctuating increases and decreases in measurement across both trials and participants.

    - Examples: 1) If *reaction-time/latency measurements* include the *speed/duration* of the responses, as is typically the case (see earlier discussion of this), then there would be some relatively *constant error* added. There would also be *variable error*, which would result from any random fluctuations in the accuracy of the timer, as well as chance variations in the attention, perceptual processing, and motor behavior of the research participants.

      2) When you measure your *weight* on a *scale,* you will ordinarily get slightly different values if you step on and off the scale several times. This represents *variable error*. Moreover, if the scale is not correctly *calibrated* to zero when there is no one standing on it, then there would be *constant error,* in addition to the variable error, each time a measurement were taken.

  - IN SUMMARY: All measures can be thought of in terms of components:

    Measure = True Score + Constant Error + Variable Error

— Determinants of reliability
  - Relative magnitudes of the true score and variable error
    ∞ *Measurement consistency* and thus *reliability* is greater to the degree that the *variable error subcomponent* of measurement is *small* relative to the true score component.
  - Sample size of a set of measurements
    ∞ Reliability also increases as more measurements are *averaged* together, because *variable error* reflects *chance/random* increases and decreases in measurement across both trials and individuals, and thus, statistically, variable error *averages toward zero as sample size increases* (see Chapter 11 under "Sampling Reliability").

— Maximization of reliability

•   Importance

∞   It is necessary to maximize reliability so that all measurements reflect mainly the *true score,* i.e., the *independent-variable effects* in experiments.

•   Variability associated with the *error score component* — i.e., *extraneous variable effects* causing measurement error — tends to *mask* (hide) or possibly confound (get *confused with*) any effects that different independent-variable conditions have on a dependent variable (covered further, along with related concepts, in Chapters 7 and 8, "Control in Experiments").

•   Testing data with inferential statistics

∞   If dependent-variable values are found to differ *significantly* for different values of some independent variable, then by definition, the dependent-variable measures differ to a *greater extent* than can likely be accounted for by just *chance variation,* i.e., experimental error.

•   *Sufficient reliability* of a dependent-variable measure is thus indicated by the attainment of *statistical significance.*

∞   If, on the other hand, significant effects are *not found,* then rather than this indicating that the independent variable does not affect the dependent variable, it might only mean that the dependent variable measure is *not sufficiently reliable,* and thus there is *masking* (known as Type II Error, and discussed in Chapter 12).

•   This *reliability issue* is an important basis for the *difficulty* of publishing studies that have only *negative results.*

∞   Note: Regardless of whether statistical significance is found, there might be *confounding* by extraneous variables that have a substantial effect on dependent-variable results.

•   This *potential for confounding,* especially by *constant error,* is a very serious problem for *validity.*

— Determinant of validity

•   Relative magnitudes of the true score and constant error

∞   *Measurement accuracy* and thus *validity,* are greater to the degree that the *constant error subcomponent* of measurement is *small* relative to the true score component (compare this with the determinants of *reliability* noted earlier).

— Consistency versus accuracy: reliability versus validity

•   *Reliability* in the form of *consistency* of measurement implies that there is also *accuracy* of measurement and thus *validity.*

∞   This represents part of the theoretical basis of the application of inferential statistics in *hypothesis testing.*

- *A key assumption,* however, is that if there is any *constant error,* it is relatively *small.*

   ∞ But, and this is very important, *if* there were a *small variable error* and yet a *large constant error,* then there would be *good reliability* in the form of consistency of measurement but nevertheless *low validity* due to the poor accuracy of measurement!

   ∞ Note: The true score, constant error, and variable error components of a *single measurement* are related respectively to the primary, secondary, and error variance components of a *set of measurements* in statistical analyses (this is covered in Chapter 7 under "Types of Variance" and is illustrated in Appendices 2 and 5).

      - *Hence the theoretical components of any measure represent a major explanatory principle not only for research but also for statistics.*

— Extraneous-variable control (elaborated in Chapters 7 and 8)

   - It is necessary to *maximize control* because, as already noted, *extraneous variables* are the source of *variable error* in the dependent variable and hence the reason for *low reliability* (inconsistency), which leads to *masking* or possible *confounding* of independent-variable effects.

   - In addition, *confounding* and thus *low validity* (inaccuracy) result from *constant error* caused by extraneous variables that have *consistent effects* and that vary *systematically* with the levels of the independent variable.

      ∞ Example: Consider two groups of research participants who are run in two *different independent-variable conditions* but using different apparatuses for each condition, which are *differently calibrated (an extraneous variable)* and which therefore give *different dependent variable measurements.*

      In cases such as this, the effects of the independent and extraneous variables on the dependent variable *could not be separated and distinguished.* Hence, there would be *confounding* and, thus, *low validity.*

      Confounding (as indicated earlier) can lead to *confusing* the extraneous-variable effects for independent-variable effects, or it can lead to *masking* of independent-variable effects.

   - Note: A very important point is that *low validity* associated with significant *constant error* can occur even if there is *high reliability,* caused by insignificant *variable error.*

      ∞ Unlike variable error and reliability problems, *constant error* and *validity problems* are not readily detected after the fact.

         - Therefore it is very important to *control constant error* by arranging for distribution of extraneous-variable effects *equivalently* across all independent-variable conditions, so as to *avoid confounding,* and then to analyze for the extraneous-variable effects so as to also *avoid masking.*

- IN SUMMARY: (see also Chapter 7, "Types of Variance" summary)
    - ∞ *Variable error* is associated with low consistency and thus low reliability, which leads to masking or possible confounding of independent-variable effects by extraneous variables.
    - ∞ *Constant error* is associated with low accuracy and thus low validity, which is due to confounding and possible masking of independent-variable effects by extraneous variables.
    - ∞ Note: both *variable error* and *constant error* can cause either extraneous variable *masking* or *confounding* of independent-variable effects on dependent variables.

Abstract Variables and Concrete Definitions

In addition to independent and dependent variables, as well as extraneous variables, there are other variables (or constructs) that are *less tangible.* Regardless of their differences, all types of variables must be clearly and unambiguously defined in terms of *observable events.* Abstract, *intervening variables* are discussed first (in relationship to the variables already covered) and then concrete, *operational definitions* are described.

1. **Intervening/Latent Variables and Hypothetical/Theoretical Constructs[19,32]**
    — Definitions
    - Intervening/latent variables (see Appendix 1, Figures A1.1 and A1.2)
        - ∞ *Abstract, unobservable or unobserved* processes or entities are called *intervening variables.*
            - In psychology this includes primarily *mental processes* — i.e., internal/organismic, central, in-the-head phenomena.
                Examples: motivation, attention, perception, emotion, learning, memory, planning, etc.
        - ∞ Intervening variables are *inferred* from behaviors that are their presumed indicants.
        - ∞ They *interconnect* concrete, *observable* independent and dependent variables, whose *relationship* they *label* (symbolize or summarize) and are also often hypothesized to *explain.*
            - *Meaning* of a given intervening/latent variable is, in fact, the particular *defined relationship* that it represents/labels between specific independent and dependent variables.
                Example: Hull's concept of *Habit Strength* $_sH_R$ is defined, other things being equal, as the *relationship* "between the number of *reinforced* repetitions of a stimulus-response sequence, and the subsequent *probability* that the stimulus will be followed by the response. Habit strength means just that relationship and

nothing else." [English, H. B., & English, A. C. (1958). *A Comprehensive Dictionary of Psychological and Psychoanalytic Terms*. New York: David McKay.]

∞ Hypothetical/theoretical constructs (scientific concepts)

∞ Either the term *hypothetical/theoretical construct* is *synonymous* with the term *intervening/latent variable* or in different usage it represents that *subset* of intervening variables in which there is the distinguishing implication that such constructs *additionally* have *independent existence* (although not fully observable) and thus that the hypothetical constructs have *surplus meaning*.

  • In other words, hypothetical/theoretical constructs are often thought of as having *measurable consequences* — explanatory, predictive properties — other than the observable phenomena and relationships that led to their being proposed and that define them.

    Example: *Learning*, which is commonly defined as a relatively permanent change in behavior as a result of experience, is believed to have *actual physical properties* (biochemical and/or anatomical changes at neuronal synapses/connections) and hence consequences and meaning in addition to the relationship that it labels/symbolizes.

— Two different interrelated levels at which scientists operate

  1) Theory, hypothesis, construct level

    ∞ *Abstract, unobservable (or not yet observed) phenomena* occur at the theory, hypothesis, and construct levels.

      • These phenomena are the *intervening/latent variable*s and *hypothetical/theoretical* constructs of hypotheses/theories.

      • They provide both *organization* for facts and *ideas* for research (which are the two functions of *theories*) by interrelating *several* independent variables on the one hand with *several* dependent variables on the other hand.

        Example: The intervening variable or hypothetical construct of anger links many independent variables with many dependent variables, such as:

        INSULTS↘   ↗ INCREASED BLOOD PRESSURE

        LOSS ↦ *ANGER* ↦ YELLING

        HARM ↗   ↘ PHYSICAL ATTACK

    ∞ Because such constructs *summarize* the effects of *several* independent variables on *several* dependent variables and usually under *many different environmental conditions* as well, they result in *parsimonious theories*.

- Note: There is *economy/efficiency of explanation,* such as in the above example where there are only six linkages necessary to explain the relationship among the three independent and three dependent variables, rather than the nine linkages (3×3) that would be necessary without the intervening variable — and this efficiency increases as the number of each variable increases.

- Note: *Science progresses* when a *single construct* can explain the *outcomes/relationships* for a *variety* of factors under a *variety* of conditions; and intervening variables or theoretical constructs provide for just such *systematized knowledge* — an important goal of science.

2) Observation Level

∞ *Concrete, observable, measurable phenomena* occur at the *observation level* (in contrast to the aforementioned theory, hypothesis, and construct levels).

- These phenomena are the *independent and dependent variables* that are *actually* manipulated and measured in research.

- They provide *data* for testing theories and hypotheses.

- Scientists must *move back and forth frequently* between the *observation and theory levels* in order to *generate* their theories from data and then to *test* their theories with additional data (note: these are the *inductive and deductive methods,* respectively, as covered earlier in Chapter 3).

∞ This shuttling back and forth requires the use of *operational definitions* (see below) of *unobservable* intervening/latent variables and hypothetical/theoretical constructs, given in terms of the *observable* independent and dependent variables that are their *presumed indicants* (empirical referents).

- In this manner, *manipulations* and *measurements* of *observables* are made possible, such that *theories* and *hypotheses* involving *unobservable* intervening variables and/or hypothetical constructs can be *empirically tested.*

∞ IN SUMMARY — *Deduction*: A theory/hypothesis with unobservable intervening variables → development of operational definitions → independent-variable manipulations and dependent variable measurements → observed test data → *Induction*: Comparisons of test data against predictions → either support for or modification of the original theory/hypothesis → further *deduction-induction cycles.*

2. Operational Definitions

— Characteristics

- Operational definitions define phenomena by specifying the *operations/procedures* used to *produce* and *manipulate* (for independent and some intervening variables) or *measure* (for dependent and some intervening variables) the concept/entity/process of interest.

∞ Example: In a study of the effects of different types of *music* on *cognitive performance*, we must specify both the operations for *producing* different forms/levels of music and the operations for *measuring* cognitive performance.

∞ Phenomena/concepts that are operationally defined are said to possess *operational meaning* — their meaning being synonymous with the corresponding *set of operations to demonstrate/obtain them* (i.e., to produce, manipulate, and/or measure them).

  • Example: *Intelligence*, an intervening variable or hypothetical construct, could be operationally defined as the score obtained on the Stanford-Binet IQ Test; and if this is how one chooses to *measure* intelligence, then in this particular instance that is its *operational meaning.*

∞ Operational definitions are not necessarily *valid definitions* — i.e., *independent, dependent, and intervening variables* are not automatically *valid* just because they have been operationally defined.

  • Example: *Intelligence* could be operationally defined in terms of the height of the forehead or the distance between the ears, but would you consider these *valid* (on the face of it) indicators/measures of intelligence?

— Functions

There are *four basic functions* of operational definitions: *to make variables empirical, repeatable, public, and clear/unambiguous.*

1) Operational definitions *transform/reduce unobservable constructs* (intervening variables) into *publicly verifiable sets of physical operations*, i.e., into *observable events* that are *public.*

∞ These definitions provide *objective meaning* to constructs by relating them to *empirical referents*, which are the very *procedures* for their production, manipulation, or measurement (a *requirement* for scientific observation).

∞ Hence operational definitions are the *mechanism* (as noted earlier) by which scientists can *move* between the level of *unobservable* intervening/latent variables or theoretical/hypothetical constructs and the level of *observable* independent and dependent variables.

  • Theories and hypotheses are often stated in the form of *abstract, conceptual independent and dependent variables* (e.g., anxiety or learning); thus these are actually *intervening variables or hypothetical constructs,* and they need to be operationally defined in order to transform them into *concrete, observable phenomena.*

    Example of such a theory/hypothesis: *Anxiety impairs judgment.* Note that this *postulate* contains only *conceptual* independent and dependent variables.

- *Full* operational definitions of *intervening variables or hypothetical constructs* would be given in terms of the *relationship* between how they're *produced* (independent variable manipulation) *and measured* (dependent-variable change), i.e., in terms of both procedures.

  Example: *Anxiety* could be operationally defined in terms of *types of threats* made to individuals, and also in terms of measured *changes in blood pressure*.

  However, when dealing with *conceptual independent or dependent variables*, the definitions are often just in terms of *either* how they are produced (for independent variables) $\overline{\text{or}}$ how they are measured (for dependent variables).

2) Operational definitions also make constructs *replicable/ repeatable*, in addition to their being made *empirical and public* as just noted (all of these are *requirements* for scientific observation).

   ∞ This results directly from operationally defining phenomena in terms of the *actual* operations that must and can be performed in order to *demonstrate/obtain* them.

3) Operational definitions additionally make constructs *clear and unambiguous*, thereby facilitating understanding and communication.

   ∞ Specifically, operational definitions *indicate which member of a class of events*, represented by a *multidimensional construct/variable* (a name or symbol for the class of events), is being referred to in a particular instance.

   - Example: *Aggressiveness* is a multidimensional construct, i.e., it has *many meanings*. Therefore we must carefully define *which form* we are referring to in any given study. Do we mean, for instance, being violent or being hard-driving and competitive?

4) Finally, operational definitions also *define* the more *concrete terms of theories and hypotheses*, i.e., the *independent and dependent-variables* (versus the intervening variables), in terms of *specific* observable events — *empirical referents*.

   ∞ Hence these, too, will be *clear and unambiguous*, as well as *empirical, replicable/repeatable, and public*.

   ∞ Example: When studying the effects of *food deprivation on problem solving for food reward*, we must specify — through operational definitions — how food deprivation is *produced and manipulated*, how problem solving is *measured*, and also how food reward is *produced* (Note: *measurement* is invariably involved too whenever producing or manipulating a variable).

- SUMMARY AND ELABORATION: Without *operationalism* — i.e., providing objective/empirical meaning to phenomena through operational definitions — there could be no *manipulation or measurement* and therefore no *experimental tests* of potential solutions to research problems!

∞ *Theories and hypotheses* whose terms *cannot* be operationally defined and thus *cannot be tested* are *scientifically meaningless*.

   • *Problems* whose terms *cannot be operationally defined* likewise *cannot be solved*, and they too are *meaningless*.

∞ In essence, operational definitions of variables are *sets of instructions* for carrying out research, e.g., experiments.

   • They are statements of the *operating procedures*.

   • They tell scientists *how to produce, manipulate, and measure* both conceptual intervening variables and concrete independent and dependent variables.

      Example: To study the effects of *daily exercise* on *improving the mind*, how would one possibly conduct such research without *operationally defining* these variables?

      Also, note how *changing the definitions* of variables might significantly *modify the study* and thereby possibly *change the findings*.

      Example: By *exercise* do we mean weight lifting, or aerobics, such as jogging or swimming? Furthermore, there are many different ways that could be devised for measuring the concept of *mind*.

— Thoroughness of definitions and complexity of constructs/variables

   • It isn't always necessary to state in research reports *complete operational definitions* for every concept/phenomena, because many have *standard operational definitions* that can be referenced.

      ∞ *New concepts, processes, and techniques*, however, require very precise and thorough operational definitions so that they have clear, empirical meaning and are repeatable and public.

      ∞ *Research problems and hypotheses* are often stated in very *general terms* in the *Introduction* of a research report, and then the terms are given *specific operational meaning* in the *Method*-section of the paper through descriptions of the materials, apparatus, and procedures of the study.

         • It should be noted that if a significant effect of the independent-variable manipulation is found, then the data *directly support* the hypothesis in just the *limited sense* that it was *operationally defined* — support for the *general hypothesis* is only indirect, i.e., a *generalization*. This is because an operational definition is only a *partial representation of a construct/variable* (e.g., anger) that represents a more *general class of events* (as pointed out earlier) — thus *no single definition provides a complete description*, and hence only *partial understanding* can typically result from a single investigation.

Circular Reasoning[26]

1. Vicious Circularity

— *Circular reasoning* consists of resting an element of one's thinking upon *another element*, which actually is *dependent on the first*.

- Example: "People who are optimistic expect the best; therefore, people who expect the best are optimistic." Does the second statement actually add anything to the first and vice versa?

— *Vicious circularity* is the term used for the form of this *erroneous reasoning* in which a *phenomenon* and its supposed *explanation* are derived from a *single set of facts* (empirical observations).

- In other words, the *answer* to a research question is based on the *question* itself, and thus we have simply gone in a *circle*.

- Such *vicious circularity* has historically occurred repeatedly in the form of *instinct and drive naming*, which can be generally described as follows:

  ∞ Why do organisms exhibit behavior "A"? Because they have drive or instinct "B." How do we know they have drive or instinct "B"? Because they exhibit behavior "A."

  - Example: Why does an anima*l ea*t? Because it's *hungry.*

    How do we know that it's *hungry*? Because it *eats.*

    Note: Eating *has not been explaine*d; we simply now have *another name* for eating behavior, i.e., hunger behavior.

- Vicious circularity occurs in *other similar forms* as well.

  ∞ Example: To ask why a man is a *hard-driving, competitive businessman* and to be answered that it is due to his *aggressive nature* is viciously circular if by aggressive nature we simply mean that the man is hard driving and competitive.

- Thus it can be seen that the *existence of a construct/concept,* even one that is adequately defined, does *not* necessarily constitute an *explanation* for the behavior or other phenomenon to which that construct is related: the construct might serve only to *differently name or describe* the behavior of interest.

  ∞ *"Nominalistic trap of tautology"* is what this result of vicious circularity is called, which refers to the *naming trap* of *needlessly repeating* an idea, statement, or word (as we have just illustrated); i.e., it is *simply renaming something* rather than actually explaining it.

— *Meaningless problems and hypotheses* are the consequences of circular reasoning (just as occurs when the variables cannot be or are not operationally defined).

- *Problems* can't be solved if *hypotheses* can't be tested due to circular reasoning that makes it impossible to disprove them.

- *To break the vicious circularity — to get out of the trap — what is needed is an operational definition of the explanatory concept that is independent of the phenomenon it is used to explain.*
  - ∞ We must appeal to *information outside the vicious circle* — only then does the concept have explanatory value.
  - ∞ Example: With regard to hypothesizing that eating is caused by *hunger*, one must operationally define hunger *independently* from eating; e.g., in terms of the amount of food in the stomach, the blood-glucose level, or the hours of food deprivation.

2. Circular Definitions

  - ∞ Closely related to *vicious circularity* are *circular definitions*.
- They result from defining a concept in terms whose definitions *refer back to the original concept*, and thus little is accomplished.
  - Example: Defining the *nervous system* as a complex, organized collection of *neurons* and then defining *neurons* as the fundamental functional unit of the *nervous system*.
    - ∞ *Dictionaries*, unfortunately, are sometimes guilty of this circularity — which of course can be very frustrating.

Review Summary

1. The three *primary elements of experimental research designs*, with respect to the three major *variables* involved in experiments, are as follows:

   a. Selection of *independent variables* and specification of their levels and the means for their purposeful, direct manipulation

   b. Determination of *extraneous variables* and specification of the techniques for their control

   c. Choice of *dependent variables* and specification of the methods for their measurement

2. *Independent variables* can be *environment/stimulus factors*, such as instructions, tasks, procedures, and external stimuli; or *organismic/participant factors*, such as anatomical, physiological, and psychological characteristics.

   These may be manipulated by *instructions*, which are associated with problems of attentiveness and interpretation; by *events*, which minimize these problems and increase realism, meaningfulness, impact, and "real world" representativeness of behavior; or by *measured selection*, which is associated with poor control of extraneous variables.

3. There must be *at least two levels/values/conditions* of an independent variable in an experiment, because *comparison* is necessary to determine any effects on a dependent variable. There are two *forms of conditions*:

   a. A *control condition* — zero or standard value — might be used to establish a *baseline* and to *control and evaluate* extraneous-variable effects

   b. One or more *experimental/treatment conditions* — non-zero or non- standard value — would be compared against each other and/or a control condition to determine the effects of the independent variable

   Variation of independent-variable levels can consist of *presence versus absence* (i.e., treatment versus control conditions), *amount* (i.e., quantitative variation in treatments), or *type* (i.e., qualitative variation of treatments). *Combinations* of these forms of variation are common.

   Although there are no theoretical or statistical limits to the *number of independent variables and levels* that can be investigated in a single study, there obviously would be practical limits. The same can be said for *dependent variables*, for which it might be advantageous to measure more than one in a study because many phenomena/constructs are multidimensional, e.g., learning, and the different measures might not be highly correlated.

4. *Dependent variables* can be *environmental/overt-behavioral responses*, such as pressing buttons, or *organismic/physiological responses*, such as nervous system activity. Combining these two categories with those for independent variables, there are three sources of variables: *environmental stimulus (S) independent variables, environmental response (R) dependent variables, and organismic events (O)*, which can be either stimulus-independent variables or response-dependent variables. There are four *types of causal relationships* among these that are commonly studied:

$$S \rightarrow R, \quad S \rightarrow O, \quad O \rightarrow R, \quad \text{and } O \rightarrow O.$$

5. There are at least six general *types of measures* for dependent, as well as independent, variables (there are alternative forms/names for each type):

   a. Magnitude/amplitude/intensity

   b. Frequency/rate/probability/proportion/number

   c. Latency/reaction time

   d. Speed/duration

   e. Accuracy

   f. Complex score/profile

6. *Nonreactive/unobtrusive measurement* is when participants are *unaware of, and thus not reacting to*, the measurement process. This is advantageous if natural behavior is desired. It can occur with or without *intervention*, i.e., the purposeful, direct manipulation of external or internal environmental variables (representing experimental vs. descriptive research, respectively).

Reactive/obtrusive measurement occurs when participants are *aware of and reacting to* the measurement process itself, which represents an *intervention*. This form of measurement is advantageous if, e.g., high motivation, and thus optimal performance are desired.

Participatory observation is the active participation of an investigator with those being observed in order to better obtain data — usually nonreactively.

7. *Validity* of dependent and independent variables refers to the *accuracy, correctness, or relevancy* of a measure or manipulation, which in turn affects the validity of a study. There are at least four types of validity:

    a. *Criterion-related validity* is obtained when the test or dependent variable accurately *predicts* behavior on some direct (criterion) measure, as assessed by computing a correlation (i.e., validity) coefficient.

    b. *Content validity* is used when the content of the measuring instrument (test or dependent variable) is a *representative sample* of the kinds of things about which conclusions are to be drawn, as judged subjectively.

    c. *Construct validity* is measured when *explanatory concepts* (constructs), which are based on the theory underlying the measure, account for performance.

    d. *Face validity* occurs when the test or dependent variable *appears* to measure what it claims to measure, i.e., it is generally accepted.

8. The *sensitivity and range* of a dependent-variable scale must be sufficient to allow the *effects* of different values of an independent variable to be *distinguished*. If dependent variable measures are too easy or difficult, then the effects of independent variable manipulations will be restricted/limited. Specifically, there will be *ceiling* (upper) or *floor* (lower) *attenuation truncation effects* when the measures are too easy or difficult, respectively.

9. *Reliability* of dependent and independent variables refers to the *consistency/stability/repeatability* of a measure or manipulation. There are three general *techniques for assessing reliability,* all of which involve *replication* of measurements and computation of a *correlation* (i.e., reliability) coefficient:

    a. *Test-retest technique* involves correlating two *successive* measures on the *same* test.

    b. *Alternate-forms technique* involves correlating two *successive* measures on two different but *equivalent* forms of the same test.

    c. *Split-half technique* involves correlating two *simultaneously* obtained measures from *equivalent* halves of the same, single test.

10. *Any measure* can be thought of as consisting of two *components*: a *true score* and an *error score*. These relate respectively to the influences of an independent-variable versus extraneous variables on a dependent variable measure. The error score component can itself be divided into two *subcomponents*: *constant error* and *variable error,* the latter resulting from chance/random factors.

11. *Reliability* is greater to the extent that *variable error* is small relative to the true score. This minimizes *masking* by extraneous variables. Reliability in the form of

consistency of measurement implies that there is also *accuracy* of measurement, and thus *validity*. But this assumes that any *constant error*, which undermines accuracy and thus validity, is relatively small. Small constant error minimizes *confounding* by extraneous variables.

12. *Operational definitions* are those that define phenomena by specifying the *operations/procedures* used to *produce* and *manipulate* (for independent and intervening variables) and/or *measure* (for dependent and intervening variables) the concept, entity, or process of interest. This provides phenomena with *operational meaning*. It is important to note, however, that operational definitions of variables are not automatically *valid* definitions.

13. There are four *functions* of operational definitions:

    a. *Transform/reduce abstract, unobservable constructs* (intervening variables) into publicly verifiable sets of physical operations, i.e., *observable events*.

    b. Make constructs *replicable/repeatable*.

    c. Make *multidimensional* constructs/variables *clear and unambiguous* by indicating which *member* of the class of events is being referred to.

    d. Define also the more *concrete* (independent and dependent) variables of theories and hypotheses in terms of specific observable events.

14. *Vicious circularity* is *circular reasoning*, which consists of resting an element of one's thinking on another element that actually is dependent on the first. Thus, the answer to a question would be based on the question itself, i.e., the phenomenon and its explanation would be derived from a *single set of facts* — empirical observations. For example; why does an animal drink? Because it is thirsty. How do we know that it is thirsty? Because it drinks. This leads to *meaningless* problems and hypotheses. *Circular definitions* are closely related to vicious circularity, because they result from defining a concept in terms whose definitions refer back to the original concept.

Review Questions

1. What are the three *primary elements of experimental research designs* with respect to the three major *variables* involved in experiments?

2. List and discuss the three general *means of variation/manipulation of independent variables,* and discuss their relative advantages and disadvantages.

3. List and describe the two *forms of independent-variable conditions/levels*, including their functions, and also the three *forms of variation of independent variables* (e.g., presence versus ...), as well as their combinations.

4. List, describe, and give an example for each of the two *categories of independent variables* and also *dependent variables,* and diagram the four *types of causal relationships* commonly studied among the three *sources* of these variables.

5. List and give an example of three of the six *types of measures* for dependent (and independent) variables, other than complex score/profile.

6. Define *nonreactive measurement versus reactive measurement*, and explain what is meant by *intervention*.

7. Define what is meant by the *validity* of variables, and describe two types other than face validity.

8. Describe *ceiling and floor attenuation effects*, and explain their causes.

9. Define what is meant by the *reliability* of variables, and describe three *techniques for measuring reliability*?

10. State the *components and subcomponents* that any *measure* can be thought to consist of, and explain how these relate to *experiments*.

11. How do *reliability* and *validity* relate to the theoretical components and subcomponents of any measure, what is the *implied relationship* between reliability and accuracy of measurement, what *assumption* is involved in this implied relationship, and why can this be a *problem*?

12. Describe what *operational definitions* are in terms of the three ways they can define phenomena.

13. What are the four *functions of operational definitions* (be sure to discuss intervening variables)?

14. Define *vicious circularity* and *circular definitions*, giving an example of each.

5

Initial and Final Phases of Research

CONTENTS

Previous chapters of this Handbook have covered a number of aspects of the principles and methods of experimental research. Before going any further, it is appropriate to now discuss *how research actually gets initiated*. In addition, at the end of this chapter the *final* step of the scientific method is covered, communicating research findings through a written research report, because of its close relationship with the writing of a research proposal, which takes place when initiating research.

Literature Surveys

1. **Definition**

    — A *literature survey* is a *thorough search and careful review* of all publications that are relevant to the research problem selected for investigation.

    • The survey should include both *research reports* and *theoretical writings* that appear in professional *journals* as well as *books*.

2. **Functions**

    There are *four major reasons* for conducting literature surveys:

    a. Surveys tell the investigator what theorizing and research *has already been done*, and what *still needs to be done*.

    • An intended study, e.g., might already have been run by *others*.

    ∞ *Replications* to confirm the findings of other investigators are often *desirable*, but they should always be *intentional*.

- Rather than simple replications, scientists more commonly review the literature because they want to learn the *specifics* of what others have done, and then *expand* on that work by *modifying* and *adding* to the body of theoretical writing and research.
  - ∞ This can involve the questions asked, hypotheses proposed, and the variables, types of participants, and methods used.

b. Surveys sometimes help the investigator to formulate *better research problems* and *hypotheses*.
- *Vague* ideas might be made more concrete.
- *Misconceptions* might be corrected.
- *Modifications* might be made that improve on an initial research problem and/or hypothesis.

c. Surveys suggest what *research apparatus* could be used, what the *extraneous variables* might be, and how best to *control* them.

d. Surveys are necessary in order to provide *summaries and references* to previous relevant studies and theories, as is required when writing research proposals and final research reports.
- Note: Although for all the above reasons a literature survey is essential *before* carrying out a research project, a *supplemental literature review* might also be necessary *after* a study is completed, especially if a considerable *period of time* has passed since the first survey, or if more information is needed in order to fully discuss and interpret any *unexpected results* of a study.

3. Mechanisms

There are a number of important aids commonly used to conduct thorough literature surveys, *five* of which are listed below. *Additional resources* are found in most research libraries, and the librarians can be of great assistance in locating and using these resources.

— *Psychological Abstracts*
- This is the *traditional* literature review resource for psychologists and other behavioral scientists.
- It is published by the American Psychological Association (APA), and is updated monthly and cumulated annually.
- *Psychological Abstracts* indexes articles from over 1300 journals as well as books and book chapters in psychology, biology, sociology, management, education, etc. — thus it is interdisciplinary.
- Each article and book chapter entry includes an abstract/summary, while book entries also contain the table of contents, and all entries contain bibliographic information for finding the item.
- Subject, author, and book title indexes are provided so that it is user-friendly, although tedious to use compared with modern, computerized search resources (see the following).

— *PsycLIT and PsycINFO*

- *PsycLIT* is a computerized, CD-ROM–based version of *Psychological Abstracts*.

- It is very fast and powerful, allowing searches by subject and author, as well as by keywords and their combinations that might be found in titles, abstracts, keyword lists, etc.

- However, CD-ROMs are updated quarterly, rather than monthly.

- *PsycINFO* is an on-line service that is the main bibliographic database for *PsycLit*, but it contains entries going all the way back to 1887 for journal articles, doctoral dissertations, and technical reports, as well as for book chapters and books.

  - ∞ PASAR (PsycINFO Assisted Search and Retrieval) will do customized searches for those not having access or not wanting to do searches themselves (it takes about 2 weeks and there is a substantial charge — contact APA).

— *Social Science Citation Index*

- This is an interdisciplinary source covering all the social sciences.

- It is used for locating *more current publications* than what the researcher has at hand.

- From any given reference on a subject, the researcher can obtain a list of more recent articles that have cited that reference, and thus that are likely to be related to the topic of interest.

- *Social Science Citation Index* is available both on paper and as a computerized data-base resource on CD-ROM.

- Subject, author, and citation searches are possible.

— *Current Contents*

- This is useful because it publishes in a timely manner, and on a weekly basis, the table of contents for journals and books.

- Author and title indexes are provided, along with author and publisher directories of addresses.

- Separate editions are published of *Current Contents* which cover the different areas of science, e.g., the *Social and Behavioral Sciences* edition for psychologists, and the *Life Sciences* edition for biologists.

- This resource is also available on diskettes for computerized database searches.

— Internet and the World Wide Web

- The *Internet* (also known simply as the Net) is an ever expanding network of networks that ties together millions of computers and tens of millions of users around the world so that they can efficiently exchange information, ideas, and opinions.

- The *World Wide Web* (also known simply as the Web, or WWW) is the graphics-intensive component of the Internet.

- Using these resources requires a personal computer with modem (preferably high-speed), as well as browser software for searching and displaying information, and an Internet service provider for connecting to the Net through a phone line or cable.

- Searching the Internet and Web resources through the use of key words can pull up electronic journals with complete articles, figures, tables, and even videos; book chapters; lists of various references; conference dates and locations; and a variety of other useful information presented in a number of ways, such as electronic bulletin boards and chat rooms.

— Publication references

- Lists of useful publication references are also found, of course, in research, theoretical, and review *articles*, as well as *books*.

  ∞ Thus, an ever expanding number of references can be located for a topic of interest by simply starting with the references found in any one article, book, or chapter on that topic, and then going to the references that those references themselves contain, and then to their references, and so on.

Research Proposal (See Appendix 3)

1. Definition

— A *research proposal* is a thorough *plan and rationale* for a scientific study (i.e, a prospectus), which is usually written out and submitted for *critical review and approval* by others with expertise in the area.

2. Functions

Written research proposals serve *four basic purposes*.

a. *Ambiguities or problems* that might exist for an intended study are *brought to light* by putting a proposal in writing, which then prompts the investigator(s) to *correct* the weaknesses.

b. *Having others critically review a research proposal*, which is best done when the proposal is put in writing, further helps to uncover any *biases* of the investigator, *overlooked points*, and *logical errors*.

c. *Obtaining research funds, space, or time* typically requires that a written research proposal be submitted to an administrator, committee, or granting agency that provides research support.

d. The eventual write-up of a study for *publication* typically utilizes much of what was written in the research proposal, e.g., regarding the study's purpose, relevant literature review, hypotheses, method, and possible theoretical and practical implications of the work.

3. Components

— These are listed and described in Appendix 3, which should be referred to at this time.

Pilot Study

1. Definition

— A *pilot study* is a *preliminary investigation* that is conducted prior to a *full-scale study.*

2. Characteristics

— Pilot studies vary greatly in *extent* and *complexity.*

- They can be limited to a *simple tryout* of an *element* of the research, e.g., the apparatus, instructions, or experimental tasks.

- More commonly, however, they involve a *complete run through* of all the procedural steps of a proposed study, but using only a relatively few participants.

— Pilot studies often *precede* rather than follow the formal *writing of a research proposal.*

- In fact, given their functions (see next), pilot studies might actually be *required* before submitting a proposal to get *funding* for research.

3. Functions

There are *seven primary reasons* for conducting pilot studies.

a. *Experimental procedures* (e.g., instructions and tasks) plus the *apparatus and materials* can be *checked* — either completely or in part.

- *Flaws or weakness* in the instructions (e.g., insufficient clarity) and in the tasks (e.g., they might be too easy or too difficult), as well as *malfunctioning* of the equipment and *inadequacies* of the materials, are likely to be revealed so that they can be corrected before the full study is carried out.

   ∞ *Feedback from pilot participants* can be very helpful in this regard, and therefore a *set of questions* should be prepared in advance, and participants should also be encouraged to make any *additional comments* they want to after studies are run.

b. *Practice* is provided for researchers on the *procedures* involved with independent variable manipulations and dependent variable measures, including the use of materials and any needed apparatus.

c. *Independent variable levels* can be evaluated to determine those that appear to be *most effective* in producing differential effects on the dependent variables (i.e., maximizing the primary/experimental variance, which is covered in Chapter 7).

d. *Dependent variable measures* can be evaluated for those that are *most sensitive* to manipulations of the independent variables, and for having a *sufficient range* to avoid ceiling and floor truncation effects (which were discussed earlier in Chapter 4).

e. *Extraneous variables* can be *uncovered* that need to be dealt with, and various *control procedures* can be evaluated for their *effectiveness* (as discussed in Chapters 7 and 8).

f. *Data variability* (error variance) can be *estimated* to determine the sample size needed in order to obtain *sufficient reliability* of dependent variable effects (as discussed in several chapters).

g. *Results* that will likely be found in the main study can be *estimated*.

- *Modifications* of the research hypothesis(es) might thereby be *suggested*.

- *Other research problems/questions* might become apparent that could be *substituted* for the original problem, or be pursued later.

- *Potential contributions of the study* could be *evaluated* to determine whether it would probably be *worthwhile* to actually go ahead with the full study.

Serendipity

1. **Definition**

   — *Serendipity* is the art, process, or fact of *finding something of value* while looking for something else.

2. **Implications for Research**

   — For the scientist, serendipity means being alert to and recognizing the potential importance of *unusual or unexpected observations* — and then following up on them.

   - Such accidental observations can lead to *fruitful hypotheses* for testing, which might be more interesting than the original ones.

     ∞ Many *important discoveries* have been made serendipitously.

       - Example: James Olds and Peter Milner unexpectedly discovered the *pleasure/reward centers of the brain*, located in the limbic system, while looking for more information about the reticular activating system and its role in drive and motivation [Olds, J., & Milner, P. (1954). Positive reinforcement produced by electrical stimulation of the septal area and other regions of the rat brain. *Journal of Comparative and Physiological Psychology, 47*, 419-428.]

   - Note, however, that the *large majority* of unusual or unexpected occurrences have *little scientific significance* — they are often just distracting *extraneous variable effects* that should be controlled.

∞ *Overemphasis* on the role of accidents in scientific discovery is thus *misleading*, and can lead to *reduced productivity*.

- Scientists, therefore, should not enter their laboratories thinking that serendipity can take the place of the *systematic application of the scientific method* for making meaningful discoveries.

- *Following up on every unexpected occurrence* during research could consume most of an experimenter's time and energy, while providing little benefit.

∞ *Experience and good judgment* clearly are important factors in evaluating the potential significance of unusual events.

Research Report

1. Definition

— A *research report* is a *presentation* of the purpose, background, hypotheses, method, results, and conclusions of a scientific study.

- The research report is best done in *writing*, but additionally or instead a report is often delivered *orally* or with *posters* at a conference or other meeting of some professional organization.

2. Functions

There are *five general reasons* for preparing research reports.

a. *Communication to other scientists*, etc., is necessary in order that the data and their interpretation become part of the *systematic body of knowledge of science*, rather than remaining just *private knowledge*.

b. *Historical functions* are served by research reports, i.e., identifying *who was first* to accomplish something and *when*.

c. *Judging the quality* of research by others, and conducting further experiments for *verification and elaboration* of the findings, requires research reports that can be carefully studied and built upon.

d. *Unsuccessful approaches* to research problems are sometimes noted in reports, and these methods can then be *avoided* by others.

e. *Unintended repetition* of studies can be *minimized* by research reports that get published, and are thus made widely available.

3. Qualities to Strive For When Writing Research Reports

There are *five criteria/characteristics* of good report writing.

a. Accuracy

- Scientifically and ethically it is an absolute necessity that research reports be *factual*, and also as *precise* as possible.

b. Completeness
- Sufficient information must be provided in scientific reports to permit *replication* of a study, and thus *verification* of findings.

c. Clarity
- *Ambiguity* must be avoided in all scientific communication.
- *Precision* in the choice and use of words is therefore required.
- *Orderly* presentation of ideas is necessary for their *logical development* so as to facilitate *comprehension*.
- *Transitional markers* should be used so that the writing flows *smoothly* and without impediments to *understanding*.

d. Conciseness
- *Succinctness* is important to minimize *reading time*, and also to hold down *publication costs*.
- *Constraints*, however, are obviously imposed on brevity by the *preceding requirements* of accuracy, completeness, and clarity.

e. Readability
- Exploring research reports should be *enjoyable* and *painless*.
- Writers must therefore *do more than just document facts*.
- Research reports should not be dull and formal just to make them appear more scientific, rather they should be written in an *interesting and compelling* fashion so that they will be read.
- Readers should feel *invited* to dig in, and *encouraged* to continue, by the *clarity and persuasiveness* of the report's presentation of ideas, methodology, findings, and conclusions.
- *Restrictions*, of course, are placed on a writer's *literary freedom* by the other criteria of good report writing; but, although readability might be the *least important quality* to achieve, it is nevertheless *very desirable*.

4. **Publication Manual of the American Psychological Association**[4]

— The *primary guide* in psychology for preparation of research reports, doctoral dissertations, master's theses, and term papers is the *Publication Manual of the American Psychological Association*.

- Because of its quality and comprehensiveness, this APA style manual, which has gone through several revisions, is used in *many disciplines* in addition to psychology.
- *Copies* can be purchased from most university bookstores, or by writing to the American Psychological Association Order Department, P.O. Box 92984, Washington, D.C., 20090-2984.

— With the informative purpose of research reports so paramount, and with the economy of time and money so pressing, *important writing conventions* have evolved that are contained in this style manual.

— Chapters are provided in the latest edition on the following topics:

1) Content and Organization of a Manuscript

2) Expressing Ideas and Reducing Bias in Language

3) APA Editorial Style (this is the most extensive chapter)

4) Reference List

5) Manuscript Preparation and Sample Papers

6) Material Other than Journal Articles

7) Manuscript Acceptance and Production

8) Journals Program of the American Psychological Association

— Note: The *APA Publication Manual* contains such a wealth of information on effective writing (Chapter Two) and editorial style (Chapter Three), that no attempt will be made to duplicate this material here (summaries of this information are available on the Internet, but the manual itself should be studied for details).

— What follows is an outline (built around Chapter One of the manual) on the *content and organization* of typed research papers, as well as a few of the more important elements of the *APA editorial style*.

5. Parts of a Research Paper[4]

Research papers are written with *nine standard components*, which make the reports easier to follow and aid the reader in finding specific information. Each major part of the paper, except the title page and the introduction (which begins with a repeat of the title), is identified by a *heading* that is capitalized and centered left-to-right. Certain parts of the paper have subsections, whose headings are capitalized, placed flush left, and underlined. Some research papers have more than just these two levels of headings (see the *APA Publication Manual* for specifics).

a. Title page

1) Title and running head

∞ *Titles* should provide a concise (10–12 words), self-explanatory *summary label* for a paper — avoid abbreviations.

- Example: "Effects of Choline Supplements, Given Orally and Daily, on Recall Memory of Alzheimer's Patients."

- *Research problems/questions* should be specified in the title, usually by stating the specific *independent-dependent variable relationships* that were studied.

- *Participant populations*, if not humans in general, should also be indicated (e.g., children, schizophrenics, or rats).

- *Research methodology* should be concisely noted as well.

- *Unnecessary words and phrases* should not be included, e.g., "The" and other articles at the beginning of a title, or "An Investigation of" or "A Study of" as a title lead-in.

∞ *Running heads* are an *abbreviated title* (≤ 50 letters, spaces, and punctuation) that is printed at the top of all pages of an article, *when published*, in order to identify it for the readers.

- *When typing* the manuscript to be submitted, however, the running head is placed flush left near the top of only the title page, just below the *manuscript page header*.

- *Manuscript page headers* are what are placed at the top right of each page when typing, in case the sheets get separated, and should include just the first two to three words of the title, along with the page number (Note: *when the article is published*, the *running head* becomes the page header).

2) Authors and Institutional Affiliations

∞ *Authors* listed on the title page should include all individuals who played a *substantial role* in the research.

∞ The names are usually listed in the *order of importance* of the researchers' contributions, although in some non-APA journals the names are simply listed *alphabetically* (center the names just below the centered title of the paper).

∞ Give the first name, middle initial, and last name of each author; but *omit titles*, such as Dr. or Ph.D.

∞ *Institutional affiliations* indicate the workplace of the primary author(s), which is usually where the study was conducted (center this just below the names of the authors).

b. Abstract

- The *Abstract* is a one-paragraph, *concise yet comprehensive, self-contained summary* covering the main points of a study — but not containing information that isn't in the body of the paper.

∞ *Length* of an abstract should not exceed 120 words (the paragraph is not indented, and to conserve characters all *numbers* should be typed as *digits*, except those that begin a sentence — these rules are less complex than those for numbers in the body of a paper).

- Statements about the following should be placed in the abstract:

∞ *Research problem(s)* investigated;

∞ *Research method*, including participant characteristics (number, species, sex, and age), apparatus/materials, and procedures;

∞ *Results*, including statistical significance levels;

∞ *Conclusions*, especially regarding whether or not there was support for the *research hypotheses*, and also noting the theoretical or practical implications of the study's findings.

- Usually it is *easiest* and *best* to write the abstract *after the rest of the paper has been written*, in order to ensure that all of the most important information is summarized.

- The abstract is usually the *first contact*, other than the title, that a reader has with an article when doing a *literature search*, and it often determines whether or not the paper as a whole will be read — in this sense it is the most important part of the paper.

- *Key words* for indexing purposes are imbedded within the abstract (or in many non-APA journals they are listed separately just below the abstract).

c. Introduction

- The *Introduction section* is the part of the paper that lets the readers know 1) the *reasons* for conducting the research, 2) how the research relates to *prior work* — the scientific literature, 3) what the investigators *hypothesized* they would find, and 4) the general nature of the *research design*.

  ∞ *Reasons (goals/purpose)* for doing the study should be given near the beginning of the Introduction section.

    - *Research problems (questions)*, whose solutions were the reasons for the study, should be stated clearly, along with an explanation of the theoretical and/or practical importance of finding the solutions/answers.

  ∞ *Scientific literature* related to the research problems should be *reviewed* in a *logical sequence* (e.g. chronologically, or from the more general and less relevant to the more specific and relevant), and it should be made very apparent how the study is logically derived from (connected with) earlier work.

  ∞ *Research hypotheses* should be stated clearly, preferably in an "if ___ then ___" or *mathematical form* that specifies the predicted relationships between *independent and dependent variables*, as well as any *conceptual (intervening) variables*.

    - *Rationales* for all the hypotheses, i.e. their *logical bases*, should also be presented.

  ∞ *Research design* used in the study should be described in at least a general manner.

    - *Details* about the design, as well as *operational definitions* of the independent and dependent variables, can be saved for the "Method" section (see below).

d. Method

- The *Method section* is the part of the paper that describes in detail *what* was done in the study and *how* it was done.

  ∞ It is important that sufficient information be provided so that the research methods of the study can be *evaluated* for their *quality and appropriateness*, and hence the data for its *validity and reliability*, and so that others can attempt to *replicate* the study.

- This section of the paper is typically *divided into subsections* (usually three) for ease of reading and clarity:

1) Participants or Subjects

    ∞ *Species* that served as research participants/subjects, and the *total number used*, should be indicated along with sufficient information about their *characteristics* to permit comparisons with other studies, as well as possible attempts to replicate the research.

        - *Human participants* would be described in terms of the number of each sex, ethnic/racial group, and national origin, as well as ages, education levels, pathologies, etc.

        - *Nonhuman animal subjects* would be described in terms of their genus, species, strain, and supplier; their sex, age, and weight characteristics; and any pathologies.

    ∞ *Ethical standards* that were adhered to, both *professional and statutory,* should be indicated for all species when applicable.

        - *Housing, care, and handling* of nonhuman animal subjects should be described sufficiently to permit evaluation, comparison with other studies, and possible replication.

    ∞ *Selection* of the research participants should be explained.

        - *Sampling procedures* that were used should be described, since their *validity and reliability* determine the *representativeness* of a sample with respect to a population, and thus what *generalizations* of results are appropriate.

        - *Compensation* that was given to humans for participation in the research (monetary, etc.) should be specified.

    ∞ *Assignment* of participants to the different conditions (independent variable levels) should be described in terms of *procedures used* and the *number* assigned to each condition.

        - The *number* of individuals in each condition that *did not complete* the study should be specified, along with the apparent *reason(s)* for the *attrition*, and the probable *representativeness* of those that quit, and hence of those that remained.

2) Apparatus and materials

    ∞ *Equipment* and other *materials* (e.g., stimuli, paper-and-pencil tests, and simulated naturalistic environments) should be *described*, and the *reasons* for their selection explained.

        - *Specialized equipment* should be identified by the manufacturer's name and model number, as well as an address (note: long detailed descriptions, if any, belong in an *appendix*).

        - *Specialized tests* would be similarly identified.

- *Illustrations* can be helpful when describing complex or custom-made equipment (possibly placed in an *appendix*).

∞ *Laboratory or field settings* where the study was conducted should have all their pertinent *features* described, and in sufficient detail to permit replication.

  - *Laboratory studies* would specify the crucial characteristics of the *room* where the study was run; e.g., its size, lighting, windows, sound attenuation, and special furnishings.

  - *Field studies* would specify the crucial characteristics of the *environment* (outdoors for most animals, but possibly indoors for humans) in which the observations were made; e.g., buildings, vehicles, terrain, vegetation, other organisms, and weather.

3) Procedure

∞ *What was done* by the *researchers* and the *participants* from their very first encounter to the final contact should be reported *step-by-step*, and in enough detail for replication.

  - *Independent variables* should be *operationally defined* by specifying how they were *produced and manipulated. Instructions, tasks,* and *stimulus delivery* should all be described.

  - *Dependent variables* should be *operationally defined* by specifying how they were *measured. Response observation and recording techniques* should be carefully described.

  - *Extraneous variables* should be specified, along with the techniques by which they were *controlled*.

  - *Ethical standards* adhered to, and the *means* by which they were *upheld*, should be stipulated.

  - *Design* that was used in the study should be characterized in a *concise but detailed* manner.

    Example: The study used a $2 \times 2 \times 3$ mixed factorial design, with repeated measures on one factor (each factor would also be identified by name, and each level by its value; see Chapters 9 and 10 for discussions of research designs). If the design were complex and the description long, the design might be placed in a *separate subsection* (usually the last under "Method").

e. Results

- The *Results section* is the part of the paper that presents the *findings* of the study *objectively*, i.e., without explanation or interpretation.

∞ *Conclusions* involving evaluations and inferences, including theoretical and applied implications of the findings, are presented *later* in the *Discussion section* (see below).

- A *brief summary* of the most important results should be given at the outset, and then all the data should be presented in *sufficient detail* to justify the *conclusions* presented later.

  ∞ *Major findings* with respect to the research problems and hypotheses should be presented first.

    - *Statistical analyses* that were used to test each of the *research hypotheses* would be stated, along with whether in each case the data analysis *supported or failed to support* the hypothesis.

  ∞ *Secondary findings* should then be presented, e.g., those regarding unexpected results, specific comparisons among multiple levels of the independent variables, higher-order interactions (covered later in Chapter 10), and also any findings not directly related to the research hypotheses.

    - Note: *All relevant findings* must be presented in the Results Section of the report, including those that *run counter* to the research hypotheses.

- *Descriptive and inferential statistics* are presented in detail after the brief summary, and consist of *mathematical summarizations of the data*, and *analyses of the data's statistical significance* (see Chapters 11 and 12 for discussions of statistics).

  ∞ *Descriptive statistics* must always be provided to indicate both the *size and direction of effects*.

    - *Central tendency* measures (mean, median, or mode) should be accompanied by *variability* measures (e.g., standard deviation, standard error of the mean, variance, or range).

    - *Raw data* should not be presented, except in single-case studies or to provide some individual scores as examples.

  ∞ *Inferential statistics* should include the computed *numerical value* of a statistical significance test (e.g., analyses of variance, or *F*-test; *t*-test; chi-square test), the *degrees of freedom* in the analysis, and the associated table value of the *probability* that chance alone could have produced results as great or more extreme than those found in the study.

    - It is useful to report both *a posteriori probabilities* (the exact values obtained in the study, e.g., $p < .03$) as well as the *a priori probability* (the one selected beforehand as being required for rejecting the *null hypothesis* that chance factors alone are the cause of any obtained differences in dependent variable values under the different independent variable conditions).

        Note that the *a priori probability* is called the *alpha level*, and it is *conventionally* set at either .05 or .01, which also equals the probability for falsely rejecting the null hypothesis, i.e., making a *Type I Error* (discussed in Chapter 12).

- *Statistical significance tests* are influenced by sample size, and are measures of the *reliability of effect*, not the magnitude of effect. The importance of *sample size* and the varying *statistical power* of different analyses should be considered when testing hypotheses, and also when testing the assumptions underlying statistical models.

- *Magnitude-of-effect measures*, which do not depend on sample size, determine, e.g., the *proportion of variance* on the dependent variable that is accounted for by independent variables — and thus they are measures of *practical importance*, or strength of association, which should be reported in addition to tests of statistical significance, or reliability (see Chapter 12).

- *Tables and figures* should be utilized where they *facilitate comprehension*, but not where a few sentences in the text would do just as well.

  ∞ *As few* tables and figures should be used as are necessary, since tables are more expensive to publish than text, and figures are even more expensive than tables.

  ∞ Tables and figures serve to *supplement* the text, they cannot do the entire job of communication on their own.

  ∞ *Tables* are best for presenting means, standard deviations, and correlations; for illustrating *main effects* of independent variables; and for summarizing inferential statistical analyses.

  ∞ *Figures* are best for illustrating *interaction effects* of independent variables and for making general comparisons, but they are *less precise than tables* (Note: in addition to the usual graphs of data, figures also include drawings and photos of apparatus, stimuli, physiological recordings, etc.).

  ∞ *Table titles and figure captions* should provide enough information for the tables and figures to *stand on their own*, but the *main points* must still be described in the *text*, along with any additional explanation that facilitates understanding.

    - *Avoid repeating* in the text, however, *all* the data that are in the tables and figures (which clearly would be needless *redundancy*) — just provide a summary in the text that calls attention to the *major features* in the tables and figures.

  ∞ *Refer* to each table and figure as near to the relevant material in the text as possible (illustrative materials should never be included in a paper without referring to them).

    - Note: *Tables followed by figures* should be placed at the *very end of a submitted manuscript* (absolutely last) to facilitate processing and typesetting by the publisher, who will reposition them at the correct locations *within the body of the text* when the article is eventually *printed* (submit each table and figure on a separate sheet).

∞ *Figure captions* should contain the figure number, title, and a concise clarifying description/explanation.

- Note: For some publishers (e.g., APA), figure captions must all be submitted *together on a separate page* that is placed *before* the set of figures, rather than each caption being on the individual figure pages themselves.

∞ *Legends* are integral parts of *figures* that explain the *symbols* that are used; if not placed within the figures, then symbols, along with *abbreviations and units of measurement*, should be clarified within figure *captions*.

f. Discussion

- The *Discussion section* is the part of the paper that contains the *interpretation and evaluation of results*, particularly as they pertain to the research problems and hypotheses that were presented in the Introduction.

  ∞ *Inferences and conclusions*, are drawn, including *generalizations* of the results to *conditions* as well as *categories of participants* not included in the study.

    - *Relatively free rein* is permitted in this section, but remember that the focus must be on the *most important findings* regarding the research problems and hypotheses.

    - *Speculations* are appropriate in this section, but they must be *clearly identified* as such, they should be kept *modest and concise*, and their *logical* relationship to *empirical data or theories* should be spelled out.

  ∞ *Key issues* are: 1) Did the data support the *research hypotheses*? 2) How has the study helped to solve the *research problem(s)*? 3) What are the *overall conclusions and implications* (theoretical and applied) that can be drawn from the study, i.e., what has the study *contributed*?

    - Readers have a right to *clear, unambiguous, and direct answers* to these and related questions.

- *Begin* the Discussion section with a *straightforward statement* of whether or not the results support the *research hypotheses*, along with any necessary *qualifications and explanations*, such as those regarding unusual findings.

- *Similarities and differences* between the results of this research and those of *related studies*, especially the ones cited in the Introduction, should be covered next — with differences *explained*.

  ∞ This *integration* helps in the *interpretation* of the findings, as well as in the *evaluation* of the *validity* of results.

- *Interpret the results* with respect to the *research problem(s)*, pointing out the *theoretical and practical implications* of what was found in the study.

- *Shortcomings* of the study, e.g., *limitations of the methodology*, should also be discussed if they might have had a *significant influence* on the results;

however, *do not dwell* on the *flaws*, and unless there is a very good documented reason to think so, don't use them to try to *explain away* (rationalize) *negative findings*.

∞ *Nonsignificant statistical results* should be accepted for what they are, the *failure* to demonstrate in the study a reliable effect or association between variables.

- *Conclude* the Discussion section, if appropriate, by examining the likelihood of possible *alternative explanations* of the results (e.g., "Due to the inadequately controlled extraneous variable of . . . , it is possible that . . . ."); then briefly state ways for *improving* upon the study; and finally, note any ideas for *expanding* on the research in future investigations.

∞ When relevant, specify how *extraneous variables* might be better controlled, what potentially worthwhile modifications to *independent or dependent variables* might be made, and possible beneficial changes in the *participant population(s)*.

- Note: If the discussion is *brief and straightforward*, then it may be *combined with the Results section*, which would therefore be titled "Results and Discussion," or "Results and Conclusions."

g. References

- The *references* consists of a listing of *all* the information sources *referred to* in the text of a paper — and *no others*.

∞ Examples: journal, magazine, newspaper, and newsletter articles; brochures; books and book chapters; technical and government reports; proceedings of meetings and symposia; doctoral dissertations and master's theses; audiovisual and electronic media (CD-ROM, Internet, etc.); unpublished works, and publications of limited circulation.

- *Purpose* of the references is to *document all statements* made about the scientific literature and other sources.

∞ References are not meant to be exhaustive — instead they should be *chosen judiciously* because of their *particular relevance* to the research being reported.

- *Format* for typing the references is specified by the journal to which the manuscript is submitted, and of course it is necessary that the references be *complete and accurate*.

h. Appendix

- This is the location for including *detailed material* that would be *disruptive* if placed in the main text of a paper, but which is *important* because it helps the reader to understand, interpret, evaluate, or replicate the research.

∞ Examples: a listing or reproduction of all the *stimulus materials* used in a study, a copy of a *newly designed test*, a detailed description of a complex piece of *specialized equipment*, or the lines of code for a *computer program* written especially for the research.

- *Do not include* material that has been *already published*, or that is otherwise *readily available* — instead, refer the reader to the appropriate source(s).
- Note: Appendices are *optional*, and thus frequently not included.

i. Author note

- When submitting a manuscript for publication, the following should be provided on a separate sheet of paper:
  - ∞ *Departmental affiliation* of each author;
  - ∞ *Financial support* received for the study (e.g., grants);
  - ∞ *Acknowledgements* of the professional contributions of others to the research (e.g., assistance in running the study, analyzing the data, or critiquing the paper), and also for special help in preparing the manuscript (e.g., typing the text, or producing the figures and tables);
  - ∞ *Name and address* (including E-mail when available) of the person that readers may contact for *additional information* about the study;
  - ∞ *Disclosures*, if applicable, of certain other information:
    - *Special basis for the research*, such as a master's thesis or doctoral dissertation;
    - *Previous presentation of the results*, such as at a meeting, in a chapter of a book, as a part of another paper, or in a nonprofessional publication;
    - *Large-scale multidisciplinary project* of which the study is a component;
    - *Conflicts of interest*, such as relationships with governments, industries, corporations, business firms, or individuals who could benefit from the published research.
- Note: In APA style, it is *after* any appendices and the author note that any *footnotes, tables, figure captions, and figures* are placed in the manuscript — and in that order with all pages, except the figures, numbered consecutively.

Review Summary

1. *Literature surveys* are *reviews* of the literature that entail a *thorough search and careful study* of all publications that are relevant to the research problem selected for investigation. There are four *functions*:

   a. Helps in *formulating* better research problems and hypotheses

   b. Tells the investigator whether or not the research is *needed*

   c. Other studies provide numerous *suggestions* as to research apparatus, possible extraneous variables, and control methods

    d. Previous relevant studies and theories must be *summarized and referenced* in research proposals as well as final research reports

2. There are several *mechanisms/aids* for conducting literature surveys:

    a. *References* — a list of sources that have been referred to in research, theoretical, and review articles, as well as in books

    b. *Psychological Abstracts* — a monthly publication of the American Psychological Association containing a summary/abstract of articles and book chapters, as well as a table of contents for books, which is indexed by subject, author, and book title

    c. *PsycLIT* and *PsycINFO* — a computerized, CD-ROM version and an on-line version, respectively, of *Psychological Abstracts*, with fast and powerful engines for searching by subject, author, and keywords

    d. *Social Science Citation Index* — an interdisciplinary listing of publications that have cited a particular *earlier* publication, which is available both in paper and on CD-ROM, with subject, author, and citation searches possible

    e. *Current Contents* — a timely publication containing copies of the table of contents for journals and books, with separate editions covering the different sciences, and which is available both in paper and on computer diskettes, with author and title indexes

    f. *Internet* and the *World Wide Web (WWW)* — a network of computer networks, interconnecting computers around the world by modems and phone lines for the efficient exchange of information, ideas, and opinions, the WWW being the easy-to-use, graphics-intensive segment

3. *Research proposals* are thorough written *plans and rationales* for scientific studies. They have four *functions*:

    a. *Ambiguities or problems* are made more apparent

    b. *Biases, overlooked points, and logical errors* can be uncovered by others who read them

    c. *Research funds, space, or time* usually require a submitted proposal

    d. *Published reports* utilize much of the information assembled for research proposals (e.g., the literature review, purpose, and methods)

4. *Pilot studies* are *preliminary investigations* conducted prior to a full-scale study. They vary considerably in *extent and complexity*. Their *functions* can include:

    a. *Checking* the experimental procedures, apparatus, and/or materials

    b. *Practicing* the procedures and/or the operation of equipment

    c. *Evaluating* the probable effectiveness of various levels of *independent variables*

    d. *Evaluating* different possible *dependent variable measures* for sensitivity and adequate range

    e. *Uncovering extraneous variables* and trying out *control procedures*

    f. *Estimating data variability* and thus the necessary *sample size*

    g. *Estimating results*

5. *Serendipity* is the art, process, or fact of *finding something of value* while one is looking for something else. For the scientist it means being *alert to and recognizing* the potential importance of *unusual or unexpected observations* — and *following up* on them.

   This has led to many *important discoveries*. However, it should be noted that the *large majority* of unusual or unexpected occurrences have little scientific significance, and thus are distractions.

6. *Research reports* contain the purposes, hypotheses, methods, results, and conclusions of a scientific study — presented in written form, or orally at a conference or other professional meeting. They have five *functions*:

   a. *Communication* to others so that the research becomes part of the *systematic body of knowledge of science*

   b. *Historical identification* of who did it first and when

   c. *Judgment* by others of the *quality of research*, and follow up for *verification and elaboration*

   d. *Avoidance* by others of approaches found to be *unsuccessful*

   e. *Minimization* by others of unintended *repetition*

7. There are five *qualities* to strive for when writing research reports:

   a. *Accuracy* is absolutely required — there's nothing without it

   b. *Completeness* is necessary for possible replication and thus verification of results

   c. *Clarity* is critical for avoiding ambiguity and facilitating comprehension of the research logic

   d. *Conciseness* is needed for economy of reading time and publication expense

   e. *Readability,* in the form of interesting and compelling writing, is important so that the reader is encouraged to continue to the end

8. *Research papers* typically consist of about nine standard *major parts*:

   a. *Title page* — containing a title (concise, self-explanatory summary label), running head (abbreviated title), authors, and institutional affiliations

   b. *Abstract* — a concise but comprehensive self-contained summary

   c. *Introduction* — which states why the research was done (research problem), what was expected to be found (hypotheses), and the general design of the study, along with a literature review/survey

   d. *Method* — which describes in detail what was done and how it was done, including participants, apparatus and materials, and procedure

   e. *Results* — which objectively present the findings in the form of descriptive and inferential statistics, usually with tables and/or figures

   f. *Discussion* — which contains the interpretation and evaluation of results, i.e., inferences and conclusions, including generalizations

   g. *References* — which consist of a listing of all the sources referred to in the text of a paper, and only those sources, using a specified format

h. *Appendix* — which is the location for detailed material that is important, but which would be disruptive if placed in the main text of the paper

i. *Author note* — which identifies the department affiliation of each author, as well as the name and address of the person that readers may contact for additional information, specifies sources of financial support, provides acknowledgements of the contributions of others, and discloses any special relationship of the study to other research, prior presentation of the results, and possible conflicts of interest

Review Questions

1. List and briefly discuss four *reasons* that a *literature survey* is an important phase of research.

2. List and briefly describe five *aids* for conducting *literature surveys*.

3. Describe what a *research proposal* is and list four important *functions*.

4. Describe *pilot studies* and list seven *functions*.

5. Define *serendipity* and state its *implications for research*.

6. State five *functions of research reports*.

7. List the five *qualities* to strive for when writing *research reports*, and explain why they are important.

8. List and describe the nine standard *major parts* that typically make up a *research paper*.

Part Two

Research Ethics and Experimental Control

6

Ethics of Research

CONTENTS

The topics of *research ethics* and *experimental control* are placed together in Part Two of this Handbook because they are highly interrelated. When designing the study and developing procedures to control extraneous variables, as well as to manipulate independent variables and measure dependent variables, researchers must always give serious consideration to all the ethical issues.

Non-Participant Ethical Issues[14]

There are several ethical issues in research that do not directly relate to the use of participants in a study, but that nevertheless are very important.

1. Data Collection and Analysis

— Great care must always be taken to ensure that research data and analyses are *accurate and complete*, and that *alternative hypotheses/explanations* for the findings are acknowledged, e.g., inadequately controlled extraneous variables (see Chapters 4 and 8).

  • *Falsifying data* or *selectively reporting* only the data that confirms the research hypotheses is *fraud* and is totally unethical (it is also usually *uncovered* through later research by others).

    ∞ Example: The most famous case is probably that of *Sir Cyril Burt*, who did research on *intelligence and its genetic basis*. He reported that the

IQ scores of identical twins who were reared apart were highly similar. However, a careful examination of the data uncovered the fact that several correlation coefficients for different sets of twins were exactly the same out to three decimal places, which would be unbelievably unlikely, and subsequently it was determined that Burt must have fabricated his data to support beliefs regarding the relationship between social class and intelligence.

2. Publication

— *Premature* publication and *withholding* publication are *problematic*.

- *Premature publication* of results, by submitting a manuscript before a study is fully completed, might *unethically* be done in an attempt to *beat out other investigators* who are working on the same research problem; e.g., a vaccine or cure for cancer.

- *Withholding publication* of results, on the other hand, might be done in an attempt to maximize gains from the *commercialization* of research findings before others can apply the same data; e.g., regarding an effective treatment for dyslexia (reading difficulty).

  ∞ Note: The *degree* to which withholding publication would be *unethical* is dependent on whether the research is done by a scientist while working in a *public as opposed to a private/commercial institution*, and by whether *funding* for the research (space, equipment, and salaries) is public/charitable or private.

— *Duplication* of the publication of findings is also *usually unacceptable*.

- *Exceptions* to this proscription are cases where the multiple publications of the same data represent *different degrees of sophistication* and/or are aimed at *different audiences*.

  ∞ Examples of when multiple publication is usually considered *acceptable*:

    1) Publication in both a professional journal and in a nonprofessional magazine;

    2) Publication in both the proceedings of a conference and in a professional, *peer-reviewed* journal.

  ∞ Note: In all cases *reference* to the earlier publication should be made in the later publication, not only for *informative purposes* but so that it will *not* be viewed as being deceptive.

3. Crediting Others

— Significant work done by others involved in the research should be recognized through *acknowledgement or co-authorship* in the report, depending on how substantial their contributions were to the study.

- This includes *acknowledging* organizations and individuals that provided *financial or material support* for the research.

- *Authorship*, however, is a *privilege* belonging only to those who make a *significant contribution* to a research study in terms of its conceptualization, design, execution, analysis, or interpretation.

  ∞ Example: *Student lab assistants* would ordinarily receive *an acknowledgement*, unless their contribution was substantial.

4. Conflicts of Interest

— Any *associations* with businesses, government organizations, etc., that might be considered to create a potential *bias* in one's research or reporting of results should be clearly *identified* when communicating research findings in professional journals, conferences, and magazine or newspaper articles.

- This provides others with important information for drawing their own *conclusions* about the possibility of *undesirable influences*.

- Example: When reporting the results of a major study indicating that a large percentage of American men and women experienced *sexual dysfunction*, the *Journal of the American Medical Association (JAMA*, February 1999) failed to note that the authors of the study had been paid to review clinical trial data on *Viagra* before this drug for alleviating *impotency* was submitted for government approval. Moreover, in an Associated Press story, the lead author said that the study "gives us a base for explaining why we had this enormous response to Viagra." It would seem obvious that there was a conflict of interest.

5. Military Applications

— *Personal philosophy* regarding the morality of the use or threat of force to protect or extend a country's interests, as well as the *nature and degree of force*, are involved in deciding whether it is *unethical* as opposed to a *patriotic duty* to conduct research on military issues.

- *Violence* and the resulting *harm* to life and property are obvious concerns regarding the applications of military research.

- *Secrecy and restricted dissemination of findings*, presumably for national security reasons, are also troubling issues for science.

- On the other hand, the *defense of life, property, commerce, national borders, and ideals* are certainly important as well.

- Questions to ponder: Would you personally feel ethical conducting research that enhances our country's *nuclear, chemical, or germ warfare* capabilities, or for that matter participating in any *military research* at all — even if it would ostensibly only be used for *defense* rather than *offense*?

Human Participant Ethical Issues[1,2,11]

There are *two major sets of ethical issues* pertaining to humans: 1) *How research participants are obtained and the information provided*, and 2) *How research participants are treated during and after a study.* After discussion of these, some additional issues will be covered, concluding with the ethical principles of the American Psychological Association (See Appendix 4).

1. How Research Participants Are Obtained and the Information Provided

— This first major set of ethical issues involves the related topics of *coercion, concealment, deception, and informed consent.*

— Coercion

  • Any form of *duress* wrongly denies individuals *complete freedom of choice* regarding their *participation* in research, which includes the right to *withdraw* after beginning to take part in a study.

    ∞ *Power relationships* are a problematic issue — especially when using students, employees, patients, children, prisoners, and military personnel, over whom one has *authority/influence.*

      • Freedom of choice can be very *limited* in such cases.

    ∞ *Monetary and other compensation* also can be coercive when it is *substantial* relative to the participants' needs.

  • Although great urgency often exists to obtain individuals for research, nevertheless *participation should be voluntary, and there must be the right to withdraw without penalty.*

    ∞ When research participation is a *course requirement* (e.g., in introductory psychology classes) or is an opportunity for *extra credit*, an *equitable alternative* must be provided (e.g., writing a brief paper summarizing some research).

      • In addition, the research and alternatives should be of *educational value* to student participants, e.g., informing them about the nature of scientific research.

— Concealment

  • Sometimes participants are *not made aware* of certain aspects of a study, or do not even know that they're taking part in a study.

    ∞ Examples: 1) Observing the *purchasing behavior* of people at shopping malls in response to manipulations, such as shopping *incentives* (This would not seem to be a serious ethical problem since there is *little if any invasion of privacy).*

    2) Observing the *shoplifting behavior* of people in department store dressing rooms — by using a *hidden camera* — to measure the *deterrence effect* of employees counting the number of items an individual takes into the rooms (This is a serious ethical problem due *to extreme invasion of privacy).*

— Deception

  • Even when people are aware of participating in a study, they are often given *false information* (i.e., lied to) about its *purpose and/or what will occur.*

    ∞ Example: *A potentially beneficial medical treatment* might be withheld from some individuals who volunteered to receive it, without them being informed about this, in order that these research participants could serve as a *control group*.

- Note: This might be *justified* if there were a *shortage* of some drug, and control group participants were *randomly selected* and were placed on a *waiting list* to receive the treatment when more of the drug became available — assuming, of course, that it was found to be effective.

- However, it would be *more ethical* if instead of deception a form of *partial concealment* were used, in which all participants would be *told* at the outset that some of them would be randomly selected to *first* serve as controls, but without being informed who had been selected to do so.

- Naturally, the use of control groups is a *particularly serious ethical problem* when the treatment is for a disorder that could lead to *death*, such as AIDS, cancer, or clinical depression. Note, however, that it's sometimes possible that the experimental treatment not only could be ineffective, but *harmful* — thus the control participants might actually be better off.

- *Concealment and deception* are sometimes *combined*.
    - ∞ Participants might be *unaware* that they are taking part in a study, and they also might be *given false information*.
        - Example: *Posing* as a fellow student and observing the willingness of others to take part in *playing a prank* on an instructor when *falsely* told either that the instructor is a homosexual, an atheist, or just a demanding teacher.

— Informed consent

- This is the *integrating theme* of all preceding participant issues.

- An individual's decision to take part in an investigation should be made on the basis of *sufficient and accurate information* about the purpose/goal of the research, what will happen in the study, what they will be called upon to do, and anything else that might reasonably be expected to influence *willingness* to participate.
    - ∞ This should be done in *writing* using *layperson's terminology*, and the form should be *signed* and *dated* by the individual if she or he agrees to participate, and by the investigator as a witness (see example in Appendix 4).
    - ∞ *Consent forms* are now commonly integrated with written *instructions* about a study that are shown to participants.
    - ∞ Potential participants *must be informed about*:
        - The study's *purpose and procedures*
        - The *place, time*, and expected *duration* of the study

- The *right to decline* to participate or to *withdraw* from the study at any time without penalty

- *Potential risks* of discomfort or harm, if any, and their likelihood; as well as what steps will be taken to prevent or minimize them, and what corrective steps will be taken

- *Possible benefits* of participation, as well as any *alternatives* that might also benefit the individual, if the study is of a clinical nature

- *Compensation*, if any, that will be given for participation, as well as when and how it will be provided

- *Limitations*, if any, regarding the *confidentiality* of the data obtained; and the steps that will be taken to continually maintain the participant's *right to privacy*

- *Anything else* that might possibly be *relevant* or that the investigator is *asked about*

∞ *Special safeguards* are needed when *children* and individuals with *mental or communication impairments* are to be used, since they have less capacity to give *truly* informed consent.

  - In these cases, the informed consent information should be *adjusted* to a level they might understand, and *assent* should be obtained from both the *potential participant* and an individual who has *legal responsibility* for him/her.

- *Problems* arise for those studies that *could not be conducted* if potential participants were informed about *all* the significant aspects of the research.

  ∞ Example: see Milgram's Obedience Studies, presented later.

  ∞ *Ethical issue* is whether in such cases the *potential scientific, educational, or applied value* of the study *justifies* some *deception or concealment* (by commission or omission); and even if it does, whether *alternative research procedures* couldn't be used that would *not require* such undesirable practices, but which would still achieve the same ends.

    - *Special safeguards* are required for the protection of the participants' *welfare and dignity* whenever deception or concealment is used, even though considered justifiable.

    - *Debriefing* is required after the use of deception or concealment in research. Specifically, participants should be *fully informed* about the research, given an *explanation/justification* for deception or concealment, and any *stressful after-effects* should be detected and corrected (debriefings are further discussed later in this chapter).

2. How Research Participants Are Treated During and After a Study

— This second major set of ethical issues involves the *welfare of participants during the research* and the *rectification of any undesirable consequences following the study.*

— Welfare of participants

- Investigators must *protect* individuals from any *physical and mental discomfort, stress, and harm* that might result from their participation in the research.

- *Noxious physical or psychological stimuli*, however, can be an essential and sometimes *justified* part of research designs.

  ∞ Example: Volunteers who are fully informed might be given *electric shocks* to study the effects of *aversive stimulation* on undesirable behaviors, such as smoking cigarettes.

— Undesirable consequences

- Any harm that is done to participants must always be *detected and corrected*, i.e., *eliminated* (see Debriefing below).

- Procedures likely to cause *serious or lasting harm* should not be used, except under the following *special circumstances*:

  ∞ If the *potential benefit* of the research is *exceptionally great*, and indeed *imperative*, and also *fully informed, voluntary consent* is obtained from all the research participants.

    - Example: Searching for a prevention or cure to *criminal sexual behavior*, such as pedophilia, incest, or rape.

  ∞ If *avoiding* the use of some research procedure involves the possibility of exposing the *fully informed volunteers* to even *greater harm* (in which case the decision is more clear-cut than for the preceding circumstances, but rarely simple).

    - Example: Deciding to try a new, risky, experimental treatment for *clinically depressed, suicidal patients,* when it is unlikely that a potentially safer treatment will be discovered soon enough to avoid their death.

- Note: Deception or concealment and physical or mental stress/harm can be *interrelated ethical issues* in research studies.

- Examples: 1) To determine possible effects on task performance, *deceitful* test-result reports about *personality* characteristics, such as a lack of masculinity or femininity, could be given to a random half of the participants. This study clearly might cause *mental distress* and even lasting harm, and therefore it is *very unlikely to be ethically justifiable.*

  2) To determine how police cadets respond to *stressful situations* when they have versus when they have not received *special training*, they could be exposed to *shouts and cries for help* made by hired actors that are in a locked room. The cadets would be told that a *deranged killer* has taken hostages. This research would probably be considered *justifiable*, because the ability of police officers to perform under stress in the real world can determine the life or death of others.

Milgram's Obedience Studies: Major Example of Several Ethical Concerns

Purpose and Design

- These were studies about blind obedience to an authority figure, conducted in the early 1960s. [1] Milgram, S. (1963). Behavioral study of obedience. *Journal of Abnormal and Social Psychology, 67,* 371-378. 2) Milgram, S. (1964). Group pressure and action against a person. *Journal of Abnormal and Social Psychology, 69,* 137-143. 3) Milgram, S. (1965). Some conditions of obedience and disobedience to authority. *Human Relations, 18,* 57-76.]

- Volunteers were recruited to participate in research at Yale University, *allegedly in a scientific study of memory and learning.*

- A *volunteer* reports to Milgram's lab and meets a *scientist* dressed in a lab coat and also an individual introduced as another volunteer, but who is actually an *actor serving as a confederate* of the experimenter.

- The scientist *falsely tells the volunteer* that this is a study of the *effects of punishment on learning and memory.*

  - ∞ One person is to serve as a *teacher* and the other as a *learner.*

    - The supposedly random assignment process is *rigged,* and the *real volunteer is always selected to be the teacher,* while the confederate is always selected to be the learner.

- *Electrodes* are attached to the confederate, and the real volunteer is positioned in front of an *impressive looking shock machine* with levers that the participant is *falsely told* will deliver shocks to the learner when pressed.

  - ∞ The *sequence of levers* are labeled from 15 volts up to 450 volts and are also marked "Slight Shock," "Moderate Shock," etcetera, up to "Danger: Severe Shock."

- The *confederate* is told to *learn a series of word pairs,* and is then given a test to see if he remembers the paired associates (which word goes with which).

  - ∞ The *volunteer teacher* is to *deliver a shock as punishment* every time the learner (confederate) makes a mistake, with the level of shock being increased for each successive error.

    - The *confederate learner repeatedly makes mistakes,* and as the *supposed shock* gets more intense the confederate begins *screaming in apparent pain,* and eventually yells that he *wants to stop the study.*

    - The *volunteers* were found to become *quite visibly upset* by the pain the confederate was presumably experiencing.

      - * *If the volunteer teacher wanted to quit,* the scientist would say that he could quit, but would nevertheless urge him to continue by using a *series of prods.* The investigator would *stress the importance of continuing the study,* and if necessary would go so far as to *assume full responsibility for any problems.*

Results

- In the experimental condition just described, 25 of 40 volunteers continued to deliver shocks all the way up to the limit of 450 volts.
  - Thus, in over 60% of the cases the *authority of the experimenter overrode the personal conscience of the participants.*
    - This was even true at the high levels of shock where screaming had stopped and the volunteer actually expressed the *belief that the "learner" had died.*

Scientific and Applied Value of the Research

- Milgram's findings received much publicity and have *challenged beliefs about our ability to defy authority,* even when it appears appropriate to do so.
  - His findings have important implications for the real world.

Examples:

1) Behavior exhibited by citizens and soldiers in *Nazi Germany*
2) Behavior exhibited by American soldiers at Vietnam's *My Lai massacre*
3) Behavior exhibited by employees who know that their employer is *cheating customers*
4) Behavior exhibited by students asked to do something *unethical* by a professor when helping with research

Ethical Issues

- Milgram's studies are very troubling: *Was the research justified,* i.e., did the benefits/values outweigh the costs?
  - *Stress* was experienced by volunteer participants while delivering intense shocks to an obviously unwilling learner.
    - *Films of the volunteers* show them protesting, sweating, even laughing nervously while delivering the shocks.
  - *Deception* was practiced and thus there wasn't *informed consent.*
    - *Volunteers had agreed* to participate in a study of learning and memory, but they actually took part in a study on obedience to authority.
      - ∞ *How were they to know* that they would be called upon to deliver supposedly high intensity, painful shocks to another person, possibly to the point that they would believe the person had died?
      - ∞ *How were they to know* that the learner was actually a research confederate and was not receiving any shocks?
    - *Deception and lack of informed consent,* however, were *apparently necessary* in order to carry out this study.

∞ Would not knowledge by the volunteers that the research was a study of obedience have *altered/distorted their behavior*? Would not most of the participants probably have tried to show that they wouldn't be obedient, since this would be the *socially acceptable behavior*?

∞ Would not the *participant sample* have been *biased* (non- representative) by being told in advance that the experiment was going to involve inflicting pain? That is, *wouldn't only certain types of people have consented to participate*? Would not this have created problems for generalizing the results? Would not critics have been able to say that the obedient behavior in Milgram's experiments occurred simply because the participants were *cruel, inhumsane sadists in the first place*?

— *Long-range consequences* of the deception were possible.

- Those who obeyed the experimenter might feel *continuing remorse* and begin to see themselves as *cruel and inhumane*.

— *Thorough debriefings* were used to detect and remove any negative effects of deception and stress, but was this adequate?

- Volunteers who were obedient were *told that their behavior was normal* and that they had acted no differently than most other participants.

- They were *made aware of the strong situational pressure* that had been applied.

- They were *assured no shocks* had actually been delivered.

- There was a *friendly reconciliation* with the confederate.

- Efforts were made to *reduce any tension* the participants felt.

- Later the participants were mailed a report of the research findings and *asked about their reactions to the experiment*.

  ∞ *Responses indicated* that 84% were glad they had participated and only 1% were sorry. Furthermore, 74% said they had benefited from the experience.

- Participants were *interviewed by a psychiatrist* a year later.

  ∞ *No ill effects* of participation could be detected.

Conclusions

— It could be that the *negative effects* of deception and stress were adequately dealt with, and that given the *importance of Milgram's work* the deception and stress were *justified*; however, there are still some *troubling concerns*. [Baumrind, D. (1985). Research using intentional deception: Ethical issues revisited. *American Psychologist, 40,* 165-174.]

- *Was it logical* for Milgram to use the self-reported, post-study judgments of overly acquiescent ("destructively obedient") participants to establish the ethical propriety of his experiments?

- Was the study *methodologically sound* and thus *justified*?

∞ Since *people were paid* for participating in Milgram's research, and were recruited on that basis, might not some of their obedience have reflected a *sense of fair play, or employee loyalty,* rather than blind obedience?

∞ Were not Milgram's directives to reluctant participants *incongruous and bizarre,* thereby *confusing* them?

∞ Were not Milgram's orders legitimized by the *laboratory setting,* thus permitting participants to resolve incongruity by trusting the investigator?

- *To sum up these concerns*: Was Milgram's research actually *relevant to the real world,* i.e., would it permit *valid inferences/generalizations* to real-life situations? Hence, was the research truly justified?

- Finally, and perhaps most interestingly, *were not all the necessary data actually provided by the confederate actors*?

  ∞ Didn't they themselves demonstrate conclusively, by obeying Milgram's instructions to inflict suffering upon the volunteers, that when asked to do so by an authority figure, normal, well-intentioned individuals will hurt innocent people?

    - Therefore, were not Milgram's deceitful and stressful manipulations actually *unnecessary,* as well as *invalid,* and hence thoroughly *unjustified*? Well, *not necessarily.* To get meaningful results, Milgram would still have needed the deceived and stressed volunteer "teachers" in order to get data from the research confederate actors.

---

3. Implications for Society of Using Deception in Research[17]

— *An important ethical question* is whether researchers, simply for the purpose of experimentation, have the right to add to the many *anxieties* that already exist in life.

- Are not research participants just like other human beings whose *dignity* should be preserved in our interrelationships with them?

- Do researchers really have the right to treat participants with less *respect* than they would give to any other fellow human: to lie, trick, mislead, conceal, and even coerce?

— If it is considered acceptable to use deceit for the *advancement of scientific knowledge,* then why not for anything else that is valued?

- Once deception is justified for any purpose, *where does it stop,* and *who decides* what merits the sacrifice of truth?

- Example: Do *political leaders* have the right — perhaps the responsibility — to deceive the populace to achieve some goal?

— Society places great *trust* in the *honesty of scientists* and in the *respectability* of their activities.

- Does not the use of deception undermine society's *confidence* in science, and ultimately in the *morality* of all humankind?

- Do not the cumulative effects of deception *alienate* humans from each other and thus from society?

- Does not deception ultimately make us *distrustful* of everything?

  ∞ Is it not *illogical*, in fact, to expect participants to believe the *justifications* researchers later give when during *debriefing* they confess that they lied earlier — particularly if participants are young, and thus inexperienced in the complexities and compromises common to everyday adult life?

— On the other hand, it must be kept in mind that the *potential benefits* for society of *certain research* might be so great as to outweigh the *expected costs of compromised ethics*, assuming that the data could not be obtained in a *more ethical fashion* (institutional review boards are responsible for such decisions, as discussed later in this chapter).

4. **Methodological Implications for Researchers of Using Deception**

— Scientists who assume that participants are *naive* and *unaware* that they are being *deceived* might actually be naive themselves.

  - It is becoming *increasingly difficult* to find naive participants.

— Participants are often *uncertain* as to what an experiment is actually about, because, due to direct and vicarious experiences with deception in research, they frequently *suspect* that the research is *not* what the investigator claims to be the study's purpose.

  - Obviously, suspicions by individuals that they are being deceived in some fashion can *undermine the intended effects* of the artifice and *influence a study's results*, even if the participants have not correctly inferred the true nature of the research.

    ∞ Hence the *value of using deception might be lost*, even when it is true that valid results could not be obtained if the participants were aware of certain facts about the study.

— Just as the prophylactic use of antibiotics and pesticides can reduce their effectiveness when later needed, so can the *repeated use of deception and concealment* eventually reduce their usefulness.

  - Moreover, even in studies where deception or concealment is *not used*, their *expected use* can distort the behavior of participants.

  - IN SUMMARY: *Scientists have an ethical responsibility toward science, each other, and society not to use deception and concealment unnecessarily.*

— Debriefing

  - This is a standard *postexperimental conference/interview* between the experimenter and the research participants that has a number of functions.

  1) Debriefing can provide *important information* about:

    a) Whether the research participants *suspected* any deception;

    b) What the participants *thought* the study was really about.

2) Debriefing can also *help determine*:

    a) *Effectiveness* of the independent variable manipulation(s);

    b) *Strategies* that the participants employed on the tasks.

3) Debriefing should *always* be used, *after obtaining* some or all of the information just noted, in order to:

    a) *Undo and explain/justify* any *deception* that was employed;

    b) *Detect and remove* any *negative effects* of participation;

    c) *Educate* the participants about psychology and its research;

    d) *Give* individuals a sense of *satisfaction* for participating by explaining the expected value of the research.

4) Debriefing is *often* also used to:

    a) *Convince* participants *not to discuss* the experiment with anyone else until they are told that the entire study is completed;

    b) *Notify* participants about how they can receive *additional information* about the research and its results once the study has been concluded.

5. Alternatives to Deception

It is not always necessary to use deception, there are at least *three options, and a compromise: role-playing studies, simulation studies, honest experiments, and assumed-consent research.*

a. Role-playing studies[17]

- Instead of running a deceptive experiment, in these studies a situation would be *described* to the research participants by the investigator, and the participants then would be asked to indicate how, in all probability, they or others *would behave.*

- This is *not usually satisfactory,* however, as an alternative to deception.

    ∞ *Participants would not be involved deeply enough* in the *real situation* to get accurate/valid results, i.e., their behavior would likely be *different* when placed in the actual situation.

        • People are not very good at *judging/predicting* what their behavior or the behavior of others would be under *imagined circumstances.*

        Example: If Milgram's research design were simply described to people, how many do you think would have predicted that they or others would be completely obedient? (Even experts underestimated.)

    ∞ *Demand characteristics* are often a potential problem.

        • When a complete description of some situation is given to participants, it might make the experimenter's hypothesis *obvious,* which would tend to *"demand"* that the subjects act accordingly (discussed further in Chapter 7 under "General Sources of Variance"). This relates to the bias of individuals toward making *socially desirable/acceptable* behavioral responses.

- Examples:

    1) In role-playing studies of *altruistic behavior*, wouldn't most people say that they would *help* others who were in need, especially when the described need was great? Yet in actuality, wouldn't many individuals try to *ignore* others when help is required, particularly if additional people are around who might provide the necessary assistance?

    2) What would most people say they would do when asked to imagine that they had found a wallet on the ground full of lots of money or even just a little?

b. Simulation studies[17]

  - These studies entail *role-playing* that takes place in a *simulation of the real-world situation* of interest.

    ∞ This creates a *high degree of involvement*, and thus produces *more accurate data* than would just imaginary role-playing.

        Example: The *Inter-Nation Simulation*, in which participants role-play being leaders of nations, while researchers observe the processes of negotiation and problem solving as a function of the particular situation or conditions.

    ∞ *Ethically unacceptable consequences*, however, might occur even in simulation studies.

      - Psychological and physical stress or harm can be as great or greater than would be found in deception experiments.

        Examples: 1) A newspaper article (12-8-91) reported that a couple in Salt Lake City had been accused of beating a roommate to death with a claw hammer as he slept following a fight over a *Monopoly game* — which is a simulation of the world of high finance.

        2) A well known *simulated prison study*, which was conducted in the basement of the psychology building at Stanford University using college students who played the roles of prisoners and guards, had to be ended after only 6 of the planned 14 days because of the cruelty seen in the "guards," and the incredible levels of stress observed in the "prisoners." [Zimbardo, P. G. (1973). The psychological power and pathology of imprisonment. In E. Aronson & R. Helmreich (Eds.), *Social Psychology*. New York: Van Nostrand.]

c. Honest experiments[29]

   There are *four honest experiment strategies*, and they don't involve any role-playing: *fully-informed participant studies, explicit behavioral-change studies, field experiments, and naturally occurring event studies.*

1) Fully informed participant studies
   - ∞ *No deception* would be used about the purpose of the research in these studies.

     Example: Individuals could be told that they will be participating in an experiment about the *effects of different types of music* on cognitive performance — and that is exactly what the study would be about.

2) Explicit behavioral-change studies
   - ∞ With *full knowledge*, participants would volunteer to have some *maladaptive behavior* modified as part of these studies.
     - Example: Experiments have been run on the relative effectiveness of *weight-reduction programs* that use either dieting, exercise, counseling, or some combination.

3) Field experiments
   - ∞ Manipulations would take place in some *natural context* in these studies, i.e., they would involve behavior in public places in everyday situations, with *no deception* employed.
     - Example: Studying the effectiveness of *conspicuous observation of driving behavior* on increasing lawfulness, such as a greater likelihood of full stops at stop signs.
   - ∞ *Ethical problems*, however, can exist.
     - Researchers would usually *conceal* the fact that an experiment is taking place (as in the example, but this is not a real problem since the behavior observed is public).
     - Participants' *time* might be taken up (not really an issue in this example, since the time spent making a full stop is to everyone's advantage, and it is the proper behavior).
     - *Stress* might be produced in participants (this an issue in our example, since having someone stare at you as you drive toward a stop sign could be stressful because you wouldn't be certain what the individual was up to).

4) Naturally occurring event studies
   - ∞ Nature and society sometimes provide *nonsystematic/random events* that can be studied for their influence behavior.
     - Example: What are the effects on behavior when people *win lotteries* for enormous amounts of money?

d. Assumed-consent research[6]
   - Assumed consent is a *compromise approach* that could be used *when deception is the only means to obtain valid data,* and the potential value of the research *justifies* the use of deception.
   - *Complete and accurate* description of the experimental procedures and their purpose would be given to a *pilot sample of potential participants*.

∞ These individuals would be drawn from the same *population* as would the sample for the ensuing actual experiment.

- The pilot sample — after having been *fully informed* — would then be asked *whether they would consent* to participate in the study.

- If the vast majority of individuals (e.g., ≥95%) indicated that they would give their consent, then *consent would be assumed* for the people who are later selected to *actually participate*, but who could not be given an opportunity for fully informed consent due to the adverse effects it would have on the study's validity.

6. Confidentiality

— *Confidentiality* and also the *planning process* (see below) are important ethical issues that are in addition to the *two major sets of ethical issues* (with their components) that have been covered.

— A significant ethical concern is that information obtained about research participants must be kept *anonymous*, unless otherwise agreed to in advance, since investigators have an obligation to *protect the privacy* of individuals that participate in their research projects.

- *Data sheets* should be *coded* with *numbers and/or letters* that indicate the different participants and the conditions they were in, but the *names* of individuals would *not* be on the data sheets.

  ∞ Just in case a participant might have to be contacted later for additional information, a *separate sheet* listing the names of participants with their corresponding code numbers/letters could be kept until the data analyses were completed — but then this *cross-referencing form* should be destroyed.

- *Ethical dilemmas* can arise, however, when it appears necessary to violate the *principle of confidentiality* in order to uphold the *principle of protection from harm*.

  ∞ Information obtained during the course of a study might indicate that an individual is *likely to harm* himself/herself, or possibly even someone else, in which case there would be an obligation to *disclose the information to others* (e.g., a counselor) who could help *prevent* the harm from occurring.

7. Planning Process

— Before any research begins, it is necessary to determine the *ethical acceptability* of a study (as has been noted earlier).

- This involves careful *evaluation and weighing* of the risks to participants, including possible effects of deception, *versus* the potential scientific, educational, or applied value of the research.

- Furthermore, whenever there are problematic issues, there is an *obligation to seek the advice of others* in order to ensure the rights and welfare of research participants (see Institutional Review Board below).

8. **American Psychological Association Ethical Principles[1,2] (See Appendix 4)**

— *Ethical Principles in the Conduct of Research with Human Participants* was first published by the American Psychological Association in 1973 after 3 years of deliberation.

  • *Review* of these principles was later initiated in 1978 and, after much consideration, a revised, refined set was published in 1982.

  • *Ten principles* are enumerated (see Appendix 4) that are the basis of all the ethical issues that have been discussed here.

— *Ethical Principles of Psychologists and Code of Conduct* was later published by the American Psychological Association in 1992.

  • These principles cover *animal research* as well as *human research*, and also *clinical/counseling practice and teaching*.

  • The reader is encouraged to review this lengthy document on the Internet at www.apa.org/ethics/code.html, but the APA publications contained in Appendix 4 probably provide a *more useful* set of "Ethical Principles in the Conduct of Research with Human Participants," as well as a *more extensive* set of "Guidelines for Ethical Conduct in the Care and Use of Animals."

9. **Institutional Review Board (IRB)**

— This is a *research review committee* for studies involving human participants, which must be established when an institution receives money from the federal government, and it serves to evaluate the *ethical propriety* of proposed studies, in addition to ensuring that governmental and institutional *laws and regulations* are adhered to.

  • A detailed *research proposal* must be submitted, accompanied by a completed *application* with answers to a number of questions.

— Note: There are *no magical rules* that help make decisions simple for the review committee as to whether or not a given study is ethical.

  • *Every study has to be evaluated in terms of*:
    a) Whether there are any *ethical problems* with the procedures;
    b) Whether *more ethical alternative procedures* are available;
    c) Whether the *importance of the study* is such that any ethical compromises are *justified* (i.e., a cost-benefit analysis is needed).
    ∞ Note: Research proposals are *likely to be rejected* (versus modified) only when studies involve great amounts of physical or psychological stress, and there is no informed consent.

Animal Participant Ethical Issues

The topic of *animal research ethics* is one that is important, controversial, and of considerable interest to many members of society.

1. Facts Versus Myths About Animal Research[27,28]

— *Animal rights groups* have raised society's consciousness about the care and use of animals, but many have also disseminated a great deal of misinformation about: 1) the *treatment* of animals in research, 2) the *necessity* for using animals, and 3) the *significant role* that animals have played in advancing scientific, medical, and veterinary knowledge and practice.

- It has been pointed out that the *message* of the more radical groups not only reflects opposition to research with animals, but in fact is actually *antiscience in general.*

- Some individuals and groups have even gone *beyond debate and political action.* They have broken into laboratories and snatched research animals, and they have vandalized, destroyed, or stolen expensive equipment, and sometimes years of hard-gained data.

  ∞ It should be noted that occasionally these *criminal acts* have significantly *held up* developments that would have *benefited animals*, such as the improvement of facilities and procedures for the care and use of animals.

- *Throughout the past century,* contrary to the claims of many in the animal rights movement, *nearly every major advance* that has been made in understanding the *physiology of the body and its nervous system* and in *medical science* has been the result of using nonhuman animals in research — including development of antibiotics, vaccines, therapeutic drugs, and surgical procedures.

  ∞ Examples: Due to research made possible by using animals, smallpox as been eradicated; other terrible diseases such as polio and rheumatic fever have been virtually eliminated; tuberculosis, diphtheria, whooping cough, typhoid fever, scarlet fever, dysentery, cholera, and tetanus are much less common; effective drugs have been developed for treating schizophrenia, depression, anxiety, pain, high blood pressure, diabetes, and Parkinsonism; and angioplasty, heart-pacemaker implants, and organ-transplant surgery are now common procedures for saving lives.

- *Today,* as well, advances continue to be made in the prevention and treatment of *debilitating and deadly disorders* — again based on research that is very dependent on the use of animals.

  ∞ Examples: Progress is being made in solving the profound problems of AIDS, cancer, diabetes, arthritis, strokes, psychoses, mental retardation, and Alzheimer's disease.

- *Psychological experiments* with animals led many years ago to a better understanding of incentive and motivational mechanisms, and to the discoveries of *classical and instrumental conditioning,* along with the *learning principles* of reinforcement, punishment, extinction, generalization, and discrimination — all of which are applicable to human learning and behavior modification.

∞ As a result, the field of *behavioral medicine* was established and several *behavior therapy techniques* were developed, such as shaping behavior by successive approximations, modeling, counterconditioning, aversive conditioning, systematic desensitization, cognitive restructuring, and biofeedback — which have all been very important for helping individuals with a variety of problems.

- Examples: Neuromuscular disorders impeding control of the body, and causing urinary and fecal incontinence; bed wetting (enuresis); strokes and head injuries affecting language; alcoholism and other drug abuses; phobias, depression, migraine headaches, hypertension, smoking, and obesity; compulsive disorders such as bulimia or anorexia nervosa; and self-injurious behavior in autistic children and adults.

- It should be noted that these *applied, clinical benefits* of the earlier *basic research* on learning were largely *unforeseen*, which is often the case, and this should therefore be taken into account whenever evaluating whether a particular study has sufficient justification based on some analysis of *cost-benefit expectations*.

- Animal research has also *experimentally demonstrated* the important relationships that exist between *psychological stress* and *physical health*, at least some of which involve suppression of the *immune system*.

  ∞ Examples: Cancer, high blood pressure, heart attacks, strokes, infections, ulcers, and kidney failure can all be affected by psychological stress.

- Animal research is being used for *many other purposes* today.

  ∞ Examples: To learn about the *nervous system mechanisms* of *psychoactive drugs* that produce addiction, pain relief, and anxiety reduction; to study *environmental toxicant effects* and the *effects on fetuses* of alcohol and other drugs used by pregnant mothers; to investigate the *effects on youths* of early psychosocial deprivation and other childhood abuse; and to better understand the causes and means of controlling *aggressive behavior*, which is all too common in our society.

- Importantly, it should not be overlooked that animal research isn't just done to help humans — it *benefits nonhuman animals* as well.

  ∞ Examples: 1) The *veterinary treatments* that pets receive for their disorders, such as rabies, canine parvovirus, distemper, heartworm, and feline leukemia, are all due to research conducted with animals (as well as sometimes with humans), 2) Research on the *reproductive behavior* of animals has made it possible not only to help humans but also to breed endangered species in captivity so that they can be saved from extinction and reintroduced into the wild, 3) Research on *learned taste aversion* in rats has led to nonlethal and humane means of keeping crows away from crops, and coyotes away from sheep — thereby saving the lives of many animals.

- It is also important to point out that more than 85% of the animals used in research are *rats, mice, and other rodents* that are specially bred for this purpose; and hence only a minority of research animals are cats, dogs, birds, or nonhuman primates.

- Furthermore, far fewer animals are employed in research, education, and testing than are used for other purposes — for example, less than 1% of the number that are *killed for food*.

- Moreover, only about 6% of research animals are exposed to *painful procedures*, and these are usually *neither severe nor long-lasting*, and are permitted only because to relieve the animals of pain would defeat the purpose of finding ways to alleviate the *chronic pain* experienced by millions of humans (and costing tens of billions of dollars in health care and lost work every year).

- Finally, it should be noted that in the case of *psychologists*, less than 10% of the research involves the use of nonhuman animals in *any way*.

— Animals are used in research for *several reasons*.

1) They are *interesting* and form an *important part of the world* we live in, hence we want to better understand and care for them, and thereby perhaps also better understand and help ourselves.

    ∞ Examples: How can we design better living conditions and in other ways enhance the well-being of captive animals? What are the cognitive abilities of other primates; how does this relate to the possible evolution of human intelligence, consciousness, and language; and can this knowledge be used to help humans with brain dysfunctions?

2) *Ethics prohibits* the initial testing on *humans* of unknown drugs, as well as other dangerous, irreversible manipulations.

    ∞ Examples: New forms of chemical or radiation therapy for cancer, and destruction of areas of the brain as treatments for uncontrollable violence or intractable epilepsy.

3) Animals can serve as *highly controlled models* of physiology, behavior, and disorders of humans and other species; and they commonly have *shorter life cycles*, thereby facilitating the study of genetics as well as the processes and diseases of aging.

    ∞ Examples: Studies of the genetic vulnerability and mechanisms of drug addiction and cardiovascular disease.

— *Philosophical justification* exists for using nonhuman animals to facilitate the advancement of science and the betterment of health.

1) Nonhuman animals *also exploit* other species for their needs.

2) Most humans readily accept the use of nonhuman species for *food, clothing, and labor* — so why not for research, testing, training, and instruction?

3) Humans *kill* enormous numbers of animals because they are *troublesome*. Millions of pounds of *poison* are used to eliminate rats annually, and

tens of millions of cats and dogs are *abandoned* and eventually *destroyed* every year.

∞ For every dog or cat obtained for research from animal *shelters and pounds*, 100 others are killed in shelters and pounds because they do not find homes.

∞ If these animals were not used for research, then there would be *no benefit* arising from this unfortunate situation.

∞ Also, other animals would then have to be specially bred for research, resulting in the *unnecessary* death of even more animals, which clearly would be perverse.

∞ In addition, the cost of research would be driven up, with a resulting decrease in funds available for productive use.

4) Most importantly, and sometimes overlooked, the *discomfort and loss of life* for some research animals, mostly rodents, should be weighed against the *suffering and death* to which millions of humans would be condemned if animal research were curtailed.

∞ Consider the following: If a *choice* had to be made between the well-being or life of a human child versus that of a loved household pet, what would be *your decision*?

∞ It must also be kept in mind that there is far more abuse, neglect, and cruelty to many *farm and even pet animals* (who are cooped up, abandoned, and worse) than to research animals.

• Hence isn't the emphasis of most animal rights groups *grossly misplaced*?

— Regardless of philosophical position, no one should dispute the proposition — which is supported by essentially every scientist — that animal research participants deserve to be treated as *humanely* as possible, and with *respect*.

• Since nonhuman animals cannot provide *informed consent or withdraw* from research, additional protection must be provided to ensure that the research is *meaningful and productive*, and to minimize any *physical or psychological discomfort, stress, and harm*, unless justified by the probable benefits of the research.

• Moreover, the following additional principles should be followed:

a) The *lowest level species* should be used that could provide *valid data* for generalization to humans, if that's the goal;

b) As *few animals* should be used as could provide *reliable data*;

c) *Alternatives*, such as *in vitro biological systems* (e.g., cell and tissue cultures), *computer simulations*, or *mathematical models*, should be substituted for animals whenever possible.

• But note that research using animals would *first* be necessary in order to *develop* and *validate* alternatives.

- Also, the alternatives to using intact functioning organisms would seem to be of *limited usefulness* for studying the development and prevention of many *diseases and disorders* involving either the *brain and mind* or *multiple aspects of the body*, e.g., depression and arthritis.

- Moreover, any therapy that is even remotely *potentially dangerous* must eventually be tested on *whole animals*, with all their organ systems, before being tried on humans — thus "alternatives" are really only *adjuncts* to animal research, not *substitutes*.

- It should be obvious that because of the costs relative to the funds that are available, there are *strong economic pressures* on scientists *against* using higher species or greater numbers of animals in research than are necessary, and *in favor* of protecting the wellbeing of animal participants, and of using alternatives/adjuncts whenever it is humanly possible to do so.

2. Guidelines for Ethical Conduct in the Care and Use of Animals[3]

— Many *professional organizations*, such as the American Psychological Association, the Society for Neuroscience, and the Federation of Societies of Experimental Biology, have excellent standards and guidelines for the humane care and use of animals to which its members must adhere.

— The *American Psychological Association* revised and expanded its *Guidelines* in 1992–1993 and 1996 to address a number of issues, including justification of research, personnel, care and housing; acquisition of animals; experimental procedures; field research; and educational use (See Appendix 4, second portion, for the complete Guidelines).

— In addition, the *Animal Welfare Act* (a federal law) sets forth standards, as does the *Animal Welfare Policy* of the U.S. Public Health Service, which provides detailed recommendations in the publication titled *Guide for the Care and Use of Laboratory Animals*. The *U.S. Department of Agriculture* is responsible for enforcing the Act and conducting periodic unannounced inspections of facilities.

3. Institutional Animal Care and Use Committee (IACUC)

— When an institution receives money from the Federal Government, it is required to establish an Animal Care and Use Committee that serves to ensure the *welfare and humane treatment* of nonhuman animals used in research, testing, training, and instruction, and also to enforce governmental and institutional *laws and regulations* regarding the care and use of animals.

- Such committees have *diversity*, in that they must consist of at least five members, and include at least one veterinarian with appropriate training and experience, one scientist experienced in research with animals, one member whose primary concerns are in a nonscientific area (e.g., ethicist, lawyer, or clergyman), and one individual from the community who is not otherwise affiliated with the institution either directly or through family.

- The IACUC reviews each research protocol, and studies may not begin, or animals even be obtained, unless the committee has approved the protocol for adherence to Federal and institutional guidelines for the appropriate, humane care and use of animals.
  - ∞ In addition, periodic inspections of the institution's animal care and use facilities must be conducted, and careful records maintained of all the committee's work.
- When reviewing the research protocol, the IACUC considers not only the rationale for the research, and the conditions of animal care during the study, but also the rationale for the number of animals that will be used, whether this number is appropriate for the proper interpretation of results, and additionally whether the research is unnecessarily duplicative of previous studies.
  - ∞ Perhaps most importantly, the IACUC must explicitly consider the issues of pain and suffering that might be involved in the research, and suitable approaches for minimizing and alleviating any pain or suffering that might occur.
- Finally, it should be noted that because of the *combined knowledge* of its members, these committees also serve to increase the *quality of research* through the *thoughtful recommendations* made to investigators.
  - ∞ Naturally, in most studies the data obtained will be *more reliable and valid* if the animal participants are given the best possible care, and if unnecessary discomfort and stress are minimized.

Review Summary

1. There are a number of *non-participant ethical issues* in research:
   a. Care must be taken to ensure that data are *accurate and complete*, and that alternative hypotheses/explanations are acknowledged;
   b. *Credit* for significant work done by others involved in the research should be recognized through *co-authorship or acknowledgement* in the report;
   c. *Premature publication* and *withholding publication* are unacceptable, as is usually *duplication* of the publication of findings;
   d. Whether it is unethical or a patriotic duty to conduct research having *military applications* is a matter of personal philosophy, as well as the nature and degree of force to which the research might lead.

2. Two major sets of *ethical issues* pertain to *human participants*. The first is *how participants are obtained* and the *information provided* to them. This involves four concerns:

a. *Coercion*, where participants are denied complete freedom of choice about participation and withdrawal (which is particularly problematic when there is a power relationship of the researcher over the potential participants);

b. *Concealment*, where participants are not made aware of certain or all aspects of a study;

c. *Deception*, where participants are deceived about the purpose of a study and/ or what occurs

d. *Informed consent* (the integrating theme and opposite of the preceding concerns), where individuals' decisions to take part in an investigation are made on the basis of accurate and adequate information about the purpose of the research, what will happen in the study, what they will be called upon to do, and anything else that might reasonably be expected to influence their willingness to participate.

3. The second major set of ethical issues pertaining to human participants deals with *how participants are treated during and after a study.* For the *welfare* of participants they must be protected from *physical and mental discomfort, stress, and harm* that might result from their participation in the research. However, *noxious physical or psychological stimuli* might be an essential part of a study. In that case, any *undesirable consequences* must be *detected and corrected* afterward. Moreover, except under very special circumstances, procedures likely to cause *serious or lasting harm* should not be used.

4. The use of *deception* in research has a number of *implications for society*:

a. Do researchers have the right to add to the many *anxieties* that already exist in life?

b. Should not research participants be treated with *dignity and respect*?

c. If it is considered acceptable to use deceit for the *advancement of science*, then why not for anything else that is valued?

d. Moreover, society places great *trust* in the *honesty* of scientists and in the *respectability* of their activities — does not the use of deception undermine society's *confidence* in science and in the *morality* of all humankind, therefore making us *distrustful* of everything and *alienating* humans from each other and thus society?

5. The use of *deception* also has *methodological implications for researchers.* It is becoming *increasingly difficult to find naive participants.* Because of direct and vicarious experiences, participants frequently *suspect* that experiments are not about what the researcher *claims.* This can *undermine the intended effects* of deception and *influence* a study's results. Hence the *value of using deception might be lost.*

6. *Debriefing* is a standard *postexperimental conference/interview* between the experimenter and research participants. It is *required* after the use of *deception or concealment*, but regardless, it has a number of functions:

a. It can provide important information about whether the participants *suspected* deception and what they *thought* the study was about;

    b. It can also help determine the *effectiveness* of independent variable manipulations, and the *strategies* that participants employed on tasks;

    c. It should always be used to *undo and explain* any deception that was employed, *detect and remove* any negative effects of participation, *educate* the participants about psychology and research, and give participants a sense of satisfaction for participating;

    d. It is also frequently used to *convince* participants not to discuss the experiment with others, and to *notify* participants how they can receive information about the results of the research when the study is concluded.

7. There are several *alternatives to using deception*:

    a. *Role-playing*, where some situation is *described* to the research participants by the experimenter, and then the participants are asked to indicate how they or others *would behave*;

    b. *Simulation studies*, which involve role playing that takes place in a *simulation* of a real-world situation;

    c. *Honest experiments*, which include *four strategies* — participants being fully informed, explicit behavioral change programs, field experiments, or naturally occurring events;

    d. *Assumed consent*, where a complete and accurate description of the experimental procedure and its purpose would be given to a *sample of potential participants*; and if they consented to participate, this would imply consent for other, incompletely informed participants who are later selected to actually participate in the study.

8. *Confidentiality* involves keeping information obtained about research participants *anonymous*, unless otherwise agreed to in advance, in order to *protect the participants' privacy.* However, ethical dilemmas can arise when it appears necessary to violate the confidentiality principle in order to uphold the *principle of protection from harm*, i.e., if information obtained during the study indicates that participants are likely to injure themselves or others.

9. Before any research begins, a *planning process* is necessary that carefully *evaluates and weighs* the risks to participants against the potential scientific, educational, or applied value of the research. This invariably involves an *Institutional Review Board (IRB)* that judges the acceptability of research proposals in the context of various laws, regulations, and established ethical principles, such as those of the American Psychological Association.

10. Throughout the past century, contrary to the claims of many in the animal rights movement, *nearly every advance* that has been made in the *medical and veterinary sciences*, and a great many in the *behavioral sciences*, has been the result of *utilizing* animals in research. This is also true today. These advances include, e.g., the development of antibiotics, vaccines, treatment drugs, surgical procedures, and behavior therapy techniques, as well as progress in understanding how the brain operates, the mechanisms of drug effects, and the causes and treatment of migraine headaches, epilepsy, and antisocial behavior.

11. More than 85% of the animals used in research are *rodents specially bred for this purpose*, thus only a minority or research animals are cats, dogs, birds, or non-human primates. Furthermore, the number of animals used in research is less than 1% of the number that are *killed for food*. Moreover, only about 6% of research animals are exposed to *painful procedures*, usually *neither severe nor long-lasting*, and this is done to find ways to alleviate the *chronic pain* experienced by millions of humans.

12. Animals are used in research for *several reasons*: a) They are *interesting* and form an *important part of the world* we live in; b) *Ethics* prohibits the initial testing on humans of unknown drugs and dangerous, irreversible manipulations; and c) Animals can serve as *highly controlled models* of physiology, behavior, and disorders of humans and other species, and they commonly have shorter life cycles, facilitating genetics and aging studies.

13. There are a number of *philosophical justifications* for using nonhuman animals to facilitate the advancement of science and the betterment of health, e.g., we also use animals for food, clothing, and labor. Nevertheless, animals should always be treated as *humanely* as possible and with *respect*. Their use in research should be *meaningful and productive*, and *physical or psychological discomfort, stress, and harm* should be minimized. Professional organizations and the Government provide guidelines, and an *Institutional Animal Care and Use Committee (IACUC)* is responsible for ensuring the welfare and humane treatment of animals, and seeing to it that laws and regulations are followed when funds are provided by the Federal Government. In addition, the following three principles should be followed:

    a. The *lowest* species should be used that could provide *valid data*;

    b. As *few participants* should be used as could provide *reliable data*;

    c. *Alternatives*, such as *in vitro* biological systems, computer simulations, or mathematical models, should be substituted for animals *whenever possible*.

Review Questions

1. Describe three *non-participant issues* that are of ethical concern in research, besides military applications.

2. List and describe the four *components* of the major set of *human participant ethical issues* that deal with *how participants are obtained* and the *information provided*, and give an *example of a study* (you can invent one) that illustrates violation of two of the ethical components (identify them).

3. List and describe the four *points* of the major set of *human-participant ethical issues* dealing with *how participants are treated during and after a study.*

4. Discuss four *implications* for society of the use of *deception* in research.

5. Discuss the *methodological implications* for researchers of using *deception*.

6. Define *debriefing*, and give at least seven *functions*.

7. Describe three *alternatives to using deception*, including assumed consent.

8. Describe the *confidentiality principle* and how this relates to the *principle of protection from harm*.

9. With respect to ethics, briefly describe the *planning process* that is necessary before any research begins.

10. Describe at least five ways in which *animal research* has benefited *humans* as well as *animals*.

11. Put into perspective the *numbers and types of animals* used for research, and the experience of *pain*.

12. What are three general *reasons for using animals* in research?

13. In addition to treating animal subjects/particpants as humanely as possible and with respect, what are three *principles* that should be followed when considering their use in research?

7

Control In Experiments — Basic Principles and Some Techniques

CONTENTS

Experimental control is an extensive subject, and therefore *two chapters* are devoted to this very important issue. *This chapter* covers the topics of extraneous variables and variance, maximizing the primary variance produced by independent variables, and six techniques for controlling the secondary variance caused by extraneous variables. *The next chapter* will cover five more techniques for controlling secondary variance, and conclude with a discussion of minimizing error variance due to extraneous variables.

Two Meanings of Experimental Control

When thinking about control, the fact is sometimes overlooked that *independent variables*, in addition to *extraneous variables*, are controlled in experiments (as well as in some descriptive research).

1. Control Over Independent Variables

— Experimenters *decide/control which independent variables* will be studied, and *what values* of independent variables will be used.

— Moreover, experimenters *produce/control* the independent variable *values* by *purposely and directly manipulating* the independent variables in the manner that *they decide upon.*

- Purposeful and *direct* manipulation is the *essential feature* that distinguishes an experiment from descriptive research, which instead uses naturalistic observation with purposeful measured selection from already available values as the means of manipulation, or no manipulation at all. (discussed in Chapter 2).

- But note that to *manipulate* an independent variable, whether purposely and directly or by measured selection, is to *control* it.

2. Control Over Extraneous Variables

— Extraneous variables are *extra/unwanted variables* operating in the experimental situation, which like independent variables might affect the dependent variable(s) (covered earlier in Chapter 2).

— Reasons for controlling extraneous variables

There are *two* very important reasons for exerting control: to minimize *masking* and to minimize *confounding*.

1) Masking

   ∞ This is when extraneous variables *hide/obscure* the effects of an independent variable on a dependent variable, often by causing substantial variability themselves in the data.

   - Example: It would be hard to detect the differential effectiveness of various *rehabilitation programs* (the independent variable) on *alcohol abuse* (the dependent variable) when there is simultaneously a lot of variation among the research participants in their *daily exposure to other people using alcohol* (an extraneous variable).

   - Such variability of dependent variable scores due to *extraneous variables* is seen in the *variation* of scores *within* the conditions of the independent variable, where the level is constant and thus could not be the source of variability in the dependent variable measure.

   - These extraneous variable effects lead to *decreased reliability* (consistency) of any apparent *independent variable* effects — which are *implied* by differences in the mean dependent variable scores *between* the conditions.

   - Stated differently, *confidence* in real *independent variable* effects (measured by the variation *between* conditions) is reduced by *extraneous variable* effects (measured by the variation *within* conditions) since the extraneous variables could themselves be the *entire source* of dependent variable differences in mean scores *between conditions*. This comparison of between-conditions variance (variation) to within-conditions variance is what is evaluated by *tests of statistical significance*.

     Example: In general, chance *individual differences* among participants, an *extraneous variable*, might produce dependent variable effects, seen as *within-conditions variation* (variance), that equal or exceed the *between-conditions variation* caused by purposeful and direct manipulation of some *independent variable* — hence greatly masking the independent variable's effects.

To make this more concrete, consider a situation where there are only two conditions with just four participants in each; and where under the *Treatment Condition* the dependent variable scores for the participants are 80, 70, 60, and 50% correct, with a mean of 65%; and where for the Control Condition the scores for the participants are 70, 60, 50, and 40% correct, with a *mean of 55%.*

Note the large amount of *variability/variance* among the scores *within each condition* of the independent variable (hence due to extraneous variable effects) compared to the difference in the *mean* dependent variable scores *between the conditions* (also due, at least partly, to extraneous variable effects, as well as *hopefully* to the desired independent variable effects).

Only when the *between-conditions variance* is *much larger* than the *within-conditions variance* (which is not the case in this example) can we have *sufficient confidence* that there is an effect of the independent variable in addition to the effects of extraneous variables. When this is not the case, it is declared that there is *not* a significant independent variable effect.

∞ *Errors of omission* are caused by masking.

- This is the *failure to declare an effect* of an independent variable when, in actual fact, *there is an effect* (referred to as a Type II/Beta Error — discussed in Chapter 12, along with the Type I/Alpha Error).

2) Confounding

∞ This is when extraneous and independent variables are *systematically and inextricably intertwined (mixed together)*, so that their effects on a dependent variable cannot be separated/distinguished, hence causing *confusion and errors.*

- If experimental manipulation of an independent variable *appeared to produce an effect*, the investigator *could not be certain* (given confounding) that the results were *actually* due to the independent variable, rather than an extraneous variable (or possibly both) — this is called an *internal validity problem* (see Chapter 9).

- *Confounding* occurs when variation of one or more *extraneous variables* (e.g., age, gender, time of day, apparatus, experimenter) is *systematically associated with* (related to) variation of an *independent variable* — such extraneous variables are often referred to as *confounding variables.*

Example: If *all participants* for one independent variable condition were run in the *morning*, and *all those* for the other independent variable condition in the *afternoon*, then there would be confusion as to whether any detected *dependent variable effects* (differences in mean scores) were actually due to the different *independent variable conditions*, or due to differences in the *extraneous variable* of time of day.

∞ *Errors of commission* are caused by confounding.

- This is the *mistaken declaration of an effect* of an independent variable when there is *none* — the dependent variable differences under different independent variable conditions being due only to a *confounding extraneous variable* (similar to Type I/ Alpha Error, where *chance* effects are mistaken for an independent variable effect).

  Note: In contrast to *complete confounding*, which has just been discussed, *partial confounding* occurs when participants in the various conditions are different or are handled or measured differently only *on average*, rather than consistently; this *commonly occurs* to varying degrees by *chance*, e.g., due to random assignment of participants (randomization is covered in Chapter 8).

∞ *Errors of omission* also might be caused by confounding.

∞ This would be a *failure to detect an effect* of an independent variable when, in fact, *there is an effect*: a confounding variable/factor simply *canceled out* the independent variable effect on the dependent variable.

  Example: Using one *apparatus* for all the participants in one condition, and a different apparatus for all the participants in another condition, might result — if the two apparatuses do not operate identically — in an *increase* of dependent variable values for the independent variable condition associated with lower dependent variable values, and vice versa, thereby *negating* the independent variable effects.

- Although the *consequences* are the same as for masking, and therefore it is referred to as *masking*, the cause is a *validity problem, not reliability* (see earlier).

- IN SUMMARY: *Either chance or systematic events* can lead to extraneous variable effects that either *hide* (i.e., mask) or are *mistaken* for (i.e., confounded with) independent variable effects.

  ∞ This was also noted in the Chapter 4 summary at the end of "Accuracy and Consistency of Variables," and is expanded upon later in this chapter under "Summary of Primary, Secondary, and Error Variance."

  ∞ Example illustrating all possibilities: If there are two independent variable conditions and two experimenters, and one experimenter runs all the research participants in one of the conditions, and the other experimenter runs all the participants in the other condition, then there would be *complete confounding* of the independent variable conditions with the extraneous variable of experimenter differences.

- This could result in either an *error of commission,* where an effect of the independent variable is claimed when there is really only an effect of the extraneous variable; or an *error of omission* could occur, where there is an effect of the independent variable, but it is *masked* by the extraneous variable because the value of the *extraneous variable* leading to higher dependent variable scores is paired with the value of the *independent variable* associated with lower dependent variable scores, and vice versa.

∞ Continuing the example: If instead the research participants, and the conditions they are in, are *randomly assigned* to one or the other experimenter, in order to *avoid systematically associating* the different values of the independent variable with the different values of the extraneous variable of experimenter differences, it is nevertheless possible by *chance* to end up having *partial confounding,* with the associated potential for an *error of commission or omission.*

- *Partial confounding* could be *avoided,* however, by purposefully *balancing* the extraneous variable of experimenter differences across the different levels of the independent variable, rather than leaving it *only* up to chance randomization; i.e., each experimenter would be randomly assigned an *equal number* of participants in each of the independent variable conditions (this technique of Balancing Extraneous Variables is further discussed later in this chapter).

- *Masking* of the effects of the independent variable by the effects of the extraneous variable would still be a potential problem, however, since the different experimenters running participants under each independent variable condition would likely lead to *within- conditions variance,* which would reduce the reliability of any apparent independent variable effects suggested by differences in the mean dependent variable scores *between conditions.*

  Controlling for this masking would be possible if the variance associated with the secondary, extraneous variable (experimenter, in this case) were *analyzed,* and hence *accounted for* (this technique is discussed in Chapter 8 under "Treatment Groups and Systematizing Extraneous Variables").

- Note that *confounding and masking are not completely parallel concepts,* because, although each is both a process and a result, *masking* can be due to confounding, but not vice versa.

Forms of Extraneous Variables

Appendix 5 summarizes Chapters 7 and 8. It should be noted that the extraneous variables of one experiment can, in many cases, be correlational or experimental *independent variables* of other studies, depending on what the investigator is most interested in studying at the time. The following descriptions of *12 types of extraneous variables* are elaborated in Chapters 9 and 10, "Designs for Experiments," in which control is a central issue.

1. Environment

— *Physical and social conditions of the experimental setting*, other than independent variables, that might affect a participant's performance, are referred to as *environmental* extraneous variables.

  • They are similar to the extraneous variable of *contemporary history* (see below), but in contrast, environmental extraneous variables occur specifically during the *time of treatment and/or measurement*, rather than during a pretest-posttest interval.

2. Contemporary History

— *External environmental events* (physical and social), other than the independent variable(s), that occur during the interval of time *between any pretest and posttest measurements* of a dependent variable, represent *contemporary history* extraneous variables.

  • Note that the *longer* the testing interval and the *less isolated* the participants are from the environment, then the more likely would be this extraneous variable problem.

3. Instrumentation

— Change that occurs over time in measurement of a dependent variable due to *variation in mechanical or human observer factors* is referred to as an *instrumentation* extraneous variable.

  • Variation might occur in the *calibration of measuring and/or recording instruments*.

  • Variation also might occur in the *observer's skill*, due to practice; or the *measurement criteria*, due to experience; or the *fatigue level*, due to exertion.

4. Statistical Regression

— Change attributed to the tendency of participants with *extremely high or low test scores* to *move toward the mean* when measured again represents the extraneous variable of *statistical regression*.

  • It is due to *unreliability* of the test instrument or the participants, which partially accounts for extreme scores on the first measure (see Chapter 9 for an extended explanation).

5. Maturation

— Changes due to *conditions internal to an individual* that vary as a function of the *passage of time*, i.e., not due to particular environmental events (hence not learning effects), represent *maturation* extraneous variables.

• *Developmental processes* of growth and aging that lead to anatomical, physiological, cognitive, and behavioral changes are examples of maturation.

• *Biorhythmic processes* (e.g., circadian, menstrual, and circannual cycles) are also considered varieties of maturational change.

• *Short-term processes*, such as becoming hungry or thirsty, are technically referred to as maturational effects as well.

6. Selection

— Use of *differential selection procedures* for the participants assigned to the different comparison groups (independent variable levels) represents the extraneous variable of *selection*.

• This leads to *individual participant/subject differences* between conditions.

• In addition to *direct effects* on dependent variables, *interaction effects* can occur between selection and other extraneous variables such as contemporary history, maturation, attrition, or testing.

∞ *Interactions* are when the effects of one variable are *influenced by*, and thus *dependent upon*, another variable (see Chapter 10 under "Factorial Designs").

7. Participant Attrition/Experimental Mortality

— *Loss of participants* from the various comparison groups that occurs between multiple testing or training sessions after an experiment has begun, and which might be different for the different groups, is referred to as the extraneous variable of *participant attrition*.

• When participants have to be present on more than one occasion, they sometimes fail to return (but usually not due to death, in spite of the word mortality).

8. Sequence/Order

— Participation by individuals in *more than one condition*, for comparison of independent variable effects, can produce extraneous variable consequences known as *sequence/order effects*.

• Experiencing one condition can affect performance under a subsequent condition due to *practice and/or fatigue*, as well as due to changes in *motivation and/or attention*.

9. Testing

— Being *tested more than once* in a *single condition* can produce extraneous variable influences referred to as *testing effects*.

• Experiencing an initial test, e.g., a *pretreatment test*, can affect the scores of a subsequent test, e.g., a *posttreatment test* (this is similar to sequence/order effects for different conditions).

- Influence on dependent variables can be either *direct,* or it can be *indirect* by producing *sensitivity or resistance* to the experimental treatment — in which case it would be an *interaction effect* (see Chapter 9; e.g., "One-Group Pretest-Posttest Design").

10. Participant Bias

— *Expectancy, attitude,* or *motive* of participants represents the extraneous variable called *participant bias.*

- These *three aspects* of bias are illustrated respectively by the *placebo effect* (getting better due to belief in the treatment), the *Hawthorne effect* (doing better due to knowledge or belief that one is being observed and participating in an experiment), and the *motive for positive self-presentation* (behaving in a way that presents oneself in the most positive manner).

11. Participant Sophistication

— *Familiarity* of participants with research studies and/or experimental procedures, e.g., deception (all of which can affect participant bias) is the extraneous variable of *participant sophistication.*

- This can influence participants' *interpretation* of instructions, etc.

12. Experimenter Features

— *Expectancy/bias, performance, appearance, and personality* of the researcher(s) represent the extraneous variable called experimenter *features.*

- These can affect *participant bias,* and therefore their *behavior.*

- Moreover, experimenter expectancy or performance can also affect the data *directly* (this and other aspects of experimenter factors are elaborated later in this chapter).

- This extraneous variable category *also includes* the undesired variation and effects of *procedures, instructions, and apparatus* not already noted under the preceding forms of extraneous variables, such as those of testing and instrumentation.

Concept of Variance and Good Experimental Designs

As an introduction to the material that follows, it is useful now to clarify what is meant by *variance* and why it is a very important concept.

1. Variance (See Appendix 2)

— Definition

- *Variance* is simply a *statistical measure of variation* in a set of observed scores, such as for the *dependent variable* of a study (see Chapters 8 and 11 for additional details).

— Relation to statistics and reliability

- *Analyses of variance* are tests of *statistical significance* that involve comparisons of *between-conditions* variance against *within-conditions* variance of a dependent variable, thereby providing a measure of *reliability* of independent variable effects (see earlier discussion under "Control Over Extraneous Variables").

— Relation to control

- *Experimental control* of independent and extraneous variables is best understood as the *control of variance*.

— Relation to research

- *Experimentation* can be defined as the *controlled study of variances*, i.e., the variance presumably produced by the independent variable(s) as opposed to extraneous variables.

2. Good Experimental Designs

— Goals in terms of variance

- *Maximizing* the variance of the dependent variable produced by manipulation of the *independent variable(s)* is the first goal of good experimental designs.

- *Minimizing*, or otherwise *controlling*, the variance of the dependent variable produced by *extraneous variables* is the second goal of good experimental designs (for more details, see later in this chapter under "Minimax Principles of Variance").

— Benefits of achieving the goals

- *Masking and confounding* are minimized when the goals are met.
    - ∞ As a result, the experimenter is *more likely to determine* — both reliably and accurately — the *relationship*, if any, that exists between the independent and dependent variables.

General Sources of Variance

There are *three* general sources of variance: *participant factors, environmental factors, and experimenter factors.* These three sources encompass the 12 specific types of *extraneous variables* discussed earlier. It is important to note, however, that these three general sources of variance apply to *independent variables* as well as to extraneous variables.

1. Participant Factors

— Variations in *characteristics of organisms* are participant factors.

- Specifically, they consist of differences in the participants' *genetic traits* (e.g., intelligence), *past experiences* (e.g., learning), or *present state* (e.g., anatomy, physiology, emotion, motivation, or cognition) that serve as independent or extraneous variables.

∞ Examples: gender, shyness, physical abuse, hormone deficiencies or excesses, fear, racism, and other biases.

2. Environmental Factors

— Variations in the *physical and social surroundings* of a study are environmental factors.

• Examples: room size, light intensity, temperature level, time of day, and the presence of other individuals and their behavior.

3. Experimenter Factors

— Variations in *experimenter characteristics* as well as *experimental procedures, instructions, apparatus, tasks, and conditions* are called experimenter factors.

• These are a *subset* of environmental factors that deserve separate consideration because of their special relationship to any experiment.

— *Experimental contamination* is what the result of this source of variation is called when it is both *undesired and uncontrolled*.

• As for environmental and participant factors, elements from this category can be *manipulated* as independent variables, or they can be *controlled* as extraneous variables.

• *Control* of some factors (e.g., experimenter behavior) is more difficult when there is *more than one experimenter*, but the *generalizability* of findings would be greater.

— Experimenter characteristics

There are *four categories* of experimenter characteristics: *appearance, personality, performance, and expectancy/bias.*

• Experimenter characteristics are a subset of *experimenter factors* that are important but nevertheless often overlooked/neglected as a source of *extraneous variance*, in which case, as already noted, they can lead to what is called *experimental contamination*.

• Experimenter effect

∞ The experimenter effect is any *variance* that is caused by experimenter *characteristics*, which includes the following:

1) Appearance

∞ *Age, race, gender, clothing, grooming,* and *hygiene* of an experimenter all represent aspects of appearance.

• *Human research participants* can have their *motivation and behavior* affected by any combination of these.

Example: A mature experimenter, wearing a lab coat and having neatly trimmed hair, might elicit more respect and thus cooperation from research participants than would a youthful, slovenly dressed, unbathed researcher.

• *Higher nonhuman participants,* particularly primates, can also be affected by appearance, e.g., a change in clothing or odor of a researcher with whom they are familiar.

2)   Personality

∞   *Pleasantness and cordiality versus unpleasantness or belligerency* are important aspects of experimenter personality.

- Both *human and higher nonhuman animals* can have their *motivation and behavior* affected by *nonverbal communication,* such as raising one's voice or scowling.

3)   Performance

∞   *Skilled versus inept research behavior* is an influential characteristic of experimenter performance.

- *Human participants* can be affected by the *perceived efficiency* of the experimenter.
- *Animal participants/subjects* can be affected by the *gentle versus rough handling* that they receive.

∞   *Accuracy and completeness* of recording and analyzing data are also experimenter performance characteristics.

- Obviously these are *extremely critical* variables affecting the research results and the conclusions drawn.

4)   Expectancy/bias

∞   *Anticipation or desire* of the experimenter with regard to the outcome/ results of the study (i.e., the research hypotheses) represents the expectancy/bias of the experimenter.

∞   Experimenter bias effect (Pygmalion Effect)

- This is the *influence* of the experimenter's bias/expectancy, which can be a *self-fulfilling prophesy.*
- Bias can affect the *experimenter's performance* such that the expectancy is transmitted to the research participants, thereby-causing *participant bias;* or the participants could be *subtly rewarded* for behaving in a manner that *supports* the experimenter's expectancy.

  This influence can thus act indirectly via the motivation of participants, or it can have rather direct effects on the *behavior* of participants.

- Bias can also affect the experimenter's *recording* of the participants' behavior by influencing his/her *selective attention and memory;* it can affect the *interpretation* of behavior if the data are open to *multiple explanations.*

  This relates to the *first requirement for scientific observations*: they must be empirical, i.e., objective and thus uninfluenced by the expectations and desires of the investigator.

∞   Bias affects *human participants* through *what the experimenter says* when reading instructions, describing how to execute a task, or giving feedback; as well as through *nonverbal communication* during these procedures — i.e., by *body language,* etc., consisting of the following components:

- *Body posture and gestures* (e.g., sitting erect versus slumped, and with the hands open versus clenched);
- *Facial expressions* (e.g., smiling versus frowning);
- *Tone of voice* (e.g., changes in pitch and amplitude).
- Example: *Elementary-school children* took a standard intelligence test, which *teachers* had been led to believe would predict intellectual blooming. The researchers then chose at *random* 20% of the pupils in each classroom and told the teachers that the test had predicted unusual intellectual gains for these children. Eight months later the children were retested. The 20% that had been randomly singled out were found to have done significantly better than the rest of the children, especially in tests of reasoning. [Hall, J.A., Rosenthal, R., Archer, D., Dimatteo, M.R., & Rogers, P. L. (1978, May). Decoding wordless messages. *Human Nature,* pp. 68-70.]

∞ Bias affects *animal participants* by the manner in which they are *handled,* by *prodding* them, and by the *cuing* of correct responses, such as through the use of *body language.*

- Example: In a *T-maze study* (where the animal starts at the base of the t-shaped maze, and when reaching the crossing section must decide whether to turn right or left to receive a reward goal) if the researcher stands next to the end of *a T-maze* where the goal is, then he/she would likely *cue the correct turn,* as the animals would probably have a learned tendency to approach the investigator as a result of prior associations of humans with rewards, such as food, water, and petting.
- *Non-verbal communication,* such as facial expressions and body gestures, is of particular concern when working with *nonhuman primates,* as opposed to other animals, because they have more of the same *body language* as humans due to a common evolutionary history.

∞ Demand characteristics

- *Attributes* that an object or event has for a particular organism which cause the organism to *tend to behave* in a certain way are referred to as *demand characteristics.* Note that this is only a *descriptive concept,* it is not explanatory, i.e., it does not provide the cause.

    Example: A *full moon* seems to *demand* romantic behavior from young lovers, but what's the cause?

- This concept applies to *experimenter bias effects,* in that when a participant gains knowledge about the investigator's research hypothesis, this knowledge tends to *demand* that they behave in a manner which supports the *hypothesis* — resulting in a *participant bias effect.*

Example: In *sensory deprivation experiments*, the hallucinations and bizarre behaviors that occur might actually be due more to the overly cautious treatment of participants, the screening for mental and physical disorders, the awesome release forms, and particularly the presence of a panic release button. [Orne, M.T. (1962). On the social psychology of the psychological experiment: With particular reference to demand characteristics and their implications. *American Psychologist*, 17, 776-783.]

- Demand characteristic effects are commonly *controlled* through the use of *deception or concealment*, if possible.

Types of Variance

There are *three types of variance: primary* (or experimental), *secondary* (or extraneous), and *error* (or random). The figures in Appendix 2 and Appendix 5 illustrate and clarify the nature of these three different types of variance. It should be noted that each of these types of variance can be due to any of the three *general sources* of variances just covered: participant factors, environmental factors, or experimenter factors.

1. Primary/Experimental Variance

— *Systematic/consistent variation* of the dependent variable produced by manipulation of an *independent variable* is called the *primary/experimental variance*.

— This variance relates to the *true score component* of measures (See Appendix 2).

— Hence this variance is *desired* by the experimenter — indeed, it's the *primary reason* for conducting an *experiment* (that's why it is called the *primary/experimental* variance).

- Primary/experimental variance should be *maximized*.
  - ∞ In this manner the *likelihood* of determining the effect of the independent variable would be increased.

2. Secondary/Extraneous Variance

— *Systematic/consistent variation* of the dependent variable produced by *extraneous variables*, i.e., by factors other than the independent variable(s) of the study, is called *secondary/extraneous variance*.

- Extraneous variables that produce secondary variance are sometimes referred to as *secondary variables*.

— This variance relates to the *constant error score component* of a measure, and thus mainly to *validity*.

- Note: Sources of *systematic/consistent* variation can have some *inconsistent* effects as well. Moreover, as we shall see, besides possibly causing *confounding* they can also produce *masking* — depending on the form of *experimental control* that is employed (see below).

— This is *undesirable variance.*

- Secondary/extraneous variance should be *controlled* by making the extraneous variable effects *equivalent* under all the independent variable conditions, and evaluating the effects.

  ∞ *Confounding* can result if this variance is not controlled.

    - Examples: Consider the effects of running the participants in each of the different independent variable conditions at different times of the day, or with different experimenters, or using different apparatus.

  ∞ *Masking*, it should be noted, is possible in addition to or instead of confounding if experimental control is less than optimal, e.g., if control only involves randomization (see discussion earlier and also later under "Making Equivalent the Secondary Variance").

3. Error/Random Variance

— *Unpredictable/inconsistent variation* of the dependent variable produced by *extraneous variables* is called *error/random variance.*

— This variance relates to the *variable error score component* of a measure, and thus mainly to *reliability.*

— This is *undesirable variance.*

- Error/random variance should be *controlled* by being *minimized.*

  ∞ *Masking* could occur if this variance is too large (elaborated later in Chapter 8 under "Minimizing the Error Variance").

    - *Type II/Beta Error* would be the result of this masking (discussed later in Chapter 12 under "Potential Errors During Statistical Decision Making").

    - Examples of *error variance sources* are: inconsistent experimental procedures, unreliable measurement techniques, and forms of individual differences among participants that are currently unmeasurable and unpredictable.

- Random/variable error is *self-canceling* around a mean of zero as the number of measures increases, i.e., it *averages toward zero.*

  ∞ *Confounding* thus should not be a big problem with error variance, which represents inconsistent/random effects.

  ∞ *Partial confounding*, however, and thus dependent variable differences between conditions that might be *mistaken* for independent variable effects, can result *when chance variations have not averaged/ summed to zero.*

    - Thus the need for *statistical significance tests*, which measure the *probability* of this potential error.

    - *Type I/Alpha Error* would be the consequence of such a mistake (covered later along with "Type II/Beta Error").

4. **Summary of Primary, Secondary, and Error Variance**
   (See figures in Appendices 2 and 5)

   — *Primary/experimental variance* is associated with the *true score component* of measurements, and thus *independent variable effects*.

   — *Secondary/extraneous variance* is associated with the *constant error component* of measurements due to *extraneous variables having consistent effects*, thus leading to *reduced validity* due to *confounding* by extraneous variables if they are not controlled (or possibly to *reduced reliability* along with *masking* if the extraneous variables are partially rather than thoroughly controlled).

   — *Error/random variance* is associated with the *variable error component* of measurements due to *extraneous variables having inconsistent effects*, thus leading to *reduced reliability* and *Type II Error of omission*, i.e., *masking* by extraneous variables, if they are not controlled (or possibly to *reduced validity* and *Type I Error of commission* as a result of *partial confounding* due to chance).

   — CRITICAL POINTS: Both *secondary variance* and *error variance* can lead to either errors of *commission* (confounding) or errors of *omission* (masking). What differs is the *way* that they can produce these errors, the *likelihood* of producing these errors, and the *control techniques* that can be used. (See also the Chapter 4 summary at the end of the discussion of "Accuracy and Consistency of Variables.")

Minimax Principles of Variance

There are *three minimax principles* with respect to the *three types of variance,* and they represent an expanded statement of the previously noted *goals/intent of good experimental designs.* Each of these three principles is discussed in greater detail later in this and the next chapter.

1. **Maximize the Primary Variance**

   — Systematic effects on the dependent variable of manipulating the *independent variable* of a research hypothesis should be *maximized.*

     • This *increases* the probability of determining the independent variable effects, i.e., the primary/experimental variance.

2. **Make Equivalent the Secondary Variance**

   — Systematic effects on the dependent variable of *secondary, extraneous variables* should be *controlled* by making the extraneous variables *equivalent* for all the independent variable conditions/levels.

     • If not done, then the secondary/extraneous variance might be *confounded* with independent variable effects, and thus *mistaken* for them; but as noted earlier, without thorough control the secondary variance still might *mask* any independent variable effects (errors of commission and omission, respectively).

3. **Minimize the Error Variance**

— Inconsistent/random effects on the dependent variable of *extraneous variable* fluctuations should be *controlled* by *minimizing* variation of the extraneous variables.

• If not done, then the error/random variance might *mask*, or alternatively be *mistaken* for, independent variable effects (errors of omission and commission, respectively).

Maximizing the Primary Variance

Maximizing the primary variance is the *first* of the minimax principles. It is very important because an experimenter *can never completely eliminate secondary and error variance.*

1. **Purpose**

— Maximizing the primary/experimental variance allows the *effects of the independent variable* on the dependent variable to be more easily discerned from the *background variance,* i.e., from the secondary and error variance caused by *extraneous variables.*

• It's similar to *enhancing the figure-to-ground contrast* in a *photograph* so that the main subject stands out — which can be done by putting the figure in better focus than the background.

• A better analogy, which is elaborated later, is *enhancing the signal-to-noise ratio* in a communications system (e.g., a radio).

2. **Techniques**

Three approaches to maximizing the primary variance are: using either *extreme values, optimal values, or several values of the independent variable.*

a. Extreme values of the independent variable

• Employing *extreme values* is the *most obvious way* to increase the measured effect of a *quantitative independent variable* on a dependent variable.

∞ Example: To determine the importance of the prefrontal cortex of the brain for learning and memory, we might *completely remove* (rather than just partially) the prefrontal cortex from both sides of the brain in a treatment group of animal participants, while making *no lesion* at all in a control group (note however that these animals would experience all of the same surgical steps leading up to tissue removal, thus *controlling* for those extraneous variables).

b. Optimal values of the independent variable

• Employing *extreme values* of the independent variable will *not* always *maximize* the observed differential effects on a dependent variable.

∞ Thus, *extreme values* of the independent variable are not necessarily the *optimal values*.

- Example: When studying the effects of *motivation* on learning by manipulating food deprivation in rats, it is usually necessary to have *at least some deprivation at the low end of the scale* in order to get the rats to perform the research task at all, and thus to get a measure on the dependent variable. Hence we might use 1 hour of deprivation rather than 0 hours as the minimum value.

    On the other hand, *extreme deprivation at the high end of the scale* might inhibit performance too, since the rats probably would be weak, ill, or even dead if they were deprived much beyond, say, 3 days. Hence we might use 48 hours of food deprivation as the maximum value.

∞ *Choosing optimal values* can be done on the basis of previous research, a pilot study, or careful reasoning.

c. Several values of the independent variable

- Employing *several values* of the independent variable is particularly useful when there is *no existing basis for choosing optimal values,* and when for some reason (e.g., limited time or resources) a *pilot study* can't be run to suggest optimal values.

    ∞ Obviously, a *sufficiently wide range* of quantitative independent variable levels should always be chosen in order to make likely the *detection of possible effects* that exist on the dependent variable.

- Parametric research

    ∞ Studies using several values of a *quantitative* independent variable are called *parametric* (see also Chapter 10 under "Multiple Treatment Designs, Multilevel Designs").

- Advantages

    (a) *Increased probability* would occur for having *optimal values* of the independent variable among those that are chosen, and thus of being able to determine whether there is *any effect* of the independent variable on the dependent variable.

    - Note that when a *functional relationship* (an effect) actually exists, it is possible even when selecting *two very different values* of the independent variable to get the *same value* in each case for the dependent variable — hence no effect would be observed, although in fact there is an effect (see the following for an explanation).

    - Inverted U-shaped function

        An *inverted U-shaped function* is a *non-monotonic* (i.e., direction changing) *curvilinear* relationship that looks similar to an inverted U (see Figure 7.1 below).

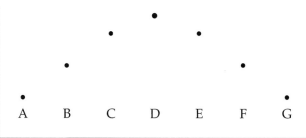

Maze Running Speed or Test
 Performance (DV)

A   B   C   D   E   F   G

Hours of Food Deprivation or
Level of Anxiety, Arousal, or Motivation
(Independent Variable)

FIGURE 7.1
Example of the inverted U-shaped function

It is a commonly found relationship in the behavioral sciences (e.g., when measuring test performance as a function of anxiety), and it is one that might yield the *same dependent variable value* if studying *only two levels* of an independent variable — and thus not *give evidence* of any of the real effects.

Note how certain pairs of different selected values of the independent variable (e.g., B and F or C and E in Figure 7.1) can result in the *same value* of the dependent variable.

Note also the *incomplete and contradictory results* that would be found if in different studies *different pairs* of independent variable values were chosen (e.g., A and D in one study versus D and G in another study).

Finally, note that if three or more values of the independent variable were chosen such that they were *distributed over the entire range of possible values* (e.g., A, D, and G), then we would observe not only an effect, but also the *non-monotonic nature* of the effect/relationship — hence the second advantage of using several values of the independent variable (see next).

(b) *More complete and accurate determination* of the nature of an effect, i.e., uncovering the *precise relationship* between the independent and dependent variables, is another possible advantage using parametric research (as illustrated above).

 • Relationships can be *graphically plotted,* and they also can be expressed as a *mathematical equation/function.*

 • *The more data points* used, the more complete and accurate the plotted or calculated relationship will be, particularly when the relationship is *non-linear,* i.e., curvilinear.

 • Note that *two points* can be used only to define a *straight line* (until shown to be incorrect through the gathering of additional data), since that is the *most parsimonious assumption* about the relationship.

Making Equivalent the Secondary Variance — Some Techniques

Making equivalent the secondary variance is the *second* of the three minimax principles. The *control techniques* that follow are listed roughly from the simplest to the more complex. Some of the techniques are an integral part of specific research designs, whereas others can be added to any of the designs. Thus several of these control techniques are further covered in Chapters 9 and 10 on "Designs for Experiments."

The topic of experimental control is a very extensive one, as noted earlier, and therefore its coverage is divided between this and the next chapter. The present chapter discusses the following six control techniques for secondary, extraneous variables: *eliminating, holding constant, balancing, matching, yoking, and using participants as their own control with counterbalancing.* In the next chapter, five additional control techniques are covered: *randomizing, pretesting, using control groups, using treatment groups with systematization, and conservatively arranging extraneous variables.*

Thus, there are *11 control techniques*. The *goal* of all but the last technique is to have *equivalence of effects* of secondary, extraneous variables across all of the independent variable conditions. This is accomplished by *eliminating differences in the average values* of the extraneous variables under the different levels of the independent variables.

1. Eliminating Extraneous Variables

— A *very desirable,* straightforward, simple control technique for secondary, extraneous variables is to just *eliminate them.*

— Environmental extraneous-variable applications

• Special research facilities

∞ *Extraneous visual and auditory stimuli* can be controlled by having windowless, sound-attenuating laboratory rooms, or specially built *isolation chambers.*

— Participant extraneous-variable applications

• Concealment or deception

∞ *Participant bias/expectancy* can be controlled by *concealing* the study or *deceiving* the participants about the research.

— Experimenter extraneous-variable applications

• Single-blind technique

∞ *Experimenter bias* can be controlled using a *single-blind procedure.*

• The *researcher who actually conducts the experiment would not be informed* (i.e., would be kept blind) as to which treatment or control condition each participant receives, and might not even be informed about the experiment's research problem and hypothesis, or even about the independent variable conditions.

Example: Half of a sample of animals might be given a specific *brain lesion* by one investigator, and then the behavioral effects would be tested by another researcher who doesn't know which animals received the treatment and which are in the control group.

- Note: Sometimes the single-blind control technique refers to a situation where the *participants* are kept blind (through deception or concealment), rather than when the researcher is kept blind.

- Double-blind technique

  ∞ Both the *experimenter and participant bias* (the latter involving *demand characteristics*) can be controlled through a *double-blind procedure.*

  - Neither the *investigator* conducting the experiment *nor the participants* in the study *would be aware* of which condition each individual receives, and might not even be informed as to the research problem and hypothesis, or even the independent variable conditions, of the study.

  - This technique is used in *drug studies* to control for *placebo effects* caused by *expectations* about therapies (discussed in Chapter 8 under "Control Group Utilization").

  - *Ethical problems* occur when research participants do not even know what treatment they *might* receive, since they could not give *informed consent* — but it might be possible to provide enough *general information* about the nature of the study to get around this problem, especially when *harmless treatments* are involved, such as vitamins.

    Example: Studying the *influence of vitamin C on colds*, while informing the participants only that they are participating in an experiment on the effects of vitamins, and not telling them which specific vitamin is being given, or what effects will be measured.

  - It is often *not possible,* however, to use this technique to keep the experimenter and participants *fully blind* as to the conditions being administered, since conditions can't always be *concealed* or made to all appear *identical* (as they could if a pill or injection were the treatment).

    Example: In a study of learning where the independent variable might be the type or amount of *verbal reinforcement* given to participants by the *experimenter,* the independent variable could not be concealed from the experimenter who must speak it, or even entirely concealed from the participants.

- Partial-blind techniques

  ∞ *Experimenter bias* would be only partly controlled when using a *partial-blind procedure.*

  - This is *used when* the double or even single-blind technique cannot be fully employed because of the *nature of the conditions* (as in the example above).

- The *researcher conducting the experiment* would be kept *uninformed* regarding the condition each participant is assigned to *throughout as many stages* of the study as possible — but not throughout all stages.

   Example: The experimenter would not know which condition a participant was in until after the individual was greeted and given all the instructions, at which time the experimenter would pull from a container of slips one which indicated the condition to be given.

- Automation
  - ∞ Experimenter-participant interactions, and thus *experimenter bias effects,* can be eliminated or minimized through *automation* employing a variety of available programming, stimulating, and recording equipment.
    - *Computers* that are powerful and relatively inexpensive, with various peripheral devices such as keyboards, joysticks, video monitors, and sound synthesizers, make it fairly easy to automate many aspects of experiments.

2. Holding Extraneous Variables Constant

— There are many instances where secondary, extraneous variables *cannot be eliminated.*

- *Participant factors* are a particularly obvious case, e.g.: gender, age, motivation, previous experience, intelligence, and health.
- *Experimenter factors* also often cannot be eliminated, e.g.: procedures, instructions, tasks, apparatus, and experimenter characteristics.
- *Environmental factors* similarly cannot always be eliminated, e.g.: temperature, atmospheric pressure, gravity, and time of day.

— *Next best thing* might be to *eliminate variation* of the extraneous variable/factor; i.e., to *hold the extraneous variable constant* for all participants, and thus for all independent variable conditions.

- *Extraneous variable effects* then should be the *same* for all the independent variable levels, unless, however, there were to an *interaction,* i.e., unless the *effects* of the extraneous variable were *dependent on (influenced by)* the values of the independent variable studied, and vice versa (this is elaborated in Chapter 10, under "Factorial Designs").
  - ∞ In such cases, the *particular value selected* for the extraneous factor would *favor* some value(s) of the independent variable over others.
  - ∞ Therefore, the results would not be *generalizable,* since the independent variable effects that were found in the study would be *relatively specific* to the *level* at which the extraneous variable happened to be held constant.
    - Example: Whether *massed versus distributed practice* results in the quickest mastery of some task might be influenced by (dependent upon) whether the participants have a *high versus low level of moti-*

vation, as well as a *high versus low level of prior experience* with the material utilized in the task. It is possible that massed practice would work best for high levels of motivation and when there are high levels of prior experience, whereas distributed practice would work best for low levels of motivation and when there are low levels of prior experience. In such cases, holding these constant would not be an appropriate form of extraneous variable control.

— Applications

- *Environmental factors* could be held constant by, e.g., using the same experimental room for all participants and thus conditions.

- *Experimenter factors* such as procedures, tasks, and instructions should be held constant, except where variations would be involved in *manipulation* of an independent variable.

  ∞ *Experimental procedures* would be practiced until routine so that the experimenter's performance would be consistent.

  ∞ *Oral instructions* would be read by the experimenter from a written script in a consistent manner.

  ∞ *Tape-recorded instructions*, however, would yield even better control over variations in voice inflections and pronunciation.

  ∞ *Written instructions* given to participants to read themselves might be best, since they could be reread if not understood.

- *Experimenter factors* such as the testing and recording apparatus might also be kept the same for all participants and conditions.

- *Experimenter characteristics* could be controlled by using just one experimenter who dressed, groomed, and behaved consistently.

- *Participant factors* could be controlled by selecting individuals who all shared one or more important characteristic, e.g., gender.

- Note: The use of the various *blind techniques* discussed earlier could actually involve holding an extraneous variable *constant* rather than eliminating it; i.e., the participants might be told *something* about the treatment, but they all would be made to think they were treated the *same*, e.g., not receiving a placebo.

3. **Balancing Extraneous Variables**

— It is often *inconvenient or even impossible to hold constant* certain secondary, extraneous variables, given the need to find a sufficient number of *participants* and to have sufficient *time* to run a study.

- Examples: Holding *age* or *gender* of research participants constant would significantly reduce the number of individuals that could participate, and holding *time of day, apparatus, or experimenter* constant would require more time to complete an experiment.

— It is also *undesirable* to hold a secondary, extraneous variable constant, even when possible, if *generalizability of results is desired* (as noted earlier).

— *Balancing a secondary, extraneous variable,* and hence its *effects,* is an *alternative* to both holding the variable constant or eliminating it.

- The different values of the extraneous variable would simply each be made to occur *an equal number of times,* and thus *to the same extent,* for all treatment and control conditions.

  ∞ *Confounding* of the independent and extraneous variables would thereby be avoided, while facilitating *generalizability.*

  - Example: *Participant factors* can be balanced by assigning to each condition the same number of individuals who have a particular value of the extraneous variable(s) of concern (such as gender, marital status, or occupation), which would not only permit generalization, but also the use of many more available participants.

TABLE 7.1

Example of Balancing a Participant Factor (i.e., Gender)

	Conditions	
Gender	**Treatment**	**Control**
Females	40 participants	40 participants
Males	80 participants	80 participants

- Note: Although there are implications for generalizability, balancing does not require that *every value* of a secondary, extraneous variable (e.g., gender) be used an equal number of times in a study (see Table 7.1), only that the number of times *any given value* occurs be *equal for all the independent variable treatment and control conditions.*

  ∞ *Several* secondary, extraneous variables can be simultaneously balanced in one experiment, and thus controlled for confounding, including those that are participant, environmental, or experimenter factors.

  - Example: Two different experimenters and participant genders can be balanced by having each experimenter run an equal number of females and males under each condition (see Table 7.2).

TABLE 7.2

Example of Simultaneously Balancing Experimenter and Participant Factors (i.e., Experimenters and their Genders as well as Participant Genders)

	Conditions	
Experimenter	**Treatment**	**Control**
John	10 females	10 females
	20 males	20 males
Jane	10 females	10 females
	20 males	20 males

∞ *Masking* of independent variable effects, it must be pointed out, would remain a possibility when *balancing* or just *randomization* (see Chapter 8) is used, as opposed to when extraneous variables are *held constant or eliminated*, since the *secondary variance* would still be present.

• *Control* of masking is possible, however, through *statistical analyses* associated with designs employing the techniques of *matching, using participants as their own control*, or *treatment groups and systematization* (all of these techniques are covered later).

4. Matching Participants on Extraneous Variables

— *Matching* is an *extended and more precise way of balancing* for a secondary, extraneous variable when: a) it is an extraneous *participant factor*, b) it is *measurable*, and c) there are *several values* of the extraneous variable, as opposed to just two or three.

• Examples: age, years of education, IQ, and severity of disorder.

• When there are only *two independent variable conditions/levels*, participants would be matched into *pairs* so that the individuals in each pair are those with the most similar (ideally identical) values on the measurable, extraneous participant variable, e.g., IQ.

∞ One member of each pair would then be assigned to each of the two conditions using *randomization* (discussed at the beginning of Chapter 8).

• In this manner, the matched extraneous participant factor would be made *equivalent* for both conditions, as should any *effect* that it might have on the dependent variable.

• Example of matching's superiority: In a verbal learning study, *intelligence* (IQ) would be a secondary, extraneous participant factor that would be distributed over independent variable conditions with *more precision by using matching as opposed to just balancing*.

Consider an idealized case where the range of IQs for the obtained participants was 80 to 150, with two individuals having IQs of 80, two at 90, and so on up through 150 (see Table 7.3).

If there were two conditions, and *balancing* were used, half the participants with IQs below the median (80 to 110) would be randomly assigned to each condition, as would half the participants with IQs above the median (120 to 150).

It is possible, with the worst possible luck (as shown in Table 7.3), to end up *by chance* assigning to one condition all the participants having the highest IQs above the median (150 and 140) as well as the highest IQs below the median (110 and 100); while the other condition is assigned all the participants with the lower IQs above the median (130 and 120) as well as the lower IQs below the median (90 or 80) — thus the mean IQs for the two conditions would be very different.

But, if *matching* were used, then the participant IQs for the two conditions would have, in this idealized case, exactly the *same distribution of values,* including the *means,* thus resulting in *greater equivalency* of this secondary, extraneous variable.

TABLE 7.3

Idealized Example of Matching versus Balancing for a Secondary, Extraneous Variable

IQ of Participants[a]	Balancing for IQ[b]		Matching for IQ[c]	
	Treatment Group	Control Group	Treatment Group	Control Group
High IQ				
150	150	130	150	150
150	150	130	140	140
140	140	120	130	130
140	140	120	120	120
130			110	110
130			100	100
120			90	90
120			80	80
Low IQ				
110	110	90		
110	110	90		
100	100	80		
100	100	80		
90				
90				
80				
80				

[a] Listed from highest to lowest and divided at *median* into high versus low IQ.

[b] Half of higher and half of lower IQ participants are placed in each group *randomly.* Due to *chance factors,* the Treatment Group might — as shown here — have higher *average IQ* than the Control Group, even though balancing was used. This is, of course, an extreme example, and would not be very likely to actually occur.

[c] *Average (mean)* and *distribution of IQs* in both groups is the same, thus matching leads to greater equivalence of secondary, extraneous variables than does simple balancing.

- When there are *three or more independent variable conditions,* the procedures used would be similar to those for just two conditions.

  ∞ Participants would simply be matched, on the basis of similarity on the measurable extraneous factor, into *triplets* or whatever number was the same as the number of conditions.

    - One member of each matched grouping would then be *randomly assigned* to each of the conditions/levels.

- Participants can be *matched for more than one extraneous variable* in a given study, and often they are matched on several.

  ∞ *Difficulty in finding matched participants,* however, increases disproportionately as the number of matching variables increases, thus excluding many potential participants.

∞ *Frequency-distribution matching* can be employed when the difficulty becomes too great for using *precision matching.*

- With this alternative, only the *frequency distribution* (mean, standard deviation or variance, etc.) of the various values of each of the participant extraneous variables would be matched for every independent variable condition.

- *Problem* is that the *combinations* of different extraneous variables might be *mismatched* in the various independent variable conditions, even when the means (etc.) for *each* extraneous variable are the same.

    Example: In one condition there might be young participants with high IQs plus old participants with low IQs, but just the opposite in another condition.

- Note: *Extraneous variables* on which the participants are matched must be *correlated with* (related to) the *dependent variable* if the *secondary and error variance* are to be reduced — thereby reducing *confounding and masking,* and thus increasing the *sensitivity/power* of the experiment (this is elaborated in Chapters 9 and 10).

    ∞ But also note that if there is *no correlation* between the participant variables used for matching and the dependent variable of the study, then these participant variables have *no effect* on that dependent variable — and thus they are *not extraneous variables,* and therefore *do not require control*!

5. Yoking Participants on Extraneous Variables

— *Matching* participants for secondary, extraneous *experimental and/or environmental* variables is the purpose of the *yoking technique,* rather than matching individuals on some extraneous *participant characteristic* (although this is an additional possibility).

— *Yoking* is used when *experimental manipulations,* themselves, *unavoidably* (and undesirably) involve the introduction of secondary, extraneous variables *along with* the independent variable; and, in addition, the other control techniques are not as satisfactory because different participants in the experimental group, *due to their own performance,* would likely experience *different exposures* to these extraneous variables, which *must be matched* for the control participants.

- *Yoking* participants together to *match* them can either be done by *physically* linking them together (as with two oxen pulling a wagon or plow — hence the name of this technique) or it can be done by using *electronic programming equipment* to link them.

- *Control participants* then would be exposed to the *same quantity,* and usually also to the *same temporal distribution,* of extraneous experimental events as the *yoked treatment participants* received.

- *Treatment* would consist of a *particular sequence or temporal relationship* of experimental events, or some *additional factor.*

- Assignment of matched participants to the treatment versus control conditions would be random (see Chapter 8).

- Example: Suppose you are studying the *reinforcing* effects of electrical stimulation of the so called *"pleasure centers"* in the brain, whenever an animal produces a desired behavior.

  - ∞ *General effects* (e.g., arousal) of stimulating the brain at just *any time* must be controlled by having a *yoked control animal* stimulated, regardless of what it is doing at the moment, every time that the paired *experimental treatment animal* is stimulated for making the appropriate behavior.

  - ∞ Thus the yoked experimental and control animals would both receive the *same amount* and the *same temporal distribution* of stimulation — the difference being that only for the experimental animals would the stimulation be *contingent/dependent upon* the animal's exhibiting the appropriate behavior, which of course is what makes the stimulation a *reinforcement treatment.*

  - ∞ Note that the yoking in this case would be by *electronic programming,* not by *physically* linking the animals.

— The *Yoked Control Group Design* (which is discussed in Chapter 10, under "Between-Participants, Related, Two-Group Designs") has as its essence this yoking of participants on extraneous experimental and/or environmental factors.

6. Using Participants as Their Own Control with Counterbalancing

— An *extreme form of matching* for secondary, extraneous participant factors, i.e., for individual differences among participants, is to use the *participants as their own control,* along with *counterbalancing.*

— Each participant sequentially receives and is measured at *all levels* of an independent variable, and thus serves as *his/her own control* for extraneous participant/subject variables.

  - In effect, each participant is *matched with himself/herself.*

— *Designs* using this control technique have two *alternative names:*

  1) Within-participants/subjects designs

     - ∞ This name indicates that comparisons between conditions are made *within each individual participant* (rather than the comparisons being made between different participants).

  2) Repeated-measures designs

     - ∞ This name indicates that measures on the dependent variable are obtained from each participant *under more than one condition* for comparison, thus requiring *repeated measurements* (note that this name is not used when just averaging repeated measures under a *single condition* in order to increase reliability).

— Advantages of this technique/design

1) *Greater reduction of secondary and error variance* occurs, and thus there is better control of both *confounding and masking.*

   ∞ *Better control of the between-conditions variance* related to *individual participant differences* naturally results from having the *same participants* in each condition, as opposed to having different ones in each condition — even if matched.

   • Specifically, *many more extraneous participant factors* are controlled by within-participants comparisons than is possible with comparisons for matched participants, who could be similar on only *some* extraneous factors.

   • Ideally, when using participants as their own control, *all participant factors* would be the same for *all independent variable levels,* because they all use the same individuals.

   However, participants do not stay the same — they vary to some degree over time in diverse ways.

   Nevertheless, the same participant should be more similar at different times than different participants would be at the same time.

2) *Fewer participants are required* here to obtain the same amount of data, since more data are gathered per individual when each participates under more than one condition; alternatively, if the *same number* participated, then *more data would be obtained.*

3) *Less total time and effort* might be required to prepare all the participants for the study (e.g., to give instructions), since fewer participants would be needed for the same amount of data.

— Disadvantage of this technique/design

   • Order/sequence effects

   ∞ Practice

   • One order effect known as *practice* leads to *enhanced performance* over successive independent variable conditions, due to the research participants becoming more skilled, i.e., due to their *learning.*

   ∞ Fatigue

   • In contrast, another order effect known as *fatigue* leads to *diminished performance* over successive conditions, due to the participants' becoming *tired or bored,* and thus also *less motivated and attentive or vigilant.*

   • *Confounding* of independent variable *levels* with the *order* of conditions, and thus with *practice and fatigue,* occurs whenever the conditions are run in the *same sequence* for *all* participants.

   ∞ This would represent *experimental contamination,* i.e., uncontrolled experimenter-factor secondary variance.

- Example: If all participants were run on a reaction-time task first to *auditory stimuli* and then to *visual stimuli,* and it was found that reaction time was fastest to the visual stimuli, it is possible that this would just be due to an *order effect of practice,* and not to the difference in stimuli, since order was *confounded* with stimulus type.

— Counterbalancing

- *Counterbalancing* is a *special type of balancing* that is used specifically to *control* for confounding by *order/sequence effects.*

  ∞ *Order of presentation* of the conditions to research participants is *systematically varied* so that it is *balanced.*

    - Which participants receive which sequence would be determined randomly (see Chapter 8).

    - *When there are two conditions*: half the participants would receive Condition A first, followed by Condition B; and the other half of the participants would receive Condition B first, followed by Condition A (see Table 7.4) This is referred to as *interparticipant counterbalancing,* as opposed to *intraparticipant counterbalancing* (see later for elaboration).

TABLE 7.4

Example of Interparticipant
Counterbalancing with
Two Conditions

	Ordinal Position	
Sequence	1	2
1	A	B
2	B	A

    - Counterbalancing is applicable for any number of conditions, but the procedure becomes more complicated when there are *more than two conditions* (covered later).

- This technique is an attempt to *distribute* the order/sequence effects of practice and fatigue *equally* to all conditions so that the influence of the effects will be *balanced,* i.e. equivalent, for all conditions, and hence the *order effects* will be *controlled.*

- Carry-over effects

  ∞ The *assumption* is made when using counterbalancing that order effects (practice and fatigue) are the *same* regardless of the *particular sequence/order* of conditions involved: i.e., that there are no *differential/ asymmetrical transfer of order effects* associated with different sequences of conditions.

    - If this assumption is not met, then the effects of order would be *incompletely balanced,* and therefore order effects would still be *confounded,* to some extent, with the independent variable conditions.

Example: In a study of the effects of *sleep deprivation* on problem-solving ability, it is quite possible that the *transfer of practice* is greater when going from the non-deprived control condition to the sleep-deprived treatment condition, than vice versa, in which case counterbalancing would not succeed at balancing the effects of order.

- Note: *Motivation and attention changes* are *order effects* that can operate independently of practice and fatigue, and they too might exhibit differential transfer.

  As is true for practice and fatigue effects, motivation or attention changes might actually be studied as an *independent variable* (see example for motivation effects later in this chapter under "Applications of Within-Participants/Subjects Design Procedures").

∞ *Carry-over effects* is the term that is used to refer to any *differential/asymmetrical transfer* of order/sequence effects.

- What it means is that the order effects are *dependent upon* — carry over from — the *particular conditions* that precede a given condition in a sequence, rather than the order effects being dependent upon just the *number of conditions* that precede a given condition in a sequence.

  Order/sequence effects thus involve not just the *ordinal position* of conditions, but also the *specific sequence* of conditions if there is differential transfer.

- Carry-over effects can be *tested for statistically*, using an analysis of variance, by making either sequence or ordinal position an additional factor in the analysis of treatment conditions, and analyzing respectively for a *sequence main effect* or a *treatment by ordinal position interaction* (see Chapter 10).

  If there is a significant *sequence main effect*, then there is good evidence of carry-over effects since all sequences include the same treatment conditions (see Table 7.5).

  Note that, in addition to evidence of an ordinal position effect in the example, i.e., the mean of 35 ≠ 40; there is also evidence here for a *sequence main effect*, and thus *carry over*, because although the two sequences are counterbalanced for treatment conditions A and B, the *sequence means* are nevertheless different, 39 ≠ 36.

  Similarly, when there is a significant *treatment by ordinal position interaction*, then by definition the effect of ordinal position *depends upon* the treatment, and hence again there would be differential transfer (i.e., carry-over effects).

TABLE 7.5

Example of Ordinal Position and Sequence Main Effects

	Ordinal Position		
Sequence	1	2	Sequence Means
1	A = 40	B = 38	39
2	B = 30	A = 42	36
Ordinal Position Means	35	40	

Note: Values shown for Treatment Conditions A and B are mean-dependent variable scores.

TABLE 7.6

Example of Treatment by Ordinal Position Interaction Showing Asymmetrical Transfer of Order/Sequence Effects

	Ordinal Position		
Treatment	1	2	Treatment Means
A	A = 40	A = 42	41
B	B = 30	B = 38	34
Ordinal Position Means	35	40	

Note: Values shown for Treatment Conditions A and B are mean-dependent variable scores.

Note that the example data in Table 7.6 are the same as in Table 7.5, only rearranged.

More importantly, note that there is evidence of a *treatment by ordinal position interaction*: The mean dependent variable scores show that Treatment Condition B gains more than Treatment Condition A in the second ordinal position (38–30 versus 42–40), thus there's *asymmetrical transfer of order/sequence effects*.

This, by the way, explains why the means were different for the two sequences in the previous table.

Finally, it should be noted that for repeated-measures designs with results like this, the *first ordinal position* would still provide useful data, i.e., data that are *unaffected by carryover* order effects.

• When there are *significant carry-over effects* (differential/asymmetrical transfer) then some experimental design other than within-participants/subjects should be used; e.g., *matched-participants/subjects*, which still partially controls for individual differences among participants, without the problems of order effects (advantages and disadvantages among different designs are discussed further in Chapters 9 and 10).

- Complete counterbalancing
  - ∞ *Complete counterbalancing* is where *every possible sequence* of conditions is used an *equal number of times,* with each research participant receiving a single sequence.
    - If N refers to the number of *conditions* (i.e., levels of the independent variable), then the number of different *sequences* that can be formed is the number of *permutations* possible with N events taken N at a time.

      This is simply equal to N factorial, symbolized as $N!$, where

      $$N! = N \times (N - 1) \times (N - 2) \ldots \times 1.$$

      Examples: If there are *two conditions*, then the number of possible sequences is $2! = 2 \times 1 = 2$, as we saw earlier; if there are *three conditions*, then the number of possible sequences is $3! = 3 \times 2 \times 1 = 6$ (if the conditions are A, B, and C, then the sequences would be: ABC, ACB, BAC, BCA, CAB, and CBA).
    - Note that the number of different possible sequences *increases rapidly* with the number of treatment conditions.

      Example: If there are four conditions, then the number of possible sequences is $4! = 4 \times 3 \times 2 \times 1 = 24$.
  - ∞ Complete counterbalancing *controls* for the effects of the *ordinal position* of conditions in the sequence because, considering all sequences taken together, and thus all participants, every condition (independent variable level) will occur *equally often* at each ordinal position.
  - ∞ Advantage
    - Complete counterbalancing *best balances* and hence *best controls* for *carry-over effects,* again because *all* possible sequences of conditions are used, and equally often.

      Nevertheless, as noted earlier, if there is differential/asymmetrical transfer of effects (carryover effects), then counterbalancing, even when complete, would not fully control for all of the order/sequence effects.
  - ∞ Disadvantage
    - *Number of participants needed might be very large,* since every possible sequence of conditions must be used an equal number of times (once, twice, etc.), and therefore the number of research participants must always be $N!$ or some whole multiple of it (e.g., $2N!$).

- Incomplete counterbalancing
  - ∞ *Incomplete counterbalancing* is used when complete counterbalancing is not practical.

- It drastically *reduces the required number of sequences*, and hence the *number of participants*, while retaining *some control properties* of complete counterbalancing.

∞ Balanced squares

- *Balanced squares* are a form of incomplete counterbalancing that retains the following *two features/requirements* of complete counter-balancing, when considering all the sequences taken together:

  (1) Every condition (value of the independent variable) occurs an *equal* number of times at each of the ordinal positions.

      This ensures that when there are no carry-over effects (asym-metrical transfer), then order effects will be controlled since *ordinal position is balanced*.

  (2) Every condition is preceded by each of the other conditions an *equal* number of times (note: this also means that every condition would be followed by each of the other conditions equally often).

      This ensures that whenever *carry-over effects* from any given condition influence only the *immediately succeeding condition*, the same degree of *imperfect control* is achieved as by complete counterbalancing.

- *Rules* for forming balanced squares when N, the number of inde-pendent variable levels, is an *even number*:

  (1) N sequences are required.

  (2) First sequence is 1, 2, N, 3, $(N - 1)$, 4, $(N - 2)$, 5, etc., until all N levels are accounted for.

  (3) Remaining sequences are obtained from the first by simply adding 1 to the independent variable levels of the preceding sequence, and substituting the value of 1 whenever a sum is greater (by 1) than N.

      Example, using alphabetical letters for the conditions, when there are four conditions (thus $N = 4 = D$) is shown in Table 7.7. Note that the two requirements (stated above) for a bal-anced square are met.

TABLE 7.7

A Balanced Square with Four
Conditions (A, B, C, D)

Sequence	Ordinal Position			
	1	**2**	**3**	**4**
1	A	B	D	C
2	B	C	A	D
3	C	D	B	A
4	D	A	C	B

- *Rules* for forming balanced squares when N, the number of inde-pendent variable levels, is an *odd number:*

(1) Two N sequences are required.

(2) First, N sequences are obtained just the same as when N is an even number.

(3) Then another set of N sequences is obtained by simply reversing the first set of N sequences.

This second, reversed set of N sequences is *necessary* to meet the *requirement* that every condition be preceded by (and thus also followed by) each of the other conditions equally often.

Obviously, *whenever possible,* the number of independent variable levels/conditions, N, should be kept *even*, thereby reducing the required number of sequences, and thus the required number of research participants.

Example of the first and second sets of N sequences, when there are five conditions (thus $N = 5 = E$) is shown in Tables 7.8 and 7.9.

TABLE 7.8

First Set of N Sequences Required for Forming a Balanced Square When Number of Conditions is an Odd Number

Sequence	Ordinal Position				
	1	2	3	4	5
1	A	B	E	C	D
2	B	C	A	D	E
3	C	D	B	E	A
4	D	E	C	A	B
5	E	A	D	B	C

TABLE 7.9

Second, Reversed Set of N Sequences, which is a Mirror Image of the First Set, Used When Number of Conditions is an Odd Number

Sequence	Ordinal Position				
	1	2	3	4	5
6	D	C	E	B	A
7	E	D	A	C	B
8	A	E	B	D	C
9	B	A	C	E	D
10	C	B	D	A	E

∞ Latin squares

- *Latin squares* are a form of incomplete counterbalancing used when the investigator is primarily concerned with *controlling for ordinal position within a sequence* and not for carryover effects.

- *Rule* for forming Latin Squares:

Set of *N* sequences is chosen (regardless of whether *N*, the number of conditions, is even or odd), usually by using a quasi-random procedure (discussed in the next chapter), such that each condition occurs *once and only once* at each ordinal position, when considering all of the sequences together.

Note that there is *no rule* that each condition has to be preceded by every other condition equally often.

Example: (more orderly than is likely to occur if using a random procedure) for when there are *five conditions* is shown in Table 7.10.

TABLE 7.10

Example of a Latin Square Form of Incomplete Counter Balancing

Sequence	Ordinal Position				
	1	2	3	4	5
1	A	B	C	D	E
2	B	C	D	E	A
3	C	D	E	A	B
4	D	E	A	B	C
5	E	A	B	C	D

• Because a Latin square must always have *at least the same number of rows as columns*, e.g., 5 × 5, the experimenter must use at least the *same number of participants* as there are conditions (minimum of five research participants are needed in the example above).

Note that when the *number of conditions is odd*, the minimum number of participants is *not* twice the number of conditions, unlike the balanced square.

If more participants are used than the number of conditions, which would be advantageous for reasons of statistical reliability, then the number of participants must always be some *whole multiple of the number of conditions* in the study (i.e., for the above example: 10 or 15, etc.).

The *same* Latin square can be used for the additional participants, which has the *advantage* of permitting a test (but not control) for significant carry-over effects, as discussed earlier — and this is sometimes referred to as a test of *square uniqueness*.

Alternatively, *different* Latin squares can be used for each successive set of participants, which has the *advantage* of increasing control over carry-over effects — for example, by expanding all the way to a Balanced Square, which in fact is also often called a *balanced Latin square*.

Although this approach does *not* permit a test for significant carry-over effects, and thus a determination of the success of partial counterbalancing; nevertheless, whenever sufficient participants are available, a balanced square makes more sense than a Latin square, since it should *better control for carry-over effects.*

- Note: Counterbalancing is useful not only for determining the order of presenting *independent variable conditions* within participants, it is also useful for determining the *order of testing/measurement* when there is *more than one dependent variable.*

  The intention would be to *balance* any order/sequence effects of measuring *multiple dependent* variables, just as for measuring the effects of multiple independent variable conditions/levels.

- Randomized counterbalancing
  - ∞ *Randomized counterbalancing* is used *when even incomplete counterbalancing is unachievable,* as when the number of participants available is smaller than the number of sequences required.
  - ∞ *Sequences to be used* are selected by a random process from the set of all *N!* possible sequences (see Chapter 9 for comparison with the "Equivalent Time-Samples Design").
    - *Any number* of random sequences may be employed.
  - ∞ Disadvantage
    - There is no assurance that the random sequences will contain the *controls* for ordinal position and carry-over effects that complete or even incomplete counterbalancing provide.

- IN SUMMARY: *Complete counterbalancing* is better than *incomplete counterbalancing,* and should be used whenever possible; but if not possible, *Balanced squares* are better than *Latin squares,* which in turn are typically better than just *randomized counterbalancing.*
  - ∞ In addition to these forms of counterbalancing, there are other *categories and terminology* that will now be described (See Appendix 7, under "Counterbalanced Designs").

- Interparticipant counterbalancing
  - ∞ *Interparticipant counterbalancing* is when counterbalancing is accomplished across/within the entire *group of participants,* rather than within each one.
    - It encompasses all the forms of counterbalancing already discussed; e.g., half the participants get sequence AB, and the other half get sequence BA.

- Intraparticipant counterbalancing: the ABBA technique
  - ∞ *Intraparticipant counterbalancing* is when an attempt is made to control for order/sequence effects within the *individual participants,* by having

each individual take part in *all* the treatment conditions, first in one sequence and then in the reverse sequence.

- When there are only *two conditions* (A and B), this would mean using *either* sequence ABBA or sequence BAAB.

- Control for order/sequence would be successful if the ordinal position effect is *linear;* i.e., if the order effect produces a *constant* increment or decrement in the score at each successive position in the sequence.

- Notice that each condition has the same *average ordinal position,* i.e., $(1 + 4)/2 = (2 + 3)/2 = 2.5$; also note that A precedes B as often as B precedes A.

- Combined intraparticipant-interparticipant counterbalancing

  ∞ *Combined intraparticipant-interparticipant counterbalancing* is when *half* the participants are run in the ABBA sequence, and the *other half* are run in the BAAB sequence.

  ∞ This controls for *non-linear ordinal position effects* that aren't controlled by intraparticipant counterbalancing alone.

  ∞ It does not, however, control for *differential carry-over effects.*

  ∞ Note: When there are *multiple presentations of each condition* for each participant, rather than just one, a useful variation on randomized counterbalancing would be the *"Block-Randomization Procedure"* (see Chapter 8), with the order of conditions randomized for each participant *within each of several blocks* of all the conditions.

    - *Block-randomization* inevitably leads to a *more equivalent distribution* of the different independent variable conditions within the total sequence, and thus *better controls for order effects,* e.g.: ACB, CBA, BCA, etc.

- Applications of within-participants/subjects design procedures

- Within-participants/subjects design procedures *must be used* if a form of *sequence/order* is in fact an *independent variable,* rather than an extraneous variable.

  ∞ Example: Studies of the effect of the independent variable of *trials* on, e.g., learning or fatigue *require* that comparisons be made within participants/subjects across the trials.

    - Note that *counterbalancing* would not be employed unless there were another element of order, one that had to be controlled as an extraneous variable; e.g., *different sets of stimuli* that were to be used on different trials.

  ∞ Another example: Studies of different *motivation effects* produced by switching from a high-reward to a low-reward situation, and vice versa — note that the independent variable would be the *order/sequence* of independent variable levels.

- Within-participants/subjects design procedures are *also required for single-participant/subject and small-N research*, where there is only one participant or a very few, and thus not enough participants for more than one group.

  ∞ *Comparisons* can then be made only *within participants*.

- In contrast, these designs are *excluded* from use if there is *significant carry-over* (differential/asymmetrical transfer) *of order effects* due to the nature of the variables.

  ∞ Examples: In *developmental studies* of the effects of the presence versus absence of social deprivation, and in studies involving different techniques for *learning* a foreign language, there would typically be *too much transfer of effects* once development or learning had occurred under one of the independent variable conditions. How could development or learning a language logically be studied once it had already occurred under a preceding condition in the same individuals?

- Finally, these designs also *cannot be used* if there are *too many conditions*, such that research participants would become so *fatigued* or lose so much *motivation* that they would *quit*.

As indicated earlier, an *additional five techniques* for controlling secondary, extraneous variance are discussed in the next chapter.

Review Summary

In addition to the material below, study the control/design techniques that have been employed in published research; i.e., read some actual studies, such as those assigned by your course instructor.

1. There are two *meanings of experimental control*:

   a. *Independent variables* and the *values* to be studied are decided upon by the experimenters, and they *produce* those values by *purposely and directly manipulating* the independent variables in a manner that they decide upon.

   b. *Extraneous variables* are *extra/unwanted variables* operating in the experimental situation, which like independent variables might affect the dependent variable(s), and thus they too must be controlled.

2. There are two *reasons for controlling extraneous variables*:

   a. *Masking* occurs when extraneous variables *hide/obscure* the effects of an independent variable on a dependent variable, usually by themselves causing substantial variability in the data. This variability leads to *decreased reliability* of any apparent *independent variable* effects, and thus *errors of omission* — failure to declare an effect of an independent variable when, in fact, there is an effect.

b. *Confounding* occurs when extraneous and independent variables are *systematically and inextricably intertwined,* so that their effects on a dependent variable cannot be distinguished. This causes *confusion,* and hence *errors of commission* — mistakenly declaring an effect of an independent variable when there is none (dependent variable differences actually being due to a confounding extraneous variable). *Errors of omission* also might be caused by confounding, if an extraneous variable *cancels out* the independent variable effect.

3. There are twelve *forms of extraneous variables.* The following summary is a copy of the list in Appendix 5 of the *Handbook.* For additional details, see the text of the *Handbook.*

Environment: *Physical and social conditions of the experimental setting,* other than the independent variable(s), that might affect the participant's performance — similar to extraneous variable of history, except not dependent on there being a pretest

History: *External environmental events,* other than an independent variable, that occur *between any pretest and posttest measurements* of a dependent variable

Instrumentation: Change that occurs over time in the measurement of a dependent variable due to *variation in mechanical or human observer factors*

Statistical Regression: Change that can be attributed to the tendency of participants with *extremely high or low scores* on a test to move toward the mean when measured again — due to unreliability of the test instrument and/or the participants

Maturation: Change due to *conditions internal to an individual* that vary as a function of the *passage of time* — anatomical, physiological, cognitive, behavioral

Selection: *Use of differential selection procedures* for participants assigned to the various comparison groups — leads to *individual participant differences* between groups

Mortality: *Participant loss* from the various comparison groups — possibly differential

Sequence/Order: *Participation in more than one condition* for comparison, leading to practice or fatigue effects, and/or possible changes in motivation or attention

Testing: *When tested more than once,* an initial test (a pretest) can affect the participant scores of a subsequent test (e.g., a posttest) either directly, or indirectly via sensitization or resistance to a treatment (similar to sequence/order effects)

Participant Bias: *Expectancy, attitude, or motive* of participants (associated, respectively, with placebo effect, Hawthorne effect, and motive for positive self-presentation)

Participant Sophistication: *Familiarity* of research participants with the field of psychology and/or its experimental procedures (relates to participant bias)

Experimenter Features: *Expectancy/bias, performance, appearance, personality* of the researcher(s); also includes undesired variation/effects of procedures, instructions, and apparatus not already noted (e.g., under testing and instrumentation)

4. *Variance* is *variation* in observed scores on a *dependent variable.* Experimentation, in fact, can be defined as the *controlled study of variances;* specifically, the variance supposedly produced by the independent variables as opposed to extraneous variables. There are *three general sources of variance,* which encompass the twelve specific types of *extraneous variables* noted earlier. These are also the three general sources of variance for *independent variables,* since the extraneous variables of one experiment can sometimes be the independent variables of other studies — it depends on what the investigator is interested in. The three general sources of variance are

  a. *Participant factors* — variation in characteristics of the organism, i.e., present state, past experience, and genetic traits

  b. *Environmental factors* — variation in the physical and social surroundings of participants

  c. *Experimenter factors* — variation in experimental procedures, instructions, apparatus, tasks, conditions, and experimenter characteristics, such as appearance (age, race, gender, clothing, and grooming/hygiene), personality, performance, and expectancy/bias

5. *Experimenter bias effect* is the influence of the experimenter's *anticipation or desire* (i.e., the research hypothesis) on the outcome/results of a study. Bias can affect the *experimenter's performance,* in terms of both *verbal and nonverbal communication* (spoken and body language), such that the *expectancy* is transmitted to human research participants. This causes *participant bias,* which motivates behavior that supports the experimenter's hypothesis/expectancy — an example of *demand characteristics.* Alternatively, human and nonhuman participants could be *cued, prodded, or subtly rewarded* for behaving in a manner that *supports* the experimenter's expectancy. Moreover, bias can also affect the *experimenter's recording and interpretation* of behavior through, e.g, *selective attention and memory.*

6. There are three *types of variance,* each of which can be due to any of the three *general sources of variances* already discussed:

  a. *Primary/experimental variance* is the *systematic/consistent variation* of the dependent variable produced by manipulation of an *independent variable.* It is related to the *true score component* of a measure, and is desired by the experimenter.

  b. *Secondary/extraneous variance* is the *systematic/consistent variation* of the dependent variable produced by *secondary, extraneous variables.* It is related to the *constant error score component* of a measure, and is undesirable.

  c. *Error/random variance* is the *unpredictable/inconsistent variation* of the dependent variable produced by *extraneous variables.* It is related to the *variable error score component* of a measure, and is undesirable.

7. There are three *minimax principles of variance* that summarize the goals/intent of *good experimental designs:* a) *Maximize* the primary variance; b) *Make equivalent* the secondary variance; and c) *Minimize* the error variance.

8. *Maximizing the primary/experimental variance* can be accomplished by one of three techniques:

a. *Extreme values* of the independent variable could be used

b. *Optimal values* of the independent variable could be used — extreme values not always being optimal

c. *Several values* of the independent variable could be used — which increases the probability of having optimal values and thus of finding any effects that might exist, and which also provides more complete and accurate determination of the effects, i.e., greater precision

9. *Making equivalent the secondary/extraneous variance* is accomplished by several techniques summarized in Appendix 5 of the *Handbook.* Of the 11 techniques, six are covered in Chapter 7. Their descriptive names are (a) *eliminating extraneous variables* (includes using the single blind, double-blind, and partial-blind techniques, as well as automation); (b) *holding extraneous variables constant;* (c) *balancing extraneous variables;* (d) *matching participants on extraneous variables;* (e) *yoking participants on extraneous variables;* and (f) *using participants as their own control with counterbalancing.*

10. There are many cases where the simple, straightforward and desirable, technique of *eliminating extraneous variables* cannot be used, e.g., with the extraneous variables of participant gender, experimenter characteristics, or environmental temperature. The next best thing might be *holding extraneous variables constant* for all participants, and thus for all independent variable conditions. This should make the extraneous variable *effects* the same for all levels of the independent variable — by eliminating any differential influences — unless there is an *interaction effect,* in which case the *results would not be generalizable.* Another problem is that it is often inconvenient or even impossible to hold constant certain secondary, extraneous variables, given the need to find a sufficient number of participants and to have sufficient time to run a study. *Balancing extraneous variables* is an alternative that does not involve the noted problems of holding extraneous variables constant or eliminating them.

11. *Matching participants on extraneous variables* is an extended and more precise way of *balancing* for an extraneous variable when it is an extraneous *participant factor* that is *measurable,* and when there are *several values* of the extraneous variable (as opposed to just two or three). In addition to *precision matching* there is *frequency-distribution matching,* which is used when participants are matched on *several variables.* Extraneous variables on which the participants are matched must be *correlated* with the dependent variable if the *secondary and error variance* are to be reduced — thereby reducing *masking* and increasing the *sensitivity/power* of the experiment.

Yoking participants on extraneous variables is a technique used specifically to *match* participants for extraneous *experimental or environmental factors,* rather than for participant characteristics. This can be done either *physically* or by *electronic programming.* A control group is exposed to the *same quantity,* and usually also *temporal distribution,* of extraneous experimental events as the treatment group. *Treatment* consists of a *particular temporal order or relationship* of experimental events, or some *additional* factor.

12. *Using participants as their own control* is an *extreme form of matching* for extraneous participant factors. Each participant sequentially receives and is measured at *all levels* of an independent variable, and thus serves as *his/her own control* for extraneous participant variables. Research designs that use this control technique are called *within-participants designs*, or *repeated-measures designs*. This technique *better controls* for the between conditions variance related to *individual participant differences*, and thus for *confounding and masking*, because comparisons between conditions are made *within* each participant. Also *fewer participants* are required, as each is measured under all the independent variable levels/conditions, i.e., *repeated* measures are taken.

13. *Confounding* by the *order/sequence effects of practice and fatigue* when using *repeated-measures/within-participants designs* has to be controlled through *counterbalancing* the *ordinal position* of conditions, either across the participants or within the individual participants. However, the effects of order would be incompletely balanced if there were *carryover effects*: i.e., *differential/asymmetrical transfer* of order effects dependent upon the *specific sequence* of conditions (not the number of conditions) that preceded any given condition.

14. *Complete counterbalancing* is where *every possible sequence* of conditions is used an *equal number of times*, with each research participant receiving a single sequence. This controls for the effects of *ordinal position*, and best balances and thus controls, although imperfectly, for *carry-over effects*. However, the *number of participants* needed might be very large, given the number of possible sequences.

    Incomplete counterbalancing drastically reduces the required number of *sequences*, and hence number of *participants*. The *Balanced Square* form of incomplete counterbalancing retains the following features of complete counterbalancing: that every condition occurs an *equal number of times* at each ordinal position, and is preceded by each of the other conditions an *equal* number of times, when considering all the selected sequences taken together. Thus *ordinal position* is balanced, and *carry-over effects* from the *immediately preceding condition* are controlled to the same imperfect degree as with complete counterbalancing. The *Latin Square* form of incomplete counterbalancing is used when one is primarily concerned with controlling for *ordinal position*, and not for *carry-over effects*. The only feature retained of complete counterbalancing is that each condition occurs equally often at each ordinal position, when considering all selected sequences together.

    Randomized counterbalancing is used when even incomplete counterbalancing is unachievable, as when the number of participants available is smaller than the number of sequences required. But the *random selection* of sequences provides *no assurance* that ordinal position and carryover effect will be as well controlled as with incomplete counterbalancing.

15. *Within-participants procedures* are *required* whenever a form of *sequence*, such as trials, is an independent variable (rather than an extraneous variable). These procedures also must be used for *single-participant and small-N research*, where there is only one participant or a very few participants, and thus not enough for more than one group. On the other hand these designs are *excluded* from use if *carryover effects* are too great due to the nature of the variables.

Review Questions

In addition to answering the following questions, also complete any control/design critique problems assigned by your course instructor; e.g., those contained in handouts or in books that you purchased or that are in the library.

1. Describe the two meanings of *experimental control*.

2. List and discuss the two *problems* that potentially can be caused by *extraneous variables*.

3. List the 12 *forms of extraneous variables*, and very briefly describe as many as possible.

4. List and describe the three *general sources of variance*.

5. Describe the *experimenter bias effect* and its mechanisms for both human and nonhuman animal subjects/participants.

6. List and define the three *types of variance* in terms of variables and their effects, and then state to which of the three *components of any measure* each type of variance is related.

7. State the three *minimax principals of variance*.

8. With respect to the preceding question, list and briefly discuss the three *techniques/approaches* to achieving the *variance goal* for good experimental designs that is related to the *independent variable*.

9. List and describe for *experimenter factors*, three specific techniques of *eliminating extraneous variables*.

10. Describe and compare the advantages and disadvantages of the three control techniques of *eliminating extraneous variables*, *holding extraneous variables constant*, and *balancing extraneous variables*.

11. Describe the control technique of *yoking participants on extraneous variables*, and discuss its relationship to, and difference from, the general technique of *matching participants on extraneous variables*.

12. Describe the technique of *using participants as their own control*, and explain its two advantages.

13. Why is *counterbalancing* used for within-participants (repeated measures) designs, and what can limit its *effectiveness*?

14. Describe *complete counterbalancing*, the two forms of *incomplete counterbalancing*, and *randomized counterbalancing*; also discuss their relative advantages and disadvantages.

15. Explain when *within-participants procedures* must be used and when they cannot be used.

8

Control in Experiments — Additional Techniques and Principles

CONTENTS

This is the *second* chapter of the Handbook on experimental control, which is an extensive and very important subject. *The previous chapter* covered the topics of extraneous variables and variance, maximizing the primary variance produced by independent variables, and six techniques for controlling the secondary variance caused by extraneous variables: *eliminating, holding constant, balancing, matching, yoking, and using participants as their own control with counterbalancing.*

The present chapter of the Handbook discusses five additional techniques for controlling secondary variance: *randomizing, pretesting, using control groups, using treatment groups with systematization, and conservatively arranging extraneous variables.* This chapter then concludes with a discussion of minimizing the error variance caused by extraneous variables.

Making Equivalent the Secondary Variance — More Techniques

As noted in Chapter 7, *making equivalent the secondary variance* is the *second* of the three *minimax principles.* As is the case for the six control techniques discussed earlier, the *goal of all five control techniques* that will now be covered, except for the last one, is to have *equivalence of effects* of the secondary, extraneous variables across all of the independent variable conditions. This is accomplished by *eliminating differential average values* of the extraneous variables under the different levels of the independent variables.

Control techniques are an integral part of research designs. Therefore, most of these techniques are further covered in Chapters 9 and 10 on "Designs for Experiments."

1. Randomizing Extraneous Variables

— Process of randomization

- *Randomization* is any *procedure* that ensures that each event in some set of events, such as the assignment of participants/subjects to either a treatment or a control condition, has an *equal probability/likelihood* of occurring.

  ∞ Examples: *Flipping a coin* or *rolling dice* to determine the assignment of participants to the different independent variable conditions.

- Randomization thus serves as a *control technique* that enhances the likelihood that secondary, extraneous variables — which for some reason cannot or should not be eliminated, held constant, balanced, etc. — will nevertheless be *equal in value and thus effect* for the different independent variable levels of a study.

  ∞ *Participants* would be *randomly assigned* to conditions, and the *conditions* would also be run in a *random sequence*.

  - *Confounding* due to *systematic influences* by extraneous participant, environmental, and experimenter factors (except bias) would be *controlled* by such randomization.

    Any *partial confounding* would be statistically less likely the *greater the number* of participants and data gathered per condition, because *random error averages toward zero* as the number of measures averaged together increases.

  - *Masking* — due to *random effects* — would still be possible, however, even if there were no partial confounding (see Chapter 7 for discussion of masking and confounding).

— Applications

1) If *known* secondary, extraneous variables are expected to be present in the experimental situation, then randomization can be used whenever it *isn't feasible* to employ a *more systematic control technique* for one of the following reasons:

   a) The extraneous variable is *not measurable.*
      - Example: The sum of *all* the relevant *previous experience* of each research participant.

   b) The extraneous variable is *too difficult to measure.*
      - Example: The current *motivational level* of each research participant.

   c) The techniques that are more systematic are *already being used extensively* to control a study's *other* extraneous variables, and thus cannot readily be used to control additional extraneous variables.

2) If *unknown* secondary, extraneous variables are expected to be present in the experimental situation, again randomization should be used, since if it isn't known what the variables are, then the *other control techniques* (e.g., balancing) *cannot* be employed — except for within-participants designs to control for participant differences.

∞ Example: Consider *subconscious* personal problems that might affect the performance of research participants.

∞ *Randomization is the only other technique available* for controlling the *unknown and/or unmeasurable sources* of secondary variance, in addition to controlling the *known and measurable sources* of secondary variance.

  • Therefore, one should *always randomize* whenever possible, even when using the other control techniques (e.g., holding extraneous variables constant) that provide *more systematic control* of known and measurable extraneous variables.

    If participants are *randomly assigned* to the different conditions, and the conditions are run in a *random sequence*, then it is likely that the values and thus effects of unknown, as well as known, extraneous factors would be *approximately equal* for all treatment and control conditions (again, the *more participants and data*, then statistically the *more equivalent* the extraneous factors are likely to be for all conditions).

    Randomization, in fact, is necessary for ensuring the *validity of inferential statistical tests*.

    Hence, even when participants are *systematically matched* on one or more secondary extraneous variables, *random assignment* of them to conditions *must* be used, as well, to control for other possible *participant factors*.

∞ In addition to extraneous participant variables, *extraneous environmental and experimenter factors* — either known or unknown, and anticipated or unanticipated — can also be controlled through randomization (as alluded to already).

  • Participants, and thus the various conditions, would simply be *run in random order* (as suggested earlier). Note that if, instead, all the participants in one condition were run first, and all those in another condition were run later, then there could be *confounding* of the independent variable conditions with *extraneous variables* that happen to *change over time*; e.g., the environment, instrumentation, and experimenter features/characteristics.

  • *Environmental extraneous variables* that would be controlled in this way are, for example, construction noises that might be greater at a certain time of the day, and odors from chemistry labs when they're in progress.

  • *Experimenter characteristics and instrumentation* that change over time are also controlled in this manner; for example, shifts in a researcher's observation skills, and the drifting of an instrument's calibration or sensitivity.

— Block-randomization procedure and the randomized-blocks design

- In the *Block-Randomization Procedure*, one participant factor is controlled using the systematic technique of *balancing*, while other participant factors are controlled through *randomization*.
  - ∞ Participants are first *blocked* — i.e., placed into groups — on the basis of some shared, measurable characteristic, such as gender; then from each block an *equal number* of participants is *randomly assigned* to each of the conditions.
    - Thus the *blocked* participant characteristic is *systematically balanced* across conditions, and due to *randomization*, it is also likely that *other* extraneous participant characteristics will be approximately equivalent for all the conditions.
  - ∞ Example: If an experiment involved one treatment and one control condition, and both *psychology and biology majors* were going to serve as participants, then the psychology majors would be *blocked* into one group and the biology majors into another, and then half the participants in each block would be *randomly assigned* to the treatment condition, and the other half to the control condition.
- *Randomized-Blocks Design* is the name for use of this procedure.
- *Matching*, which is an extended and more precise type of *balancing*, and which is used when there are several values of the extraneous variable (e.g., intelligence), can be seen as a form of this procedure, where the *blocks* are the matched participants.

— Additional combinations of control techniques

- It is typical that experiments control some of the secondary, extraneous factors (participant, environmental, and experimenter) using the *systematic techniques* of elimination, constancy, balancing, matching, etc.; then all of the other extraneous factors are controlled through the *technique of randomization*.
  - ∞ Example: The investigator might first *eliminate* noise, then *hold constant* the light level, *balance* the participants for gender, *match* them for age, and, finally, *randomly* assign the gender-blocked, age-matched participants to the various conditions — after which the conditions would be run in a random, or systematically balanced, sequence.

— Limitations

- All secondary, extraneous variables *cannot be controlled adequately* by randomization.
  - ∞ Examples are *experimenter bias and participant bias*, such as the *experimenter expectation* that the behavior of participants will support the research hypothesis, and the *participant motivation* of positive self-presentation (i.e., to behave in a way that presents the most positive image).
    - These must be *controlled by other techniques*, such as single-, double- or partial-blind procedures; automation; or disguised experiments using deception and/or concealment.

— Methods[30]

There are several methods for randomizing secondary, extraneous variables. Five common ones are *flipping a coin, drawing slips of paper, shuffling 3 ×5 index cards or test booklets, counting off, and using tables of random numbers.*

1) Flipping a Coin

∞ *Coin flips* can be used when there are *only two conditions* into which participants must be assigned.

∞ *Requirement of randomization* is met, since the events of "heads" and "tails" have *equal probability of occurring* when a coin is flipped, and thus the participants would have *equal likelihood* of being assigned to either of the two conditions that the experimenter decides "heads" and "tails" represent.

∞ *Problem*, however, is that it is *unlikely* that an *equal number* of participants would be assigned to the two conditions since, due to chance, a coin would not necessarily come up heads and tails equally often in any given sequence of flips.

• This is *undesirable* for various *statistical reasons,* but it is acceptable within limits (an unequal-*N*s analysis would have to be used, which is more complicated and statistically less powerful than an equal-*N*s analysis).

∞ *One solution:* Randomly *throw out* some of the participants from the larger group until both groups are the same size — but this is *not desirable,* since less data would be obtained.

∞ *Better solution:* Stop assigning participants to conditions randomly *once either condition had half of the total number;* all the remaining individuals would just be assigned to the other condition without further need for flipping a coin — but this is *not desirable if* people are assigned in the order that they show up, since the earlier versus later participants might be different.

∞ *Still better solution:* Flip a coin *only once for every two participants;* if the first gets assigned to Condition A, then the second would be assigned to Condition B (and vice versa) without the need for flipping the coin twice.

• *Order* in which the between-participants/subjects conditions are run could also be determined by the same coin flip results that are used to assign individuals to conditions, e.g.: B A, A B, A B, B A, B A, A B.

Using this procedure, the different conditions would be *more evenly distributed over time* than with strict randomization, and thus *extraneous environmental and experimenter factors* would more likely be equivalent across conditions, just as for the *participant factors* (note that this would not be the case for the first two assignment solutions mentioned earlier).

Block-randomization would be the outcome of this procedure, in which the experiment is run in a series of *blocks*, with all conditions occurring once in a *random order* within each block, and with participants *randomly assigned* to the different conditions.

∞ *Quasi-random assignment* is the name applied to the just described "better" and "still better" solutions, since they involve *restrictions/ constraints* placed on the randomization procedure (as is also the case for the *block-randomization procedure* described earlier).

- The purpose of such restrictions placed on randomization is to ensure *better control* over secondary, extraneous variables by *systematically* providing for their *more uniform distribution* (i.e., equivalence) across conditions.

- Note that even with such quasi-random procedures, each research participant would still be *equally likely* to be assigned to either of the two conditions — thus the outcome *would still be randomization,* i.e., each condition would still be equally likely to be assigned any of the possible combinations of one half the participants.

2) Drawing slips of paper

∞ *Drawing slips of paper* can be used for assigning participants to *any number of conditions.*

∞ *Quasi-random assignment* could be employed to ensure an equal number of participants in all conditions through the use of the following *block-randomization procedure:*

- Each between-participants condition would be written on a *separate* slip of paper that was placed in a container.

- Then one slip would be drawn to *assign* each participant/subject as they arrived, which could also determine the *order* in which the conditions were run.

- The slips would not be replaced in the container until *all* had been drawn, at which time slips would again be selected randomly to assign the next set of participants. This method is called *sampling without replacement.*

3) Shuffling 3 × 5 cards or test booklets

∞ *Shuffling cards or booklets* is an efficient, simple approach that can be used for assignment to any number of conditions *if all participants are available at the same time and place.*

∞ *3 × 5 cards* could be handed out for each participant to write down his/her name; all cards then would be collected, *shuffled,* and dealt out into as many piles as there were conditions, with an equal number of cards — and thus individuals — assigned to each condition.

- Note that all the cards for any given condition (e.g., "A") could be dealt out before assigning any to the next condition (e.g., "B"), since it is the *shuffling* that yields randomization, not the sequence in which the cards are dealt into piles (e.g., A, B, C; A, B, C; A, B, C; etc.).

- Using 3 × 5 cards would be much more *convenient and faster* than assigning participants by first preparing a list of their names, and then drawing from a container slips of paper that indicated the assigned condition for each individual (note that *especially ineffi-cient* would be to prepare as many slips of paper as participants, rather than just one slip of paper per condition).

∞ *Test booklets,* if used to manipulate an independent variable (e.g., through providing *varying instructions* for completing a task, or *different information* to influence attitudes), could simply be *shuffled* before being distributed to participants — thus randomly assigning individuals to different conditions (note that this approach would work *whether or not* the participants are all present at the same time for the study).

4) Counting off

∞ Another simple approach for assigning participants to conditions *when everyone is present at the same time and place* for a study, is to just *count off* individuals in the order in which they're *seated* in a room (e.g.: 1, 2, 3; 1, 2, 3; and so on, if there are three conditions).

- *Assumption,* which is fairly reasonable, is that the *seats* individuals take do not involve any regularly alternating participant charac-teristics corresponding to the counting off, but rather that the seating is *essentially random.*

5) Tables of random numbers

∞ *Tables of random numbers* contain the digits 0–9 arranged in a *random sequence,* i.e., each digit is *equally likely to occur* in any row and column, and there is *no systematic connection* between the location of any digit in the sequence and any other — hence the digit-occurrence events are *independent.* (See Table 8.1.)

TABLE 8.1

Sample Table of Random Numbers

91	22	76	47	51
58	13	53	01	97
03	15	87	90	11
10	42	99	66	12
64	41	16	08	11

∞ These tables *vary in appearance* due to their having different sized group-ings of numbers for convenience (e.g., in pairs, triplets, etc.), but all tables of random numbers are used in essentially the same fashion, and they are usually *accompanied by instructions and examples of applications.*

∞ *Procedure* that might be used if there are only *two conditions* is as follows:

- Prepare a *list* of the participant's names.

- Enter the table at any point in a *random fashion;* e.g., close your eyes, flip to some page, and place a pencil tip anywhere on the page.

- Read the sequence of digits in *any direction* determined ahead of time (down, up, right, or left), assigning the digits sequentially to the participants' names on the list.

- Individuals with even numbers (including 0) would be assigned to one condition, and those with odd numbers would be assigned to the other condition, after *randomly deciding* which condition would be the even or odd one.

- *Problem* with this approach, however, is that the number of individuals assigned to each condition would probably *not be equal.*

∞ *Procedure* that would assure an *equal number* of participants per condition, *regardless of the number of conditions*:

- Assign each individual a *unique number* by skipping repeated numbers when reading a sequence from a table.

- If more than 10 digits (0–9) are needed, then the digits could be read in pairs (00–99) or larger groupings.

- If there were only *two conditions,* individuals receiving numbers below the *median value* would be assigned to one condition, and those with numbers above the median would be assigned to the other, after *randomly deciding* which condition would be below or above the median.

- If there were *more than two conditions,* the numbers simply would be divided into that many sets; e.g., if there were *three conditions,* then the individuals with the lowest third of the numbers would be assigned to one condition, those with the middle third to a second condition, and those with the highest third to the last condition, after *randomly deciding* which condition went with which set of numbers.

∞ *Procedure* that is *somewhat easier*:

- List the participants and assign them *sequential numbers,* starting with the number 1 or 0, as it occurs in the tables.

- Then, e.g., if there were 20 participants and two conditions, the *table numbers* would be read in groupings of two, and the first 10 *different* participant numbers from 00–19 found in the table sequence would indicate the individuals to be assigned to one of the conditions (determined randomly); and the remaining participants would automatically be assigned to the other condition without needing to find their numbers in the table.

- • This method works just as efficiently *regardless of the number of conditions,* e.g., if there were 30 participants and three conditions, the individuals with the first 10 numbers found in the table sequence would be assigned to one condition, those with the next 10 numbers found in the table sequence would be assigned to the second condition, and the remaining 10 participants would be assigned automatically to the third condition, after *randomly deciding* which condition would be the first, etc.

- ∞ *Several other procedures* are possible, and they are described in the books that contain tables of random numbers.

- ∞ *Computer algorithms* are also available for generating "random numbers" series, which can then be used just like the published tables.

2. Pretesting Participants for Extraneous Variables

— Pretest defined

- • A *pretest* is a *dependent variable measure* that is taken *prior* to the administration of independent variable treatment and control conditions, and which can be *compared* against a *posttest* on the dependent variable taken *afterward.*

— Functions

There are three functions of pretesting participants for secondary, extraneous variables: determining *participant comparability* and making any necessary *adjustments,* enhancing *sensitivity/power* to detect independent variable effects, and determining a *baseline or trend.*

1) Participant-comparability measurement and adjustment

- ∞ *Random assignment* of individuals to conditions/groups in order to ensure *equivalent participant characteristics across all conditions* (to avoid confounding) is not *infallible.*

  - • Pretest scores on the dependent variable(s) can be used before treatments are administered to *check on the initial equivalence of groups* in the different conditions with respect to the means and variances of pertinent *participant extraneous variables.*

- ∞ Random assignment of participants *isn't always possible.*

  - • Sometimes, as in school settings, *preestablished groups* of individuals, such as entire classes, must be assigned to the different independent variable conditions.

  - • In such cases a *pretest check for comparability of groups* is even more important, since it would control for possible *confounding* by individual differences among participants that might affect the dependent variable(s).

- ∞ Pretest scores also can be used to *adjust* posttest scores.

  - • This is done by simply computing the *pretest-posttest change (difference) score* for each individual, thus taking into account *initial differences* on the dependent variable (discussed in Chapter 10, under "Nonequivalent Pretest-Posttest Control-Group Design").

- Note, however, that this control is not infallible: The participants in the different groups might vary in ways that are *not measured* by a pretest, e.g., their biases, but which nevertheless influence the effects of treatments (an interaction), and thus a study's results.

2) Sensitivity/power enhancement

∞ Use of *pretest-posttest change scores* not only controls for *confounding,* but for *masking* through reducing the within- conditions variance by adjusting for individual differences.

- This enhances the *sensitivity/power* to detect effects of the independent variable(s).

∞ *Matching* individuals across groups on the *pretest dependent variable scores* is another important method of increasing the power/sensitivity to detect effects of an independent variable (see "Pretest-Match-Posttest Design" in Chapter 10).

- This is because *initial values on* the dependent variable(s) might *influence* (interact with) the effects of treatment(s), thereby increasing the within-conditions variance.

   Example: A study with a treatment condition that tries to *reduce hostility toward some ethnic group* might find that the effectiveness of the independent variable is a function of the individual's *initial/pretest level of hostility*; e.g., the treatment might be very successful with individuals who have little hostility, but unsuccessful with extremely hostile individuals.

   This could cause *masking* unless controlled either through *matching, or blocking and balancing* on the *pretest value* and then *analyzing* the pretest value effect (see "Treatment Groups and Systematizing an Extraneous Variable" later in this chapter).

- *Matching or balancing* based on pretest values of relevant factors *other than the dependent variable* of the study, e.g., Scholastic Aptitude Test scores, or IQ scores, is *less powerful* than matching individuals based on their pretest *dependent variable* scores (see "Match by Correlated- Criterion Design" in Chapter 10).

∞ *Ceiling or floor attenuation/truncation effects* are additional influences on the sensitivity/power to detect the effects of independent variables.

- These potential truncation effects could occur for individuals having *very high or low initial values* on the dependent variable task measure.

- Pretest scores, however, provide a measure of this problem, and can be used to *adjust* for it by computing the *pretest-posttest change (difference) ratio score* for each participant (see Chapter 10, under

"Nonequivalent Pretest-Posttest Control-Group Design"). Alternatively, the dependent variable task could be *modified* before going forward with the study.

3) Baseline or trend determination

∞ A pretest dependent variable value provides a *reference level;* i.e., a *baseline* or, if multiple pretest measurements are made, a *trend* against which the posttest score(s) can be *compared* to determine what, if anything, is the *effect* of a treatment.

- *Multiple pretest measures,* as opposed to a single measure, provide *controls* for several extraneous variables that are related to *repeated measures,* and which are thus associated with pretesting itself; e.g., testing, statistical regression toward the mean, maturation, and instrumentation (discussed in Chapter 9, under "Time-Series Design").

- Another advantage is that *sampling error* is reduced by *averaging multiple measures* for each individual *before,* as well as *after,* treatments — recall that random error averages toward zero as more measures are included.

— Problems

- Pretests can *interact* with a treatment, producing *sensitization or resistance* to it (discussed in Chapters 9 and 10).

∞ Example: *Pretesting participants' opinions* might alert them to the fact that they are taking part in an *attitude change experiment,* and this knowledge could influence the effect of the independent variable, leading to results that *would not be representative* of those for a population that had not been pretested (this is a *generalization/external validity problem,* which is covered in Chapters 9 and 11).

∞ Note: Pretesting not only can produce the *indirect effect* just discussed, but also a *direct effect* on the posttest dependent variable scores (this is a *confounding/internal validity problem,* which is covered along with external validity in Chapter 9 under "One-Group Pretest-Posttest Design").

- This direct-effect problem can be *controlled,* however, by *multiple pretest measures* (noted above), or by a control group (discussed in Chapter 10, under the two "Pretest-Posttest Control-Group Designs").

— Disadvantage

- *Increased* time, effort, and money are required for pretesting.

3. Control Groups for Extraneous Variables

— Control group defined

- In relation to experimental treatment groups, a *control group* consists of a *comparable set* of participants *who do not receive an experimental treatment,* but who *in all other regards* are treated *exactly the same* as the experimental group(s).

∞ A control group can be thought of as a special treatment group that receives an *independent variable value* of *zero,* or that receives some *standard value* (i.e., one that is typical/normal).

- The *intent* is to *make equivalent the extraneous variables* for the experimental treatment and control conditions, and particularly to *hold constant* extraneous *experimenter factors*; e.g., those associated with experimenter characteristics, procedures, instructions, tasks, and apparatus; as well as *participant bias,* and extraneous aspects of the general *environment.*

— Functions

Three general functions of control groups are: determining a *baseline,* controlling for *confounding* by extraneous variables, and evaluating the *effects* of extraneous variables.

1) Baseline Determination

∞ A *control group* can be used to establish a *reference level* against which to compare the dependent variable values of the treatment/experimental group(s), and thus to *evaluate the effects,* if any, of manipulating the independent variable(s).

- Example: If an area of the brain of experimental animals were being *lesioned* (damaged) to determine possible effects on *emotional behavior,* and thereby uncover a function of this area of the brain, then a *control group* of animals would be needed in order to establish a baseline/reference level of emotional behavior for animals with normal (undamaged) brains.

2) Control of extraneous variable confounding

∞ A *control group* (or condition) is also necessary when experimental manipulations *unavoidably and undesirably* involve the *introduction* of secondary, extraneous variables *right along with* the independent variable (see placebo effects and spontaneous remission, discussed below).

- Note that if the experimental and control groups are *treated in the same manner,* except with regard to the *independent variable,* then the *extraneous variables* should be *equivalent* for the two groups; thus any difference between the groups on the *dependent variable* could logically be attributed to the *independent variable.*

- Example: In a study of the effects of *damaging* an area of the brain on some aspect of *behavioral or mental capacity,* a lesion might be produced using an electric current passed though the tip of an implanted electrode; however, it would be necessary to control for the *effects of brain surgery in general* by having a *control group* that is operated on just as the experimental group, but that does not receive damage to the structure of interest (note that this control group would also serve the function of *establishing a baseline*).

Such sham-lesion control groups (as they are called) are necessary since anesthetizing animals, drilling an opening in the top of the skull, and lowering an electrode might cause, respectively, a temporary chemical imbalance, exposure of the brain to infection, immune system reactions, and damage to other brain structures along the electrode's path.

∞ Placebo effects in therapy studies

- *Definition:* Some people who *believe* they are receiving an effective treatment for their problem(s), but who actually are not, nevertheless *show improvement.* *"Placebo"* comes from Latin, and means *"to please."*

- *Placebo control groups* are those that receive a harmless, "ineffective" treatment, such as a sugar pill, that the individuals are *led to believe* is an effective therapy.

- Placebos serve to control for any *secondary, extraneous variable participant-bias effect* that involves *expectation of improvement,* and that is *introduced along with the independent variable* of a potentially effective therapy. (Note: *evaluation* of whether in a given study there is any actual placebo effect is discussed below).

- *Ethical issues* are involved when control participants are *deceived* into thinking they will receive a real treatment; however, this problem is *diminished* if they are later given the most effective therapy (after it has been determined), which would also provide *additional data* for the study.

∞ Spontaneous remission/recovery in therapy studies

- Definition: Some people who show symptoms of mental, behavioral, or physical problems will *improve over time without receiving any treatment,* not even a placebo.

- Remission/recovery is *not truly spontaneous,* but rather is due to *maturation* and/or *contemporary history,* which of course can also account for changes over time in studies not involving *therapy* as an independent variable.

- *Time,* as an extraneous variable, and its associated extraneous factors of maturation and contemporary history, must therefore be controlled in order to avoid *experimental contamination* whenever a study involves: (1) testing individuals, (2) applying some experimental manipulation (e.g., therapy), and (3) retesting the individuals *later.*

- A *no-treatment control group* (i.e.: pretest, no treatment, then posttest) is used to control for the *experimentally introduced passage of time,* and the associated extraneous variables of maturation and contemporary history, as well as possible instrumentation and experimenter changes, etc.

- *Ethical concerns and solutions* regarding no-treatment control groups are similar to those for placebo control groups; i.e., control participants should later be given the most effective therapy, based on the study's results.

3) Evaluation of extraneous variable effects

   ∞ One versus two control groups

   - *Placebo control groups* not only control for placebo effects, but also for *spontaneous-remission effects* due to time — thus just *one control group* would be sufficient.

   - However, rather than just controlling, sometimes the *measurement/ evaluation* of extraneous variable effects is desired (the third function of using control groups).

   - *Placebo extraneous variable effects* could be *measured,* as well as controlled, if a *no-treatment control group* for spontaneous remission were also included in the design, against which performance of the placebo control group would be compared and thereby *evaluated* — hence the advantage of *two control groups.*

     Example: In a study of the effectiveness of a new drug for the treatment of bed wetting in children, one could include both a *placebo control group* and a *no-treatment control group.*

     Effectiveness of the therapy would be evaluated by comparing the measured dependent variable scores for the treatment group against those for the placebo control group, and the *placebo control effect* would be evaluated by comparing the placebo control group against the no-treatment control group.

4. Treatment Groups and Systematizing Extraneous Variables

— Functions

Although it might seem counterintuitive, treatment groups can actually serve two general *control functions,* as well as a third related function: specifically, control of extraneous variable *confounding,* control of extraneous variable *masking,* and *evaluation* of extraneous variable effects.

1) Control of extraneous variable confounding

   ∞ Adding to a study an *additional treatment group* including the *same extraneous variable(s),* then comparing effects of treatment conditions, is sometimes the best control for *confounding* by secondary extraneous variables.

   - Building on an *example* that has been used earlier: If an investigator is trying to determine the function of a particular *structure in the brain* by observing the effects of *damaging* that structure in animals, a *control problem* would be that whatever deficit occurred might not be *specific* to lesioning that *particular* brain structure. *General effects* of brain lesioning, as well as other aspects of surgery, could be the cause of any deficit.

- By adding a *second treatment group* that has a different brain structure lesioned, but otherwise is treated exactly the same as the first, and then comparing the *differences in lesion effects* — which hopefully involve unique deficits (this is known as a double dissociation of function) — the *specific brain-lesion effects* could be determined.

  Note that the *extraneous variables* of the general effects of any brain lesioning and surgery would be controlled, since these would be *made equivalent* for the two treatment groups by being *held constant*.

  Furthermore, because two different brain lesions are made, *more data* would be obtained about the functions of various brain structures than if only a *sham-lesion control group* (i.e., surgery without any lesion) or a *no-treatment control group* were used.

2) Control of extraneous variable masking and

3) Evaluation of extraneous variable effects

  ∞ Systematization of a secondary, extraneous variable

  - Both the second and third functions of treatment groups are produced by systematically *balancing a secondary, extraneous variable* in the research design, and *treating it as an independent variable* — at least in the *statistical analysis* (this requires a factorial design, and is covered in Chapter 10 under "Multiple Treatment Designs").

  - More information is thus obtained. *Secondary variance* (the *consistent effect* of an extraneous variable) is *measured/evaluated*, and as a consequence, it is also removed from the statistical error variance, thereby *enhancing detection of the primary variance* (elaborated later in this chapter under "Minimizing Error Variance").

  - *Masking* is thus controlled, in addition to *confounding* (as is the case when eliminating or holding an extraneous variable constant, but unlike when randomizing or just using balancing alone).

    Therefore, the study has greater sensitivity/power to determine any treatment effects of the independent variables of interest (i.e., the primary variance).

  - Example of both functions: The *gender* of research participants is a potential extraneous factor in many studies, but it *cannot be eliminated*; furthermore, if controlled by *holding it constant*, then the investigator would be restricted in the number of individuals available as participants, and more importantly, *generalizability* of results would be reduced.

    Balancing or randomizing for gender, on the other hand, allows more opportunity for obtaining participants, and permits greater generalizability.

However, there is a still an important problem — *masking*: Although the effects of gender would be approximately equal for the various conditions, and thus *confounding* would be controlled, the *variance* that might be produced by gender would still be *present within conditions*, and this could make it difficult to detect the independent variable effects.

But note that, after *balancing* for gender across the independent variable levels, gender could be *treated as another independent variable* in a *factorial design* (called treatments by levels), in which the levels of the experimental independent variable of interest are *likewise balanced* under the levels of gender (see earlier figure in Chapter 7).

Because *each factor or variable would be balanced under the other,* there would be *no confounding* (see Chapter 10 under Factorial Designs).

Using this approach, the effect of gender could be *evaluated*, and therefore the *secondary variance* produced by gender would be *accounted for* and hence removed from the statistical error variance.

Thus, as a result of this control technique, *masking* would be reduced, thereby enhancing *detection* of any treatment effects of the independent variable, i.e., the *primary variance*.

5. **Conservatively Arranging Extraneous Variables**[25]

— Rather than controlling for a secondary, extraneous variable by making it *equivalent* for all conditions, with the technique of *conservatively arranging* extraneous *variables* the conditions are instead *set up* so that the effect of an extraneous variable could only *reduce/weaken* (i.e., work against) the *measured effects* of an *independent variable*.

   • Hence this is a *conservative approach* — the cards are stacked against the *research hypothesis*.

   • Example: Pitting the *extraneous variable* of a *genetic predisposition* of a species to behave in a certain way (such as the predilection of cockroaches to avoid light) against a *conditioning paradigm* that is being used as an *independent variable* in an attempt to demonstrate the species' ability to learn.

      ∞ In this example, we could attempt to *train cockroaches* — in opposition to their genetically programmed behavior — to run quickly *into a lighted area* to *avoid electric shock,* which, if it worked, would demonstrate their ability to *learn* a simple brightness discrimination and evasive behavior *in spite of* a genetic extraneous variable.

— Weakness of this technique

   • *Primary variance is reduced,* rather than maximized, by the conservative arrangement of an extraneous variable.

∞ This is clearly *undesirable* since there is *less power/sensitivity* to detect an effect of the independent variable.

— Strengths and advantages of this technique

- *Confounding*, which would make it impossible to interpret results as being due to the independent variable manipulation, is *more simply controlled* through the conservative arrangement of an extraneous variable than by using additional groups and/or conditions.

- *Confidence* in the research hypothesis is even *higher* if the data support it, since there would be *less power/sensitivity* in the test of independent variable effects when using this form of control.

Minimizing the Error Variance

Minimizing the error variance is the *third* and last of the *minimax principles*. The discussion of this concept is rather technical, but it is of great theoretical and practical importance. The following material both reviews and builds on information provided earlier in this Handbook, and in turn it is very relevant to information provided in Chapters 11 and 12.

1. Review and Introduction

— *Error/random variance*, as noted before, is *unpredictable/inconsistent variation* of the dependent variable produced by extraneous variables.

- It relates to the *variable error score* component of a measure (Components of measures are reviewed below under item 2).

- Variable error is *self-canceling around a mean of zero*, i.e., it *averages toward zero* as the number of measures increases.

- Nevertheless, error variance is *undesirable,* and therefore should be *minimized.*

  ∞ *Confounding* is controlled due to *inconsistency* of effects.

  ∞ *Chance variations*, however, when they do not average to zero, can still result in *partial confounding* and thus dependent variable differences *between conditions* that might be *mistaken* for independent variable effects (this would be a Type I/Alpha Error — see Chapter 12).

  ∞ *Masking* the effect(s) of an independent variable is the other possibility if error variance is not *minimized* (this would be a Type II/Beta Error — see Chapter 12).

- Error variance *cannot be completely eliminated* (controlled) due to the typically *large number* of known and unknown extraneous variable sources of essentially random/unpredictable fluctuations, and also due to the *physical and ethical constraints* on the use of control procedures.

∞ Hence, and this is very important, there is *always some uncertainty* in the interpretation of research findings, and thus *conclusions can only be probabilistic statements* based on *statistical significance tests*.

• IN SUMMARY: Nothing is ever truly proven in science — Type I and Type II Errors due to extraneous variables are always possible, as well as other errors that result from faulty research.

— Sources of error variance

• The *sources of error variance* are the *same* as those for primary and secondary variance:

∞ Participant/subject factors

∞ Environmental factors

∞ Experimenter factors

2. **Principles Involved in Minimizing Error Variance (See Appendices 2 and 5)**

— Reliability and validity of measurement concepts revisited

• The concepts of *reliability and validity* of measurement relate to the *components* that *any measure* can be thought to consist of (discussed in Chapter 4).

∞ Components of measures:

• True score component

This is what you want in your measurements.

• Error score component

This is what you *do not want* in your measurements, and there are two parts to it: *constant error subcomponent* (*validity* is particularly affected by this error) and *variable error subcomponent* (*reliability* is particularly affected by this error).

• Note: As pointed out before (Chapter 4 and Appendices 2 and 5), the true score, constant error, and variable error components of a single measurement are related respectively to the *primary variance, secondary variance, and error variance components* of a *set of measurements*.

— Communications engineering model

• There is a *communications engineering model* that illustrates the concepts of reliability, validity, and the control of variance.

• *Two elements* must be differentiated in *communication channels* (such as radio, television, telephone, and living organisms).

∞ Signal

• This is the *information* being intentionally transmitted.

• It is the *true score component* of a measure (or perception), and it relates to the *primary variance* produced by the *independent variable* in an experiment.

∞ Noise
  - This is the *static* produced by *non-signal* factors.
  - It is the *error score component* of a measure (or perception), and it relates to the *error and secondary variance* produced by *extraneous variables.*

- Receiver's task
  ∞ Receivers must distinguish/separate the *signal* from *noise.*
    - *If noise is very great* in comparison to a signal, then it would be difficult to consistently perceive or even detect the signal; i.e., there would be *low reliability,* and *masking* might occur (Type II/Beta Error).
    - Alternatively, noise might be *misinterpreted* as, or *confused/confounded* with, a signal; i.e., there would be *low validity,* again leading to possible *error* (Type I/Alpha Error).
    - Example: When studying how the human nervous system processes information, an investigator could record the electrical activity of the brain using electrodes attached to the scalp, and try to *distinguish* the brain's electrical response to the presentation of sensory stimuli — the *event-related potential signals* — from the brain's ongoing spontaneous electrical activity and the spurious electrical activity in the recording equipment and the environment — the *electroencephalographic noise.*

      Because the electroencephalographic noise can be about the same amplitude as the event related potential signals, the noise can mask or alternatively be confused/confounded with the signals — thus it is necessary to use special techniques, such as computer averaging, to enhance the strength of the signal relative to the noise.

- Goal of receivers and transmitters
  ∞ *Maximizing the signal-to-noise ratio (S/N)* is the goal of communication receivers and transmitters.
    - This is the basic means for reducing the probability of *masking and confounding/confusion,* i.e., reducing error.

— Relation of experimentation to the communications model
  - *Major tasks of experimentation* are to *maximize* the primary variance and *minimize* the error variance, while making the secondary variance *equivalent* for all conditions: in other words, maximize the signal-to-noise ratio.
  - *In communication systems it is not possible to directly measure a signal isolated from noise.*
    ∞ If we could obtain a direct measure of just the signal, then the communication system would be *ideal/perfect,* in which case the signal-to-noise ratio concept would be *unnecessary.*

∞ In real life, however, we must contend with observations of what might be a signal *in addition to* noise, and we must compare this with observations of the expected level of noise alone, i.e.: $(S + N)/N$ rather than just S/N.

∞ *Size of the $(S + N)/N$ ratio is used to evaluate the likelihood that a signal actually exists.*

- Noise measured alone (N in the denominator) is taken as an *estimate* of the *expected noise* that would occur along with any possible signal ($S+N$ in the numerator).

- Note that the noise *is not* necessarily exactly the same under both conditions of measurement — it's an estimate.

- Most importantly, the *larger the ratio*, the greater the *probability* that there is a *signal* in addition to the noise.

- *In experiments, we similarly cannot directly measure the primary variance (PV) isolated from the error variance (EV); furthermore, secondary variance (SV) might also be present* (shown as ←SV? in the equations below, where ≡ means equivalent to).

$$(S + N)/N \equiv (PV + EV)/EV \equiv (PV + EV \leftarrow SV?)/(EV \leftarrow SV?) = BV/WV$$

∞ On the one hand, we would have an observation of *possible primary variance, plus error variance*, between conditions (PV+EV).

- This is computed from differences *between conditions* for the *average* dependent variable values, which are measured as the between-conditions, -treatments, or -groups variance (BV).

∞ In addition, we would have an observation of just the *expected error variance* (EV) present between conditions.

- This is determined from differences *within conditions* for the dependent variable values, which is measured as the within-conditions, -treatments, or -groups variance (WV).

∞ Furthermore, we also would have *secondary variance* (SV) if the secondary, extraneous variables were not *eliminated or held constant*.

- SV is *mixed inextricably* with the *primary variance* (PV ← SV + EV) if a secondary, extraneous variable is *not* controlled for *confounding* — i.e., for systematic association (or covariance) with levels of the independent variable — which should never be the case.

- SV is instead *combined* with the *error variance* (PV + EV ← SV and EV ← SV) if the secondary, extraneous variable *is* controlled for confounding but *not* for *masking* of the primary variance, as when employing the techniques of *randomization or balancing*.

- SV is, on the other hand, *separated* from *both* primary and error variance (thus just PV + EV and EV) if the secondary, extraneous variable is controlled for *both* confounding and masking, as when employing the techniques of *matching, yoking, using participants as their own control, or systematization of the extraneous variable* — all of which involve statistical analysis of secondary, extraneous variable effects (see next).

- Statistical error variance

  - ∞ *All* of the *unaccounted for variability* that occurs in the dependent variable measure constitutes what is called *statistical error variance.*

  - ∞ Thus *statistical* error variance is more inclusive than the prior nonstatistical, *research* usage of the term *error variance* (further explained below), but it too should be *minimized.*

  - ∞ This represents the *variability left over* after all *known-and- analyzed* sources of *systematic variance* have been subtracted from the *total variance* in a set of measures.

    - *Sources of systematic variance* include any effective primary, independent variable(s); secondary, extraneous variables; and any interactions among these variables (see the SS_{total} formulas and discussion below).

    - Thus *statistical error variance is reduced* when secondary, extraneous variables are *systematized* and their effects *analyzed* (techniques were discussed earlier) so that their variance is not mixed with the error variance (hence EV rather than EV ← SV in the equation).

      This controls for masking of independent variable treatment effects by the extraneous variables, in addition to controlling for confounding, since the S + N/N ratio would be greater for

      $$(PV+ EV)/EV, \text{ than for } (PV + EV \leftarrow SV)/(EV \leftarrow SV).$$

    - This contrasts with the *greater statistical error variance* that occurs when secondary, extraneous variables are *only controlled for confounding*, as when *randomization* is used, which *combines* secondary variance with the error variance (opposite of the above), thus *contributing to masking.*

      Example: Consider the difference when the *time of day* that research participants are run is controlled for *confounding* by simply *balancing* for morning and afternoon hours under each of the independent variable conditions versus the situation when additionally *analyzing* for the effects of time of day on the variance of the dependent variable — which removes this *accounted-for variance* from the *statistical error term*, and thus *reduces masking.*

- Statistical variance concepts related to significance testing

  The following information on *statistical variance concepts* is quite technical, but it lays the *mathematical foundation* for subsequent discussions (in this and later chapters) of *statistical tests of significance*. This material need not be set to memory, but the major points should be understood.

- Variance
  - ∞ *Variance* is a measure of the *variability* of scores, which for a sample would be the *sum of squares (SS) mean = SS/n*, and for a sample estimate of the population variance it would be the *mean square (MS) = SS/df*, as in the *F-ratio test* (these statistical constructs/concepts are explained next).

- Sum of squares (*SS*)
  - ∞ *Sum of squares* is a mathematical construct from which *variance* is calculated.
  - ∞ *SS* is the sum of the squared deviations of scores (*X*) from their mean (*M*), which $= \Sigma(X - M)^2$.
  - ∞ In the simplest research design, having only one independent variable, the *total sum of squares* for the dependent variable would be *partitioned* (divided up) as follows:

$$SS_{total} = SS_{between} + SS_{within} \qquad \text{which} = SS_{treatments} + SS_{error}$$

- $SS_{total} = \Sigma(X - GM)^2$ where *GM* is the grand mean of all the scores under all independent variable conditions.
- $SS_{between}$ is also called $SS_{treatments}$ and it is the *between-conditions* sum of squares, which $= \Sigma(CM - GM)^2$ where *CM* is the condition mean of the scores computed for each of the different independent variable levels.

  Note that in *more complex designs*, with more than one independent variable (i.e., factorial designs; see Chapter 10), there would be a separate sum of squares (*SS*) for each of the independent variables ($SS_{between}$), and for each of the possible interactions among the independent variables ($SS_{interaction}$).

- SS_{within} is also called SS_{error} and it is the *within-conditions* sum of squares combined for all conditions, each of which $= \Sigma(X - CM)^2$ where *X* represents the scores within an independent variable level/condition.

  This equals the sum of squares for *statistical error*, which would be reduced by the *SS* for any secondary, extraneous variable that is systematically balanced and whose effects are analyzed, as well as by the *SS* for any interaction with the independent variable (i.e., $SS_{secondary} + SS_{interaction}$), since in such a design the total sum of squares for the dependent variable would remain the same, but would be partitioned into more parts as follows:

$$SS_{total} = SS_{treatments} + SS_{secondary} + SS_{interaction} + SS_{error}$$

- *F*-ratio
  - ∞ *F-ratio* is the *analysis-of-variance* inferential statistic for testing/determining significance (elaborated below).

$$F\text{-ratio} = MS_{\text{treatments}}/MS_{\text{error}} \equiv BV/WV$$

  - ∞ Mean square (*MS*)
    - *Mean square* is a mathematical construct that represents a *sample estimate of the population variance.*
    - *MS* is the sum of squares *(SS)* divided by its degrees of freedom *(df)*, i.e., *SS/df* which $= \Sigma(X - M)^2/df$.
  - ∞ Degrees of freedom *(df)*
    - *Degrees of freedom* equal the number of *observations* on which a statistic is based, *minus* the number of *restrictions* placed on the observations' *freedom to vary* — thus *df* is simply the number of scores in a set that are free to vary.

      Example: For a set of *n* values there is one restriction, the *fixed sum or mean* of the values; thus $df = n - 1$.
- Additional statistical principles are covered later in this chapter and in Chapters 11 and 12.
— Statistical tests of significance
  - *Tests of statistical significance* measure the *probability* that, e.g., observed differences on the dependent variable for the different independent variable conditions are due to *chance alone*, i.e., *sampling error* (covered in Chapters 11 and 12). (See also Appendix 15 for a table of statistical tests for different designs and levels of measurement).
    - ∞ Conversely, they determine a *degree of confidence* that the *sampled data* reflect the *true population differences.*
      - More confidence exists at the .01 versus the .05 *level of statistical significance,* since this value represents the *probability* that dependent variable differences as *large* as those obtained (and larger) could be due to *just chance.*
  - Inferential statistics
    - ∞ *Tests of statistical significance,* are called *inferential statistics* (versus descriptive statistics), and they usually involve a *ratio,* such as in the Analysis of Variance, or *F*-test:

$$F\text{-ratio} = MS_{\text{treatments}}/MS_{\text{error}}$$

      which as noted earlier $\equiv BV/WV = (PV + EV \leftarrow SV?)/(EV \leftarrow SV?)$
    - ∞ *Likelihood (or reliability) of treatment effects* is measured by these statistical tests.

- *Magnitude (or strength) of effects* is *not* measured, however, since tests of significance, as just shown, involve a *ratio* that includes not only the effects of *treatments* (the primary variance due to manipulation of the independent variable), but also the unaccounted for effects of *extraneous variables* (statistical error variance). (For elaboration, see Chapter 12 under "Statistical Significance versus Practical Significance.")

∞ *Size of the ratio* is used to evaluate whether or not it is likely that *primary variance* — a signal — actually exists, just as for the $(S + N)/N$ ratio in communications systems (as stated earlier).

- *Statistical error variance,* or noise, measured alone *within conditions* (the denominator of the ratio) serves as an *estimate* of the statistical error variance that occurs *along with* any possible *primary variance,* or signal, measured *between conditions* (the numerator of the ratio).

   As noted earlier, by *analyzing for secondary variance* the statistical error variance would be reduced, thereby increasing the size of the ratio.

- *Larger the ratio* the greater the probability that there actually is primary variance — a *treatment effect* — in addition to error variance — or *chance variation* — since:

$$MS_{treatments}/MS_{error} \equiv BV/WV = (PV + EV \leftarrow SV?)/(EV \leftarrow SV?)$$

   If the ratio is only 1, then the between-conditions variance (BV) equals the within-conditions variance (WV), and thus there is no evidence of a treatment effect (primary variance, PV) — there is only evidence of chance variability (error variance) between conditions:

$$1 = \frac{0 \text{ treatment effect} + \text{chance variability}}{\text{chance variability}}$$

- Ratios are *interpreted* by using statistical tables to determine their *level of significance,* i.e., the statistical probability that the dependent variable differences between conditions are due only to *chance.*

   This is dependent on the *degrees of freedom (df)* that the data had to vary, which is a function of the *size of the samples.*

   The *df* for the *between-conditions variance* in the numerator equals the number of independent variable levels minus 1, since the grand mean is considered fixed (not free to vary); and the *df* for the *within- conditions variance* in the denominator equals the number of participant scores minus the number of conditions, since there is one score in each condition that cannot vary, given that the means are *fixed.*

- The *greater the degrees of freedom,* which increases with additional data *(N),* the more significant any given ratio would be — i.e., the *more reliable* the observed effects associated with manipulation of the independent variable.

- Hypothesis testing

  ∞ *Hypothesis testing* involves applying the previously noted principles and procedures of *statistics.*

    - This is for the purpose of testing *research hypotheses* about treatment effects against a *null hypothesis* that there are no effects other than those due to chance.

  ∞ *Statistical tests of significance* used in *hypothesis testing* are related to the concept of *reliability,* which implies *validity* — and all of this is related to the *control and analysis of variance* (discussed further in Chapter 12, "Hypothesis Testing and Statistical Significance").

    - Thus a number of principles, procedures, and concepts, each of which are of great theoretical and practical importance, are *highly interrelated.*

— Requirements for maximizing reliability and validity of test data

  1) *Before treatments are administered,* both the between-conditions variance and the within-conditions variance should be *minimized;* i.e., the *pretest mean* of the dependent variable scores for each condition (if actually measured) should be as *similar* as possible, as should the *individual scores* within each condition.

    ∞ This would assure the *least secondary and error variance.*

      - Note that *between-conditions variance before treatments,* when there clearly could not yet be any primary variance, would suggest that in addition to *error variance* there might be *secondary variance between conditions.*

        This would raise the possibility of partial or even complete extraneous variable confounding with the independent variable conditions (when administered), which is definitely undesirable.

  2) *After treatments are administered,* it would then be desired that the between-conditions variance (BV) should be *maximal,* while the within-conditions variance (WV) should remain *minimal.*

    ∞ This would provide evidence for the *desired primary variance* by *maximizing* the $(S + N)/N$ ratio $\equiv$

    $$(PV + EV)/EV \equiv (PV + EV \leftarrow SV?)/(EV \leftarrow SV?) = BV/WV$$

3. Techniques for Minimizing Error Variance

Eight approaches can be taken to *minimize error variance.*

a. *Individual differences* among research participants could be reduced.

- *Holding constant* hereditary (e.g., ethnic) and/or experiential characteristics of research participants would *reduce variability* in the responses of participants to independent variable conditions, as well as to extraneous environmental and experimenter factors — all of which would help to minimize error variance.

- *Generalizability* of results, however, would unfortunately also be reduced when using this technique.

  ∞ But recall that individual differences among participants that happen to be sources of *secondary/systematic variance,* as opposed to error/random variance, can be *systematized* and their variance removed from the statistical error variance in the data analyses (covered earlier), and thus their presence need not be minimized.

  b. *Environmental factors* that are randomly operating should be eliminated, held constant, or kept to a minimum.

  - Extraneous visual, auditory and other stimuli must be controlled.

  c. *Experimental procedures* should be kept consistent.

  - Random variation in the *methods* must be minimized by taking great care to control what investigators do.

  - *Holding constant* extraneous aspects of the *procedures* would reduce what is referred to as *experimental contamination.*

  d. *Measurement and analysis techniques* and the *apparatus used* should be reliable (stable/consistent) and valid (appropriate/accurate).

  - This would contribute to reducing *experimental contamination* by decreasing what is called *instrumentation/measurement error.*

  e. *Measurement observations* on participants should be *replicated* and *central tendency measures* used (i.e., mean, median, or mode).

  - This would reduce error variance by increasing the *reliability* of dependent variable measures.

    ∞ Recall that variable/random error *averages toward zero* the greater the number of measurements that are averaged together (a statistical point covered earlier).

  f. Number of participants/subjects could be increased.

  - Variable error also *averages toward zero* as *sample size* (and thus the number of measures) increases, and this is reflected in the following statistics:

    ∞ Standard error of the mean (σ_M or s_M)

    - The *standard error of the mean* is a measure of *variability* (or error) which indicates the amount that the mean obtained from a *sample* is likely to vary by *chance* from the true, *population mean.*

    - It is the *standard deviation* of a *distribution of sample means* (i.e., a sampling distribution of means), obtained from a *population* by taking repeated samples: $\sigma_M = \sigma / \sqrt{n}$ (where σ is the standard deviation of the population).

- σ_M can be *estimated* (s_M) from s, the standard deviation of a *single sample,* as follows: $s_M = s/\sqrt{n-1}$.

    Note from the formula that the standard error of the mean *decreases* with an *increase in the size of the sample (n),* and hence the amount of data that is used to determine a mean.

    Here *n-1*, as opposed to *n*, provides an *unbiased estimate,* i.e., a sample statistic whose *expected value* (long-run average) equals the value of the *population parameter* that is being estimated.

∞ Standard variance of the mean (σ_M^2 or s_M^2)

- The *standard variance of the mean* is similar to the preceding measure of *variability,* but it is for sampling *variance,* rather than standard deviation.

- It is the *variance* of a *distribution of sample means* obtained from a *population:* $\sigma_M^2 = \sigma^2/\sqrt{n}$ (where σ^2 is the variance of the population).

- σ_M^2 can be *estimated* (s_M^2) from s^2, the variance of a single sample, as follows: $s_M^2 = s^2/(n-1)$.

    Note from the formula that (as for the standard error of the mean) the standard variance of the mean *decreases* as the sample size *(n) increases* and approaches that of the population.

    Within-conditions or -treatments variance represents a *sample variance* (s²), and equals the *average* of the variances computed within all the different conditions.

    Therefore, an important point is that the *within-conditions variance* can be used to *estimate the sampling variance of the mean.*

    As noted, this variance would decrease (quite logically) as the sample size increases and approaches that of the population, and this would *reduce error* when using the dependent variable means obtained from *research samples* to estimate the *true means of populations* under the different independent variable conditions.

    Hence, as *n* and thus *df* (i.e., *n – 1*) increase, there is greater *reliability,* and therefore greater *statistical significance* (confidence) for any *differences* observed among the means *between conditions* in a study, and thus there is also greater *power/ sensitivity* to detect true effects (discussed again in Chapters 11 and 12).

g. *Within-participants/subjects or matched-participants/subjects designs* could be used.

- Such designs would reduce the error variance that is due to *extraneous participant factors,* by *statistically accounting for individual differences* among the research participants (see the earlier discussion of these control techniques, as well as the discussion of these two designs in Chapters 9 and 10).

h. *Change/difference scores* or *analysis* of *covariance* could be used.

- These are additional *statistical control techniques* for *individual differences* among participants (see Chapter 10 under "Between-Participants, Independent (Randomized), Two-Group Designs," as well as earlier in this chapter under "Pretesting Participants," for more about change/difference/gain scores).

4. **Error/Random Variance versus Secondary/Extraneous Variance**

— Some of the extraneous variance that today is considered to be *error/random/inconsistent variance* will eventually turn out to be *secondary/systematic/consistent variance*.

- The explanation for this *misclassification* has to do primarily with our current lack of knowledge, i.e., our *ignorance*, which exists for the following possible reasons:

  a) Not *wanting* to expend the necessary effort to *identify and precisely measure* certain extraneous variables and their effects, or simply not having *gotten around to it*.

  b) Not being able to *precisely enough measure* certain extraneous variables and their effects to determine whether they have systematic or inconsistent effects.

  - Examples: The entire genetic inheritance, complete nutritional status, or all the personal problems of research participants that might affect performance.

  c) Not being *able at all to measure* certain extraneous variables.

  - Example: Completely repressed traumatic events.

  d) Not being *able to even identify* some extraneous variables.

  - Sufficient research has not been done to know *all* the factors that might influence *every* dependent variable.

- *Misclassification of variance* is an important issue since, as we have seen, the *masking* effect of *secondary*, extraneous variables can be controlled through systematization and analysis of their effects, thereby *reducing the statistical error variance*.

- Considering the *concept of determinism* and the *Theory of Chaos* (both covered earlier in Chapter 1) we might question whether there is really any such thing as *chance*, since *apparent randomness* might itself be *determined*.

  ∞ Nevertheless, in the *real world*, we still must contend with the *Heisenberg Principle of Uncertainty* regarding the quantum mechanics of subatomic particles, and, of course, with the more general principle of *stochastic/probabilistic determinism*.

Review Summary

In addition to the material below, study the control/design techniques that have been employed in published research; i.e., read some actual studies, such as those assigned by your course instructor.

1. Six techniques for *making equivalent the secondary/extraneous variance* were discussed earlier in Chapter 7. An additional five techniques are covered in Chapter 8, and Appendix 5 of the *Handbook* summarizes them: (a) *randomizing* extraneous *variables*, (b) *pretesting participants for* extraneous *variables*, (c) *control groups for* extraneous *variables*, (d) *treatment groups and systematizing* extraneous *variables*, and (e) *Conservatively arranging* extraneous *variables*. As for the techniques covered earlier, the goal of all the control techniques covered here, except the last, is to have *equivalence of effects* of all the secondary, extraneous variables across all the independent variable conditions. Logically, this is accomplished through the *elimination of any differential influences* of secondary, extraneous variables.

2. *Randomizing extraneous variables* involves the process of *randomization*, which is any *procedure* (such as flipping a coin) that ensures that each event in some set of events (such as participant assignment to different conditions, and the order in which conditions are run) has an *equal probability/likelihood* of occurring. Randomization serves as a *control technique* that enhances the likelihood that secondary, extraneous variables which *cannot be systematically controlled* — e.g., because they are unknown or not readily measurable — will nevertheless be *equal in value and thus effect* for the different independent variables levels of a study.

3. *Block randomization* is a procedure used in the "*Randomized-Blocks Design*," where one participant factor is controlled using the systematic technique of *balancing*, after first *blocking* participants based on some shared and measurable characteristic (such as race), while other participant factors (such as past experiences) are controlled through *randomization*. *Matching*, which is an extended and more precise type of balancing, is a form of this procedure.

4. A number of *methods for randomizing* can be used, e.g.: flipping a coin, drawing slips of paper, shuffling 3×5 cards or test booklets, counting off, and using tables of random numbers. The latter involves tables of the digits 0–9 arranged in a random sequence. i.e., each digit equally likely to occur in any position in the table, and there is no systematic connection between the occurrence of any given digit in the sequence and any other. A *variety of procedures* are available for using *tables of random numbers*, some of which are easier and thus more efficient than others.

5. *Pretesting participants for extraneous variables* involves taking a *dependent variable measure prior* to the administration of an independent variable treatment, or control condition, against which a *posttest* taken *afterward* on the dependent variable is *compared*. Pretesting has several *functions*: a) *checking on the initial equivalence/comparability* of participants in the different conditions, and *adjusting* for differences found by computing *pretest-posttest change scores*; b) *enhancing*

sensitivity/power by *matching* participants based on their pretest scores and by *adjusting for ceiling or floor attenuation/truncation effects*; c) *determining a baseline or trend* to *control* for extraneous variables, and against which posttest scores can be *compared* to determine the effect of treatments. However, pretests can produce *direct effects* on posttest scores, as well as *indirect effects* by *interacting* with a treatment and producing *sensitization or resistance* to it.

6. *Control groups for extraneous variables* consist of one or more sets of participants that are *comparable* to the experimental treatment groups, and that are treated *exactly the same* except that they *do not receive an experimental treatment*. The intent is to *make equivalent the extraneous variables*, and particularly to hold constant extraneous experimenter factors. There are three *functions* of control groups:

   a. *Establishment of a baseline, or reference level,* against which to compare the dependent variable values of the treatment/experimental groups, and thus to evaluate the effects, if any, of manipulating the independent variables;

   b. *Controlling extraneous variable confounding* by incorporating the extraneous variables — e.g., *placebos* and the *passage of time* — into a control group when experimental manipulations unavoidably (and undesirably) involve the introduction of such secondary, extraneous variables right along with an independent variable;

   c. *Evaluation/measurement of extraneous variable effects* by, e.g., including a no-treatment-control in addition to a placebo control in order to separate out the effects of passage of time from a placebo.

7. *Treatment groups for extraneous variables* is another technique for *controlling extraneous variable confounding*. It involves adding one or more treatment groups that also incorporate the extraneous variable(s), and then comparing the effects of the different treatments. This also provides *additional data*. Moreover, through *systematization* of a secondary, extraneous variable by *balancing* it across independent variable conditions and then *analyzing* its effects, not only are its *influences evaluated*, but *masking is controlled* by removing the secondary variance from the statistical error variance (unaccounted for variance) and thereby enhancing detection of the primary variance (independent variable effects).

8. *Conservatively arranging extraneous variables* controls for them by arranging conditions so that the extraneous variable effects could only *reduce/weaken* (i.e., work against) the measured effect of the independent variable. Although this is often *simpler* than controlling for secondary, extraneous variables by making them equivalent for all conditions, the *primary variance is reduced* rather than maximized. Hence there is *less power/sensitivity* to detect an effect of the independent variable, and thus this is a *conservative approach*. For this reason, however, *confidence* in the research hypothesis would be even *higher* if the data supported it.

9. *Minimizing the error variance,* the third minimax principle, is also essential. Error/ random variance is *undesirable,* but cannot be completely eliminated (controlled), even though it is *self-canceling around a mean of zero* as the number of measures increases. Thus there is always some *uncertainty* in the interpretation of research findings — nothing is proven, only probabilistic. *Chance variations* can result in

dependent variable differences between conditions that are *mistaken* for independent variable effects. Alternatively, random variations can *mask* the effects of an independent variable.

10. Using a *communications engineering model*, the major task of experimentation is to *maximize the signal-to-noise ratio (S/N)*, i.e., the ratio of the primary variance (PV) to the error variance (EV) plus secondary variance (SV). Since it is not possible to directly measure a signal isolated from noise, it is the size of the *(S + N)/N ratio* that is used to evaluate the likelihood that a signal actually exists, i.e., that there is *primary variance* — an independent variable effect.

Note that:

$$(S + N)/N \equiv (PV + EV)/EV \equiv (PV + EV \leftarrow SV?)/(EV \leftarrow SV?) = BV/WV$$

(BV is *between-conditions variance,* and WV is *within-conditions variance.* Secondary variance can be mixed with primary variance if not controlled for confounding, or combined with error variance if not controlled for masking.)

11. *Statistical error variance* differs from the research concept of error variance. It is the *unaccounted for variability* in the dependent variable that is *left over* after all known-and-analyzed sources of systematic variance have been subtracted from the total variance. *Sources of systematic variance* include any effective primary, independent variable(s); the secondary, extraneous variables; and also any interactions among these variables. Thus statistical error variance is reduced when secondary, extraneous variables are *systematized* and their effects *analyzed* — which controls for *masking* in addition to *confounding.*

12. *Statistical tests of significance* measure the probability that observed differences between conditions are due to *chance alone,* i.e., *sampling error.* Conversely, they determine a *degree of confidence* that the sampled data reflect the true population differences. These *inferential statistics* usually involve a *ratio* such as in the *F*-test = $MS_{\text{treatments}}/MS_{\text{error}} \equiv BV/WV$. The *size of the ratio* provides a measure of the *likelihood/reliability* of treatment effects, rather than the *magnitude/strength* of effects.

Statistical error variance (noise) measured alone *within conditions* serves as an *estimate* of the statistical error variance that occurs along with any possible *primary variance* (signal) measured *between conditions.* The *larger the ratio* the greater the probability that there actually is primary variance — *a treatment effect* — in addition to error variance, i.e., chance variation.

13. *Research hypotheses* are tested against *null hypotheses* using statistical tests of significance. There are two sets of requirements for *maximizing reliability and validity of test data*:

   a. *Before treatments are administered,* both the between-conditions variance and the within-conditions variance should be minimized, thus assuring the least error and secondary variance.

   b. *After treatments are administered,* it is then desired that the between- conditions variance should be maximal, while the within-conditions variance should remain minimal, thus maximizing the

$(S + N)/N$ ratio $\equiv (PV + EV)/EV \equiv (PV+EV \leftarrow SV?)/(EV \leftarrow SV?) = BV/WV.$

14. A number of techniques for minimizing error variance are:
    a. Reduce individual participant differences
    b. Eliminate, hold constant, or minimize randomly operating environmental factors
    c. Maintain consistent experimental procedures
    d. Use reliable measurement and analysis techniques and apparatus
    e. Replicate measurements on participants and use central tendency
    f. Increase the number of participants
    g. Use within-participants or matched designs
    h. Use change/difference/gain scores, or analysis of covariance

15. Some extraneous variance that today is considered to be *error/random/ inconsistent variance* will some day actually turn out to be *secondary/systematic/consistent variance*. Such *misclassification* is primarily due to our present lack of knowledge. Indeed, one might question whether there is really any such thing as *chance*. This is an important issue since it is possible to control for the *masking* effect of *secondary*, extraneous variables through systematization and analysis of their variance effects, thereby reducing the statistical error variance.

Review Questions

In addition to answering the following questions, also complete any control/design critique problems assigned by your course instructor; e.g., those contained in handouts or in books that you purchased or that are in the Library.

1. State the *goal* of all the *control techniques for secondary, extraneous variables*, with one exception, and state how this goal is logically *accomplished.*

2. Define *randomization* and discuss what kinds of extraneous variables randomization is particularly useful for as a control technique, and why.

3. Describe the *block-randomization procedure,* name the *design* it is used in, and state the *control technique* it is related to other than randomization.

4. Describe *tables of random numbers* and a procedure for their use that is *efficient* and that should not result in the *assignment of unequal numbers of participants* to different independent variable conditions.

5. Describe the *pretesting technique,* list three of its functions, and state the problems that can result from using it.

6. Describe *control groups* and explain their three specific *functions.*

7. Explain the control technique and three functions of using *treatment groups and systematizing extraneous variables.*

8. Describe the control technique of *conservatively arranging an extraneous variable,* and list its weakness and two strengths or advantages.

9. State two reasons that *error variance* must be minimized.

10. Using a *communications engineering model,* the major task of experimentation is to do what with regard to what ratio? How would this be stated in terms of primary, secondary, and error variance?

11. What is *statistical error variance* and how can it be reduced?

12. Explain what *statistical tests of significance* measure. Also, describe the components of the *ratio* that they usually involve, and the implications of the ratio.

13. *Research hypotheses* are tested against *null hypotheses* using statistical tests of significance. What are the requirements regarding *variance* before treatments versus after treatments in order to *maximize reliability and validity of test data.*

14. List as many of the *techniques for minimizing error variance* as you can (there are about eight).

15. Discuss the possible *misclassification of error and secondary variance,* and explain why is this an important issue.

Part Three

Design and Analysis
of Experiments

9

Designs for Experiments —
Introduction and Within-Participants

CONTENTS

Research design is an extensive and important topic, and thus *two chapters* are devoted to it. *In this chapter* the general concepts, principles, and categories of "Experimental and Quasi-Experimental Designs" will be discussed, followed by detailed coverage of "Within-Participants/Subjects Designs." *The next chapter* covers both the "Between-Participants and -Subjects Designs" and the "Multiple Treatment Designs." *Within-participants designs* are discussed first because *between-participants designs* then can be viewed as expansions of these designs, providing for better control of many extraneous variables, although poorer control of individual differences among participants.

General Concepts and Principles

This chapter begins with the basics, which include the *definition* of research designs and their *components*, the *determinants for selecting* a particular design, the *evaluation criteria*, and the conventions that are used in this Handbook for *naming* research designs. It should be noted that some of the research design components outlined below are *general features* of all research designs, whereas other components *distinguish* specific designs from one another.

1. **Definition of Research Designs**

   — A research design is a *basic plan* for scientific investigation.

   • In essence, designs are *strategies* for trying to answer research questions, i.e., *solve* problems by *testing* research hypotheses.

- There are *non-experimental* as well as *experimental* designs (the distinction between "Descriptive Versus Explanatory/Experimental Research" was covered in Chapter 2).

2. Components of Experimental Research Designs

There are *six major components* of experimental research designs, with subparts to each. These are associated with *interrelated decisions* that need to be made when designing an experiment. Because the components are discussed extensively in other chapters of this Handbook, they are presented here just in the form of a detailed overview list.

a. Independent variable specifications:

- Number and choice of the *variables* to be studied;
- Number and choice of the *levels* (values) of each variable;
- Method of *selection* of the levels, i.e., systematic (a priori) versus random (see Chapter 10 under "Factorial Designs");
- Operational means, or procedures, for the purposeful and direct *manipulation* of the variables to *produce* the levels (See Chapter 4, "Variables in Research Designs").

b. Dependent variable specifications:

- Number and choice of the *variables* to be studied;
- *Scale/level* of measurement — nominal, ordinal, interval, or ratio;
- Operational means for the *measurement* of the variables (See Chapter 4, "Variables in Research Designs").

c. Extraneous variable specifications:

- Identification of the *likely variables*;
- Distinction between *systematic versus random* variance sources;
- Selection of corresponding and appropriate *control techniques* (see Chapters 7 and 8, "Control in Experiments").

d. Ethical specifications:

- Identification of *applicable ethical principles* relating to obtaining, informing, and treating research participants (see Chapter 6, "Ethics of Research," and also Appendix 4);
- Determination of whether all ethical principles will be *satisfied*;
- Justification for any principles *not satisfied*, specifically:
  - ∞ Lack of *alternative* research procedures that would avoid unethical practices while still achieving the desired ends;
  - ∞ *Cost-benefit analysis* that weighs the potential harm to participants, society, and science against the potential scientific, educational, and applied value of the study; and which finds that the *potential value* of the research is *sufficiently greater* than the *potential harm*.

e. Participant specifications:

- Choice of the *population(s)* to be studied;

- Size of *sample(s)* to be selected for sufficient statistical power;
- Method of *sampling* from the populations (see Chapter 11, "Sampling and Generalization");
- Method of *assigning* participants to independent variable levels, i.e, randomizing, matching, or placing the same individuals in all levels (note that within a single experiment the assignment method can differ for the different independent variables — see Chapter 10 under "Factorial Designs").

f. Statistical Analysis Specifications:

- Selection of the *appropriate analysis(es)*, taking into account the preceding components of designs, as well as the following —
  - ∞ Fulfillment of the various *assumptions* of the analyses;
  - ∞ Maximization of the *power/sensitivity* to detect any effects (relates to many aspects of research designs; see, e.g., Chapter 12, "Hypothesis Testing and Statistical Significance").
- Choice of the statistical significance level(s) to be used;
  - ∞ Influenced by the consequences of making *Type I errors versus Type II errors* (see discussion in Chapter 12).

3. Determinants for Selecting a Research Design

There are *three determinants* involved in selecting a research design.

a. *Problem/question(s)* being asked and the *hypotheses* being tested:

- These influence the *independent and dependent variable* considerations already noted, e.g., number and choice of the independent and dependent variables to be studied.

b. *Extraneous variable controls* and *statistical power* that are needed for obtaining reliable and valid results:

- These influence, e.g., whether the investigator will use a *within- participants versus a between-participants design* (and for the latter, whether there will be *matched or random assignment*), and what the *size of the samples* will be.

c. *Generalizability* of results that is desired with respect to individuals, independent variable values, extraneous variable conditions, etc. (see Chapter 11, "Sampling and Generalization"):

- This influences, e.g., how independent variable values will be *selected* for study, i.e., *randomly versus systematically* (see Chapter 10 under "Factorial Designs").

4. Criteria for Evaluating a Research Design

Three criteria relate logically to the preceding three *determinants* of design selection.

a. Conclusions possible

- How well can the design *solve the research problem(s)*, i.e., how well can it *answer the specific question(s)* that are asked?
- How well can the design *test the research hypothesis(es)*?

- Note: Answers to both of these questions are contingent, in part, on the *statistical tests* that can be performed on the data gathered, which is a function of the particular design used.
    - ∞ Important considerations would be having enough *power/ sensitivity* in the design and analysis to detect any effects, and also meeting the *assumptions* of the statistical test(s).

b. Internal validity[8]

- This criterion involves the question: With what *confidence* can a *cause-and-effect relationship* be inferred between manipulations of the independent variable(s) and changes in the dependent variable(s), *within the specific confines* (participants and conditions, etc.) of the given *experiment?*
    - ∞ Answers to this question are contingent upon the extent to which *confounding extraneous variables* are controlled.
        - Extraneous variables represent *alternative explanations,* i.e., *rival hypotheses* with respect to a research hypothesis, and therefore they must be *controlled.*
        - Note that *statistical significance* only indicates that the results are unlikely to be due to *chance alone,* not that the dependent variable differences are actually due to the *independent variable* studied (see Chapter 12).
            - *If controls are inadequate,* dependent variable differences could be due to *extraneous variables* that have *consistent effects* and that happen to *vary systematically* with the independent variable, i.e., that are *confounded* with it.
    - ∞ Answers to this question are also contingent upon the extent to which *extraneous variables* with *inconsistent effects* are controlled, since these can *mask/hide* the effects of an independent variable (Type II error), or alternatively their influences can be *mistaken* for independent variable effects (Type I error).
- Note: Internal validity is the *sine qua non* of any research design (Latin for "without which not"), i.e., it is something that is *absolutely indispensable.*
    - ∞ Internal validity is *essential* for a design to have *any usefulness* in that clearly there must be confidence that the dependent variable results of a study are due to the *effects of the independent variable(s),* and not extraneous variables, before even considering *generalizability* of results (see next).

c. External validity[8]

- This criterion involves the question: To what extent can the experimental results be *generalized beyond the confines* of the given specific experiment that was conducted?
    - ∞ Answers to this question are contingent upon the *representativeness* of the experiment with respect to the *target* participant populations,

settings, procedures, times, independent variable levels, and dependent variable measures (discussed in Chapter 10I, "Sampling and Generalization").

∞ Generalizability of results relates to *inductive inference,* i.e., drawing conclusions from the specific to the general, which is *never an absolute certainty.*

- Note that while it is *desirable* to select designs which are strong in *both types of validity,* unfortunately, features that increase one form of validity can *jeopardize* the other form of validity.

   ∞ Example: *Pretesting,* which is designed to *increase internal validity* by controlling for individual participant differences, lamentably can decrease *external validity* by producing *sensitivity or resistance* to the treatment(s) of the study.

      - In fact, *pretesting* can even *decrease internal validity* by producing a *direct,* rather than indirect, effect on the posttest data — but this can be controlled through the use of *multiple* pretests (see discussion later in this chapter).

5. Names of Research Designs

— Most designs, as it turns out, have *more than one name,* and the term "participants" is now often used instead of the term "subjects."

— For each of the designs discussed in this Handbook, the most *descriptive name* is listed first, and *alternative names* follow in parentheses.

- *Comparability* of naming is strived for so that *relationships* among the designs are clearer.

- Note: The designs covered in this and the next chapter, although numerous, are *not all-inclusive;* they were selected because they are common and/or they have instructional value (Appendix 7 is a summary of all the designs in this Handbook).

Pre-Experimental Design

This is an example of *what not to do* in research!

One-Group Posttest-Only Design (One-Group After-Only Design) (One-Shot Case Study)

X   O

— Description

- *One group or individual* (indicated in the schematic by the presence of only a single line) receives a *treatment* (X) and is then *observed* (O) on the dependent variable — thus, there is no pretest, only a *posttest* in this design.

— Typical uses

- Studies on the effect of some *clinical therapy/treatment* or some *educational tool/process* have used this inadequate design.

  ∞ Example: *Acupuncture* might be administered to a group of individuals, who then report any reduction in the intensity of *pain* they experience.

- *TV commercials, newspaper ads, and political claims* have also reflected use of this design, even if no research was intended.

  ∞ Examples: (1) A man gives one or more *diamonds* to a woman, and is then shown receiving *affection* from her;

  (2) A *politician* gets elected, and later there is a *drop in crime*, for which he/she claims credit.

— Weaknesses

This design has two basic weaknesses:

1) *Baseline*, or reference level data, is *lacking*.

   ∞ *Explicit data comparison* is crucial for determining causality.

   - *To measure the effect of a treatment*, you must also establish the behavior that occurs *without* the treatment, or with some *other level* (quality/quantity) of treatment.

   - In this design, there is only *implicit* comparison with other events that were probably just *casually* observed.

2) *Rival hypotheses*, or alternative explanations, are very plausible.

   ∞ *Extraneous variables* of several types (depending on the specific study) are uncontrolled for *confounding*, such as: environment, maturation, selection, and participant bias (e.g., a placebo effect).

— Conclusions

- *Causality* cannot be determined using this design.

  ∞ Thus it does not even qualify as being *quasi-experimental* (discussed below), and hence it is a *pre-experimental design*.

- *Value of this design* is mainly that it can *generate hypotheses* about causality that might be testable at a future date using *better research designs*.

  ∞ In addition, this design has didactic (educational) value, i.e., *instruction* is provided as to *what not to do and why not*.

Quasi-Experimental Designs

When researchers *cannot use experimental designs*, they can turn to quasi- experimental designs, rather than pre-experimental, as the *next best thing*.

1. Description

— *Strictest requirements* for experimental control are *not met* by quasi- experimental designs:

- *Random assignment* of participants to conditions is lacking.

- *Extraneous variables* of types in addition to individual differences among participants also are often not well controlled, e.g., contemporary history.

— But quasi-experimental designs are *better for determining causality* than the naturalistic observation techniques of *descriptive research* (which involve pre-experimental designs) for the following reasons:

- *Independent variables* are purposely and directly manipulated.

- *Pretest-posttest comparisons* are often made to control for the extraneous variables of individual differences among participants.

- *Other forms of control* also are used when possible (see later).

— Thus these designs fall somewhere *between* true experimental designs (which have excellent controls) and non-/pre-experimental designs (such as those used in descriptive research) — hence they are *quasi*-experimental.

- Example: Students in one class of an elementary school might be given an *extra recess* each day, and the number of *disruptive incidents* per day by students in that class could be compared with the number for *another class* of students who did not get the extra recess. (For a discussion of the control problems related to non-random assignment of participants to conditions, see "Nonequivalent Posttest-Only Control Group Design" in Chapter 10).

2. Causal Inferences

— *Causality* can be inferred from quasi-experimental designs to the extent that *rival hypotheses* are rendered *implausible* through the use of *controls* or *logic*.

- Incorporation of as many principles of *scientific control* as possible, under the given circumstances, is important in order to *minimize confounding* by the extraneous variables of rival hypotheses.

— *Replication* of research can be used to strengthen confidence in a causal relationship between independent and dependent variables, if *extraneous variables* do not remain constant across the replications.

- Logically, if the measured *relationship* between an independent variable and a dependent variable remains *consistent*, while the extraneous variables are *not consistent*, then the *independent variable*, rather than *confounding extraneous variables*, is likely to be responsible for the observed *dependent variable effects*.

3. Applications

— Quasi-experimental designs are useful for *exploratory or pilot studies*, and for investigating research questions when *some control* over independent and extraneous variables is feasible, but where designs incorporating *optimal* experimental control *cannot* be used — as in the following situations:

- *Undesired reactive measurement and artificiality* might result from the application of optimal experimental controls, e.g., in a lab, which would adversely affect *external validity* (generalizability).

- *Natural/field settings* (such as the work place, schools, and hospitals) might not permit the use of maximum control (e.g., random assignment of participants to conditions might not be possible in such settings when there are *pre-established collectives* of individuals that should not be disrupted).

- *Unplanned events* (such as natural disasters or war) do not allow time to set up maximum controls beforehand — but note that only *non-/pre-experimental designs* are usually possible in such cases, unless, e.g., there happened to be "pretest" data available to control for individual differences among participants.

- *Ethical constraints* (such as prohibitions against physical and psychological stress and harm) might make controlled experimental manipulation of certain independent variables unacceptable with human participants — but again, only *non-/pre- experimental designs* are usually possible in such cases.

Categories of Experimental and Quasi-Experimental Designs (See Appendices 6 and 7)

Experimental and quasi-experimental research designs can be divided into *within-participants designs* versus *between-participants designs*. An alternative, but overlapping, categorizing scheme involves *correlated-groups designs* versus *independent-groups designs*. These categories are in turn incorporated into the forms of the more complex *multiple treatment designs*. (See Appendix 15 for a table of statistical tests for different designs and levels of measurement.)

1. Within-Participants/Groups (Repeated-Measures) Designs

— Dependent variable measures in these designs are taken on exactly the *same participants/subjects* under each of the various levels of an independent variable — one of which might be zero or some standard value serving as a control condition.

— Comparisons are thus made *within* groups of participants by taking *repeated measures* on them (i.e., at least once under each condition).

- These designs can consist of either *one group or several groups,* and when there is more than one group, each group of participants receives a *different sequence* of all the conditions.

— Correlated-groups designs

- These designs include the *within-participants designs,* as well as others (see below).
  - ∞ Since the participants in the different conditions (groups) are the *same,* the dependent variable measures for individuals under the different conditions should be *correlated.*

2. Between-Participants/Groups Designs

— Dependent variable measures in these designs are taken on *different participants* under the various levels/values of an independent variable — one of which might be zero or some standard value serving as a no-treatment control condition.

— Comparisons are thus made *between* groups of participants/subjects.

— Subcategories:

1) Independent-groups designs

   - ∞ *Randomly or nonrandomly assigned participants* are used in the different groups — but in either case they are *unmatched.*

2) Related-groups designs

   - ∞ *Matched participants* are used in the different groups.

— Correlated-groups designs

- These include the *related-groups/matched-participants designs,* as well as the previously noted *within-participants designs.*
  - ∞ Since the participants in the different conditions (groups) are *matched, or identical,* the dependent variable measures for individuals under the different conditions should be *correlated.*

- As already pointed out, these designs are contrasted with the *independent-groups designs* (see Appendix 6).

3. Multiple Treatment Designs

— General forms:

1) Multilevel designs

   - ∞ These are either within-participants or between-participants designs, but with the additional characteristic that *more than two levels* of an independent variable are investigated.

2) Multifactor/factorial designs

   - ∞ These incorporate all the preceding categories of designs (including multilevel), with the additional characteristic that *more than one independent variable* (factor) is investigated.

   - ∞ Mixed designs

     - This is a type of factorial design that combines in a single study at least one independent variable whose levels are compared using *correlated groups,* and at least one other independent variable whose levels are compared using *independent groups.*

Within-Participants/Groups (Repeated-Measures) Designs

In these designs, dependent variable measures are taken on the *same participants/subjects* under all the levels of an independent variable — one of which might be zero or some standard value serving as a control condition. Comparisons are thus made *within* groups of participants by taking *repeated measures* on them (i.e., at least once under each condition). Hence every participant serves as his or her *own control* for individual differences, and *fewer participants* are needed to obtain a specific amount of data.

All the following design examples are for when there is only *one independent variable,* and just *two levels* (such as presence versus absence of a treatment). The designs are sequenced from the weakest to the strongest, and thus *quasi-experimental designs,* with their deficiencies, are presented first so that later the strengths/advantages of *experimental designs* can be better appreciated. (Appendix 7 is a summary of all the designs in this Handbook.)

1. *One-Group Pretest-Posttest Design* **(One-Group Before-After Design)**

$$O \qquad X \qquad O$$

— Description

- A *single group of participants* is observed (O) at some time *before* the onset of a treatment (X), and then these dependent variable scores are compared with those later observed (O) *during or after* the treatment.

    ∞ This design thus provides a direct measure, for each research participant, of the *change in behavior* associated with the *absence versus presence* of a treatment condition (X).

— Typical uses

- *Educational research* has utilized this design; e.g., to determine the effectiveness of a new remedial reading program for those students that are having reading difficulties.

- *Therapy research* has also used this design; e.g., to measure the effectiveness of some form of psychoanalysis for individuals with a particular mental disorder.

- *Advertising and propaganda research* has employed this design, too; e.g., to evaluate the effectiveness of ads for changing opinions about some product, politician, or political option.

— Analysis

- ANOVA (analysis of variance) or a *t*-test for related/correlated measures would be used when score data are obtained.

— Strength/advantage over one-group posttest-only design

- A *pretreatment baseline,* or reference level, is obtained for each research participant through *pretesting,* against which the posttreatment observation can be *compared* in order to evaluate whether there is a treatment effect.

— Weaknesses

There are many *uncontrolled secondary, extraneous variables* that are possible when using this design. (The extraneous variables and controls were discussed in Chapters 7 and 8).

1) Contemporary history

∞ Between the pretest and the posttest, *external environmental events* that could influence the dependent variable might occur, in addition to the experimental treatment.

- Example: The measurement of *attitudes toward capital punishment* (a dependent variable), before and then after a documentary film is broadcast on TV about death row (an independent variable), would likely be *altered* considerably by a rash of murders publicized in the news during the time between *the pretest and the posttest* (hence a contemporary history extraneous variable).

∞ Solutions

- *Shorter the interval* between observations, the less likely would be confounding by the secondary, extraneous variable of contemporary history.

- *Isolation of participants* from extraneous environmental stimuli during the pretest-posttest interval also would be a very desirable control procedure, *when it is practical.*

- *Different non-overlapping pretest-posttest periods in time* for running the different participants would be another means for minimizing the impact of any contemporary history extraneous variable events.

2) Maturation

∞ *Conditions internal to individuals* that change as a function of the passage of time, and which influence the dependent variable, are additional secondary, extraneous variables that are more likely the *longer the pretest-posttest interval.*

- *Circadian biorhythms* (~24 hr cycles) occur in our physiology (e.g., hormones) and behavior (e.g., alertness) and therefore even differences in the *time of day* that pretest versus posttest observations are made could be an important secondary, extraneous variable.

3) Participant attrition/experimental mortality

∞ *Loss of research participants* after an experiment has begun is yet another form of experimental contamination that is likely to increase with the *pretest-posttest interval.*

∞ A *non-representative subset* of the participant sample might be lost, rather than a *random, representative subset.*

- As a result, a *biased/non-representative sample* would be *left*, and this could influence the dependent variable data and make their interpretation difficult, if not impossible (see Chapter 10I, "Sampling and Generalization").

  > Example: Imagine the situation where mainly the *low-motivation or poor-ability participants* are lost. As a direct result of this, a *pretest-posttest change* in the means for the dependent variable scores might be observed whether there was any actual *effect* of the treatment, or at the very least this attrition could *influence* the observed effect of the treatment.

- *Internal validity,* which would be damaged by any *direct effect* of attrition on the change in dependent variable scores, *would not be a concern* so long as the analysis of the study appropriately did not include the *pretest data* of those who dropped out before the *posttest* was run.

- *External validity* (generalizability) would still be a concern, however, if a *non-representative subset* of the participant sample were lost, thus leaving a *biased sample,* since participant characteristics could have an *indirect effect* on the change in dependent variable scores by *influencing* the degree of treatment effectiveness.

4) Testing

- The process of collecting the *pretreatment* data might itself influence the *posttreatment* dependent variable scores.

- This *effect of testing* upon the scores of a *subsequent test* is similar to the sequence/order extraneous variable effects (practice, fatigue, attention, and motivation) that can occur when individuals participate in a study under more than one *condition* for control purposes.

- Direct effect

  > *Internal validity problems* occur when a pretest *directly* affects the posttest dependent variable scores, thus being *confounded* with the treatment.

- Indirect effect

  > *External validity problems* occur when a pretest *indirectly* affects the posttest dependent variable scores, and thus *generalizability,* by producing *sensitivity or resistance* to the treatment — which would be an interaction effect between testing and treatment (see Chapter 10 under "Factorial Designs").

- Example: In an *attitude-change study about racism,* a pretest might *influence* a reading-material treatment effect by causing sensitivity or resistance to the material, an *indirect effect* that impacts the external validity, or *generalizability,* of the findings, or the pretest might *directly* affect the posttest attitudes, thus producing an internal validity problem of confounding.

∞ Note: The *most serious shortcoming* of this research design is probably this extraneous variable of *testing.*

- *Other confounding variables* can be *controlled* to some degree, e.g., by keeping the pretest-posttest interval short and isolating participants from extraneous stimuli.

5) Participant bias

∞ *Expectancy of the participant* could lead to *placebo or Hawthorne effects,* whereby change would occur regardless of a specific treatment's actual effectiveness.

- Note that *in a one-group design* there cannot be a placebo or no-treatment control group for extraneous variables, yet some other *deception or concealment* might be possible as a participant-bias control.

6) Experimenter bias

∞ *Expectancy of the experimenter* is also a potential control problem, but this is true for *all designs,* except when investigators can be kept *blind,* or at least partially blind, as to the condition a research participant receives.

7) Instrumentation

∞ Changes over time in the calibration and operation of *measurement and recording equipment,* as well as changes in the skill, attention, and criteria of *human observers,* can produce *experimental contamination* leading to pretest- posttest differences.

∞ To minimize instrumentation extraneous variable problems:

- Equipment should be checked and calibrated frequently;
- Observers should be well practiced before they begin gathering the data that will be used for analysis;
- Test measurements should be spaced to minimize any possible fatigue of the observers.

8) Statistical regression toward the mean

∞ If participants are selected for study on the basis of their having *extremely high or low scores on a pretest,* then changes for posttest scores could simply be an *artifact* of the sample *statistically regressing* (i.e., moving) toward the *mean value of the population* on the second test — regardless of any intervening treatment.

∞ Explanation

- *Random* imperfections of the *test and/or apparatus* used for measurement are always present and cause *variation in observed participant performance,* i.e., test measurements are not perfectly reliable or correlated.

Note: The more extreme/deviant the participants' scores are from the population mean, the greater the probable *measurement error*, and thus the more likely the scores are to be different the next time.

- *Random* instability/variation of *individuals* over time leads to some participants having extreme scores on a first test for reasons that are not present to the same degree at the time of a second test (e.g., there might be changes in motivation, alertness, nutrition, health, and/or luck).

- It follows that participants who have initial/pretest scores that are *extreme*, at least partially for either of the two general reasons just given, will be *statistically likely* to have scores closer to their *true score*, and thus the *group mean*, on a subsequent posttest. This is due partly to *ceiling or floor truncation effects*, which restrict chance variation in one direction.

    Thus, participants who score at the *low end* of the scale on the first test will have, on average, a higher score on the second test; whereas participants who score at the *high end* of the scale on the first test will have, on average, a lower score on the second test; and participants who score near the *mean* on the first test will have, on average, the same score on the second test.

- Example: Studies of the effectiveness of *remedial reading programs* for students scoring *very low* on an initial test are likely to find *improved scores* for these students on a subsequent test of reading, regardless of whether there is any effect of the training, but due simply to statistical regression toward the mean.

— Conclusions

- *Quasi-experimental* is the best we can say for this design, with some researchers considering it to be just *pre-experimental*.

    ∞ There are numerous poorly controlled *extraneous variables* associated with this design, which can lead to *confounding*.

    - These represent *rival hypotheses*, i.e., alternative explanations to that of the independent variable itself having effects on the dependent variable.

    - Confidence in any *apparent* treatment effects would be dependent upon the *degree of control* exerted over the possible confounding secondary, extraneous variables.

- *Use* of this design should be *only* if there is no better approach (see the following research designs).

- *Caution* should be used when *interpreting* results of this design.

- *Replication* under varying conditions should be used to increase *confidence* in the results.

2. *Time-Series Design* (Interrupted Time-Series Design)

O   O   O . . .   X   O   O   O . . .

— Description

- *Multiple observations* (O) are made before treatment (X) on a single group of participants, and the dependent variable scores are compared with those from multiple observations (O) made during or after treatment — hence this design uses a *time-series* of observations that is *interrupted* by treatment presentation.

  ∞ This design is an *extension* of the one-group pretest-posttest design, which was just discussed.

  ∞ Importantly, a *baseline* or *trend* is established by the *pretreatment multiple observations* (there should be *at least three* in order to distinguish linear from non-linear trends).

    - Note: Observations should occur at fixed (equal) time intervals in order to facilitate a *trend analysis*.

  ∞ *Discontinuity,* or change, in the observations obtained after the onset of treatment indicates that an effect has occurred (examples below are illustrated in Figure 9.1).

    - *Shifts from a stable baseline*, i.e, an increase or decrease in dependent variable scores, represent one form of discontinuity.

        Examples are "A" and "B" in the following Figure 9.1: "A" illustrates *long-term effects*, whereas "B" illustrates *short-term effects* — such as in studies of the effectiveness of a drug therapy.

    - *Changes in the trend* of scores that are already increasing or decreasing before a treatment is another type of discontinuity; i.e., a *slope change* or an *intercept shift* (a jump or fall in scores).

        Examples are "C" and "E" in the Figure 9.1: "C" illustrates an *intercept shift,* such as a cognitive development spurt following the removal of children from abusive homes; whereas "E" illustrates a *change in slope,* such as a drug's effect in slowing the progress of some disease.

        Both an intercept shift *and* a change in slope could occur *together* in a study, but are not illustrated here.

        Note that examples "F," "G," and "H" in Figure 9.1 illustrate *no change in trend*, yet the treatment would have *appeared* to have had an effect if there had been only one pretest and one posttest (all posttest observations for "H" aren't plotted in the figure because they would have intersected with the plot for "G").

— Requirements for using this design:

1) The *temporal interval* (or number of observations) expected between the treatment introduction and the resulting effect(s) must be *specified in advance*, i.e., *predicted*.

   ∞ If not, then the design would be invalid when a change is *delayed* after a treatment, since the change could be just as likely the result of *extraneous variables (chance)* as it could be the result of the treatment.

   • Note that extraneous variable effects would be more likely to occur the *longer the interval* between the treatment and the observed change.

   • *Opportunistic capitalization* on chance extraneous variables, which *flexible timing* allows, would make any approach to the testing of statistical significance *invalid*.

        Example is "D" in Figure 9.1: Only if the delay in change were predicted, such as in studies involving drug effects, could the delayed effect illustrated here be almost as definitive as the more immediate treatment outcome illustrated in example "A".

        Pilot studies, of course, can be used to establish the predicted delay.

   ∞ *Cross-validation/replication* in order to test *causality* would be required for results involving *unpredicted delays* in treatment effects, such as can occur in *exploratory studies*.

2) *Treatments* must be *specified* before examining the outcome.

   ∞ *Post-hoc examination* of time series data to infer what event or factor was responsible for the most dramatic shift in behavior, as is done in *descriptive research,* does not involve *experimental control* — hence other, *extraneous* variables could as likely be the cause of change as the *more apparent factor.*

   • *Example: Looking for the cause(s) of intermittent increases in the incidence of* assaults *or* suicides *by trying to find, after the fact, what event(s) occurred just before the increases.*

   ∞ *Cross-validation/replication* to test *causality* would be required when treatments are not specified in advance, such as when *exploratory investigations* are conducted, and wherever possible the treatment should be *purposely and directly manipulated/produced* for better control.

— Analysis

   • ANOVA (analysis of variance) or a *t*-test for related/correlated measures would be used when score data are obtained.

   ∞ *Averaged* pretest versus posttest scores could be compared when expecting a *stable baseline and stable treatment effect.*

   • Examples are "A" and "D", but with appropriate adjustment of which *posttest* scores are averaged if a *delay* in treatment effect is *predicted*, such as for example "D".

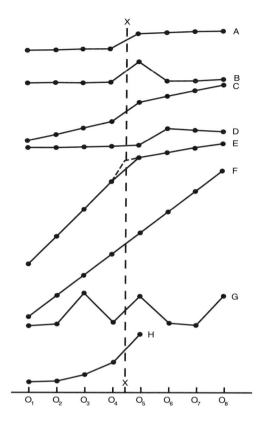

FIGURE 9.1

Some possible outcome patterns from the introduction of an experimental variable at point X into a *time series of measurements*, O_1–O_8. Except for D, the O_4–O_5 gain is the same for all time series, while the legitimacy of inferring an effect varies widely, being strongest in A and B, and totally unjustified in F, G, and H. [From *Experimental and Quasi-Experimental Designs for Research* (p. 38), by D.T. Campbell and J.C. Stanley, 1963, Chicago, Ill., Rand McNally and Co. Copyright 1966 by Houghton Mifflin College. Reprinted with permission.]

∞ *Representative* pretest versus posttest scores could be compared when expecting a *stable baseline but not a stable treatment effect*, i.e., when a short-term effect is predicted

- Example is "B", where only the first posttest score would be analyzed if, before the data are collected, the treatment effect is *predicted* to be so *short lived*.

- Trend analysis

∞ An ANOVA trend analysis would be run whenever there is *no stable baseline*.

- Examples are "C," "E," "F," "G "and "H."

- Bayesian moving average model

∞ This is a *widely used test*, but a very large number of data points are needed (perhaps 50 or more) to minimize the probability of a Type I error (see Chapter 12 for error types).

— Strengths and advantages over one-group pretest-posttest design

1) *Sampling error* can be *reduced* since *multiple* pretreatment and posttreatment measures are obtained from the participants, and *averaging* these multiple measures *controls* for *erroneous sample scores* because *random error* averages toward zero.

    ∞ *Error variance* is thereby *reduced*, and this leads to *increased sensitivity/power* to detect treatment effects.

2) *Stable pretreatment baseline or a trend* is determined by making *multiple* pretests, against which posttreatment observations can be compared — which is the *greatest strength/advantage* of this design since it *controls for several extraneous variables*.

    a) Statistical regression toward the mean

        • This is controlled in the design by the determination of *pretreatment trends*, as well as by the reduced sampling error that occurs when multiple measures are averaged.

    b) Testing

        • This is controlled for *internal validity*, since *direct effects* of tests upon the scores of subsequent tests should be reflected in the *pretreatment trend* (the external validity concern relating to indirect effects is discussed later).

            Examples: Direct effects of *practice or fatigue* that might occur as a result of pretesting.

    c) Instrumentation

        • This is likewise controlled, since it is unlikely that *equipment* changes would occur only around the time of the treatment, as opposed to also being manifested during the *pretreatment observation series*.

        • *Experimenter characteristics* such as performance changes, but not experimenter bias effects, would also be controlled by determination of any *pretreatment trend*.

    d) Maturation

        • This is similarly controlled by the *pretreatment trend* data, and the control is facilitated for this and other extraneous variables by holding constant the temporal interval between observations (as was noted earlier).

— Disadvantage

    • *Extra time and effort* are needed for making the *multiple* pretests and posttests that are part of this design.

    • Note: In this and other designs, *disadvantages* are meant to be distinguished from *weaknesses* in that disadvantages do not affect *validity*, whereas weaknesses *do* affect validity.

— Weaknesses

> There are several *uncontrolled secondary, extraneous variables* that are possible when using this design.

1) Contemporary history

> ∞ This is the *major concern* of this design, since external environmental factors that influence the dependent variable could happen to just occur *specifically* between the last pretreatment observation and the first posttreatment observation, and not also earlier.

> ∞ *Confidence* that any dependent variable results of the study indicate an independent variable effect would rest upon the *plausibility of ruling out* for that study the contemporary history extraneous variable.

> • *Shorter the interval* between the pretest and posttest observations, the less likely this problem would be (this and other points were noted for the previous design too).

> • *Isolation of participants* from extraneous environmental stimuli during the pretest-posttest interval would be a most desirable control procedure, *when it is practical.*

> • *Different non-overlapping pretest-posttest periods in time* for running different participants would also be useful for minimizing the impact of any contemporary history extraneous variable events.

2) Participant attrition/experimental mortality

> ∞ Because participants must be present for all observations made in a study, there will usually be a higher proportion of individuals lost when a time-series design is used since the design involves *multiple* pretests and posttests.

> ∞ *External validity* (generalizability) would be affected if a *systematic subset* rather than a random subset of the sample were lost, since this would leave a *non-representative sample* in the study.

> • Example: You could lose primarily *low motivation or poor performance individuals,* which would be likely to influence the treatment effect (as was noted for the previous design).

> ∞ *Internal validity* would *not* be a problem so long as *only those participants* present for *all* observations were included in the analysis — which, of course, would be the *logical* thing to do.

3) Testing-by-treatment interaction

> ∞ *Early* tests could influence *subsequent* test/observation data.

> ∞ *Internal validity* would *not* be a problem, as noted earlier, due to the control provided by determination of the *pretreatment trend or stable baseline.*

> ∞ *External validity* (generalizability) could be a problem, however, and perhaps particularly so because of the *multiple pretests,* which would be more likely to produce *sensitivity or resistance* to the treatment (i.e., interact with the treatment).

4) Participant bias

∞ If the bias effect would likely be *coincident* with the time when *treatment* is administered, such as for *placebo effects,* then participant bias *would not be controlled* by the multiple pretreatment and posttreatment observations of this design.

• However, depending on the study, it might be possible to use *deception or concealment* as a control for participant bias.

5) Experimenter bias

∞ *Expectancy of the experimenter* is also a potential control problem, but this is true for *all designs* (as noted earlier) except when observers can be kept *blind,* or at least partially blind, as to the condition a research participant receives.

— Applications

• This design is most useful for *behaviors that are naturally periodic,* such as many that occur in the workplace or at school.

∞ *Testing effects* would thereby be *minimized* because the behavior that is observed repeatedly is one that *typically occurs repeatedly,* and thus testing is less obtrusive.

∞ Examples: Work done on an assembly line, and repetitive exercises done in a classroom.

— Conclusion

• *Quasi-experimental is how* this design would be classified.

∞ It is not considered to be an experimental design primarily due to the extraneous variable problem of *contemporary history* — however, as already noted, it might be possible to satisfactorily control this variable.

3. Equivalent Time-Samples Design

$$X_1 \, O \qquad X_{0(2)} \, O \qquad X_1 \, O \qquad X_{0(2)} \, O \, ...$$

— Description

• Two samples *equivalently distributed over time* are taken of dependent variable observations (O): one sample when some *experimental treatment* has been presented (X_1), and another when it has not ($X_{0(2)}$), i.e., when either a *control condition* (X_0) or a *different experimental treatment* (X_2) has been presented.

∞ Hence, this is a *type of time-series design,* but with *repeated* introduction of the independent variable treatment(s).

∞ The *sequence of conditions* should be *randomized* as a control for *conditioned expectation,* i.e., so that there is not an effect of research participants *anticipating* conditions (note that the sequence in schematics is traditionally shown as systematic alternation, but this is just to conserve space).

- • *Block randomization* of the condition sequence, where order of conditions is randomized in successive blocks, with each condition occurring *equally often* in each block, would produce *more equivalence* of the distribution of conditions over time than would simple randomization. This would result in a *quasi-random sequence*.

  ∞ *Temporal intervals* between condition presentations should be kept *unequal*, i.e, they should be *inconsistent* (see Analysis).

  - • This prevents *temporal conditioning* that results from repeated presentation of *conditions* at fixed intervals, which could *influence* the effects of conditions.

  - • Inconsistent temporal intervals also prevent *confounding with cyclical events* in the environment and the organism (e.g., *circadian rhythms* — which are about 24 hours — and the *basic rest-activity cycle* for physiology, alertness, and behavior during wakefulness — which in humans is about 90 minutes, as are the cycles that occur during sleep).

  - • Note, however, that any of these problems would *not* affect *internal validity* when the distribution of different conditions is *randomized* over time — as it should be.

  ∞ *Multilevel and multifactor variations* exist for this design, and for the counterbalanced designs covered later.

— Analysis

- • ANOVA (analysis of variance) or a *t*-test for related/correlated measures would be used when score data are obtained.

  ∞ *Mean scores* under the *different conditions* first would be computed for each participant, and then the condition means would be compared for all participants taken together.

  - • *Unequal temporal intervals* do not pose a problem for analysis of the data, unlike in the time-series design, since treatment effects are indicated by *differences in means* under the different conditions, rather than by some *discontinuity* in the series of observations.

— Strengths and advantages over one-group pretest-posttest design

1) *Sampling error* is *reduced* since *multiple measures* are obtained from the participants under *each* of the different conditions, and *averaging* these multiple measures *controls* for *erroneous sample scores* because *random error* averages toward zero.

   ∞ *Error variance* is thereby *reduced*, and this leads to *increased sensitivity/ power* to detect treatment effects.

2) *Two (or more) different treatment conditions* can be compared, rather than studying the effects of only one treatment condition.

3) *Contemporary history* is also better controlled (see just below).

— Strengths and advantages over time-series design

1) More than one treatment condition's effects can be studied.

2) Contemporary history extraneous environmental events are better controlled by the repeated presentation of conditions, which are distributed equivalently over time.

   ∞ This both *dilutes* the effects of contemporary history events on dependent variable scores, and it spreads any such effects *more uniformly* over the different conditions.

3) *Other extraneous factors* (e.g., instrumentation, testing, and maturation) are controlled in a more straightforward fashion by the *randomized sequence* of repeated presentation of conditions and the computation of *average scores* under each condition, than they would be by the use of *statistical trend analyses*.

   ∞ *Random sequencing* makes it likely that extraneous variable effects will be *equivalent* across the various conditions.

— Disadvantage

• *Extra time and effort* are needed with this design for making *multiple* dependent variable observations under each of two (or more) conditions.

— Weakness

• Testing-by-treatment interaction

   ∞ This interaction is a more likely and hence important problem in this design than the time-series design because, as part of testing, *treatments are repeated* in addition to observations, and this added component of repetition might produce greater *sensitivity or resistance* to treatments.

   ∞ Also, if more than one experimental treatment is studied, then the *particular combination* of treatments might influence the effects observed for *specific treatments*.

   ∞ *External validity (generalizability)* is the concern for either of the above cases, not internal validity.

— Applications

• This design is most useful when treatment effects are anticipated to be *relatively transient*, or reversible, since sufficient time must be allowed *between* the presentation of conditions to avoid any possible *carry-over* of the effects of one condition to the condition(s) that follow.

   ∞ Examples: 1) *Drug studies* would typically present a problem for this design due to the rather *long durations* of many drug effects.

   2) *Transient effects* that might be studied with this design include *electrophysiological research* on the event-related-potential (ERP) responses of the brain, or on the nerve impulse frequency changes of individual neurons, produced by stimuli that vary in modality, intensity, frequency, color, pattern, motion, complexity, novelty, etc. Such research is conducted to determine the various stimulus features that

different areas of the brain and individual neurons are designed to detect and discriminate.

— Conclusion

- *Experimental* is how this design should be classified.
  - ∞ However, there are some investigators who consider this design to be just *quasi-experimental*.
    - *Problems* with the design, however, appear to be limited to only those of *external validity*.

4. Counterbalanced Designs (Cross-Over Designs)

a. Intraparticipant counterbalancing/ABBA design (just one group)

$$X_1 O \qquad X_2 O \qquad X_2 O \qquad X_1 O$$

or alternatively

$$X_2 O \qquad X_1 O \qquad X_1 O \qquad X_2 O$$

b. Interparticipant counterbalancing (more than one group)

$$X_1 O \qquad X_2 O$$

and for a second group of participants

$$X_2 O \qquad X_1 O$$

c. Intra-plus inter-participant counterbalancing (combined approach)

$$X_1 O \qquad X_2 O \qquad X_2 O \qquad X_1 O$$

and for a second group of participants

$$X_2 O \qquad X_1 O \qquad X_1 O \qquad X_2 O$$

— Description

- *Order of presentation* of conditions (X_1 and X_2) to participants is *systematically varied* so that it is *balanced,* either within each participant (intraparticipant counterbalancing), or across different participants (interparticipant counterbalancing), or by using both approaches (intra- plus interparticipant counterbalancing).
  - ∞ *Conditions* may consist only of different *treatments,* as shown in the above schematics, or they may include a *control* (X_0).
  - ∞ Dependent variable *observations* (O) are made during or after the presentation of each condition.
  - ∞ *Multilevel and multifactor variations* exist of this design.

∞ Note: *Additional details* about counterbalancing were covered earlier in Chapter 7, under "Making Equivalent the Secondary Variance, Using Participants as Their Own Control with Counterbalancing."

- Randomized counterbalancing

    ∞ When *carry-over effects* in a study would not be sufficiently controlled by using complete counterbalancing, or when there are *insufficient participants* for the necessary systematic sequences of even incomplete counterbalancing, a Randomized Counterbalanced Design might be used — thus leaving things to *chance* (See Chapter 7, as noted above).

    ∞ Note that in *contrast* to the equivalent time-samples design, the randomized counterbalanced design uses a *different* random sequence for each participant, and each condition is presented *only once*, or in a single block, per participant.

— Analysis

- ANOVA (analysis of variance) or a *t*-test for related/correlated measures would be used when score data are obtained.

— Strength

- Counterbalancing attempts to *control order/sequence effects* (practice, fatigue, motivation, attention/vigilance, etc.) by distributing them *equally* to all the conditions.

— Advantage relative to equivalent time-samples design

- *Interparticipant Counterbalancing* is advantageous over the equivalent time-samples design — which uses a randomized, *repetitive* treatment sequence — whenever the *repetition of conditions* is undesirable.

    ∞ Repetition might lead to unwanted *sensitization or resistance* to treatments, e.g., *habituation of effects*.

— Weaknesses

1) *Differential/asymmetrical transfer of order/sequence effects* across conditions is possible; i.e., *carry-over effects* might occur, which cannot be completely balanced, and thus the conditions would be *confounded* to some degree with order.

2) *Testing-by-treatment interactions* would lead to *external validity problems*, as for the equivalent time-samples design.

— Conclusion

- *Experimental* is how counterbalanced designs should be classified, if there are no serious carry-over effects.

— Single-participant and small-N designs

- When research involves just a single participant or only a very small number, it is called single-participant or small-N research.

- Such research naturally must use *within-participants designs*, such as the time-series design, equivalent time-samples design, and intraparticipant counterbalanced (ABBA) design, as well as others that have not been covered (e.g., the ABA design, where B is a treatment, and A is a control condition).

5. Evaluation of Within-Participants (Repeated-Measures) Designs

The following general evaluation of these designs is an elaboration of information given in Chapter 7, under "Using Participants as their Own Control" (see additional coverage there for the advantages, disadvantages, and also applications).

— Strengths and advantages

1) Reduced statistical error variance

∞ This is the *major benefit* of these designs.

∞ *Error variance* associated with *individual participant differences* is reduced in proportion to *twice the correlation* of the participants' scores under the different conditions.

- This correlation ($r_{1,2}$) between measures taken within participants (or from matched participants in the other form of correlated-groups designs), and the resulting decrease in error variance, is incorporated in the *statistical equations* that are used to analyze the results of such designs.

- Example: *t*-test equation for related/correlated measures

$$t = \frac{\overline{X}_1 - \overline{X}_2}{\sqrt{\frac{s_1^2}{n_1} + \frac{s_2^2}{n_2} - 2(r_{1,2})\left(\frac{s_1}{\sqrt{n_1}}\right)\left(\frac{s_2}{\sqrt{n_2}}\right)}}$$

- *Repeated measures* from participants should be *correlated positively* since individual participant characteristics normally remain fairly constant over time.

- To the extent that this is true, *individual participant differences* across conditions would be *controlled* — and hence statistical error variance would be reduced.

∞ *Masking is decreased* by the reduction in statistical error variance associated with differences among participants.

∞ *Power/sensitivity* for detecting treatment effects would therefore be *increased*.

2) Economy of participants

∞ *More data is gathered per participant,* since data for *all conditions* would be obtained from each individual.

- Hence, compared to between-participants designs, the *same amount of data* could be obtained with *fewer research participants.*

3) Reduced participant-preparation/set-up time and effort

∞ *Once prepared* (e.g., given instructions or connected to equipment) each research participant would be run under *all* of the various conditions, thus *decreasing* the total amount of time and effort required for the study.

— Applications

1) Within-participants (repeated-measures) designs are used whenever the *strengths and advantages* just noted are important.

2) They are necessary for *single-participant and small-N research*, which evaluate data from only one or a very small number of participants, as opposed to data averaged for many individuals.

3) They are also *necessary*, at least as a component in a mixed design, whenever an *order/sequence effect* (practice/learning, fatigue, vigilance/attention, or motivation) is an *independent variable*, i.e., a factor of interest (e.g., *trials* as a factor in a study of learning).

— Weaknesses and disadvantages

1) Of great concern for within-participants designs are the need and difficulty of *controlling order/sequence effects* and the other forms of *repeated-measures experimental contamination* that might exist; i.e., testing, mortality, maturation, instrumentation, and statistical regression toward the mean.

2) Yet another concern is that *demand characteristics* leading to *participant bias* are a greater problem for repeated-measures designs, since all investigated levels of the independent variable are experienced by each research participant, thus providing more information to them as to what the study is about, and hence perhaps clues as to what the experimenter hopes to find.

∞ Participants, in order to be *accommodating*, then tend to behave in ways that they *believe* the investigator wants.

— Limitations

1) You can't use some types of within-participants designs if *treatment effects are very enduring* (e.g., when testing different ways of rearing young in developmental studies, or different ways of teaching mathematics or a foreign language).

2) You can't use some types of within-participants designs if *differential-transfer/carry-over effects are significant*.

3) You can't use these designs if there are *too many conditions*, thereby causing the participants to become so fatigued and/or lose so much motivation that they would *quit* — hence resulting in *participant attrition*.

4) You can't use these designs if other forms of *repeated-measures experimental contamination*, including *demand characteristics*, are too great.

• Note: *Between-participants/groups designs* are used if any of the preceding limitations apply.

∞ *Matched-participants* forms of between-participants designs come closest to the within-participants designs in terms of *controlling for individual differences among participants*.

- Hence matched-participants designs are a good *alternative* to within-participants designs when the latter cannot be used.

Review Summary

When reading the following material, it will be useful to refer to Handbook Appendix 6 showing "Experimental and Quasi-Experimental Designs: Category–Subcategory Relationships," and Appendix 7 showing a "Schematic Summary of Designs."

1. A *research design* is a *basic plan* for scientific investigation — a *strategy* for trying to answer research questions by testing research hypotheses. Five *components of experimental research designs* are: (a) *independent variable* specifications, (b) *dependent variable* specifications, (c) *extraneous variable* specifications, (d) *participant specifications*, and (e) *statistical analysis* specifications.

2. The three major *determinants for selecting* a particular research design are: (a) *problem/question(s)* being asked and the *hypothesis(es)* being tested, e.g., how many independent variables and levels; (b) *extraneous variable controls* and *statistical power* needed; and (c) *generalizability* of results desired.

3. The three *criteria for evaluating* a research design relate logically to the preceding determinants of design selection:

   a. Conclusions possible — How well can the design *test* the research hypotheses and *solve* the research problems? This is contingent in part upon the *statistical tests* that can be performed.

   b. Internal validity — With what *confidence* can one infer a *cause-and- effect relationship* between the independent and dependent variables *within the confines* of a given specific experiment? This is contingent upon the *control* of extraneous variables.

   c. External Validity — to what extent can the experimental results be *generalized* beyond the confines of a given specific experiment? This is contingent upon the *representativeness* of the experiment.

   Note that while it is desirable to select designs strong in *both* internal and external validity, features that increase one form can *jeopardize* the other. However, internal validity is *absolutely indispensable* — without it, questions of external validity are *meaningless*.

4. *One-group posttest-only design* is where one group of individuals receives a treatment and is then observed on the dependent variable. This is a *pre- experimental design*. It is very weak because there is no *baseline,* or reference level data, and because *rival hypotheses,* or alternative explanations, are very plausible since

extraneous variables are uncontrolled for *confounding*. Thus *causality* cannot be determined using this design.

5. *Quasi-experimental designs* are those that do not meet the *strictest requirements* for experimental control. However, they are *better for determining causality* than the naturalistic observation techniques of *descriptive research*. They are used when experimental designs cannot be employed, such as might occur in *natural/field settings*, e.g., in the work place and schools. They are also used for *exploratory or pilot studies.*

6. Research designs can be divided into those that compare the effects of different levels of an independent variable *within* one or more groups of participants, and those that compare the effects *between* different groups of participants. The *between-participants/groups designs* are in turn divided into those consisting of *independent groups* of *randomly or nonrandomly* assigned participants and those consisting of *related/matched groups* of participants. Both *matched-groups designs* and *within-groups (repeated measures) designs* are forms of *correlated-groups designs*, as opposed to independent-groups designs. *Multilevel designs* are when there are more than two levels of an independent variable. Finally, *mixed designs* combine in a single *multifactor/factorial design* both *correlated groups* of participants to investigate the effects of one or more independent variables and also *independent groups* of participants for other independent variables.

7. *One-group pretest-posttest design* is where a single group of participants is observed both before and after a treatment. This has a *strength/advantage* over the one-group posttest-only design in that a *pretreatment baseline,* or reference level, is provided against which to compare the posttreatment observation. Nevertheless, it is still a very weak design since there are *many uncontrolled secondary, extraneous variables* that could be *confounding:* contemporary history, maturation, participant attrition, testing (probably the most serious shortcoming), participant bias, instrumentation, and statistical regression toward the mean. Thus, this is a *quasi-experimental design* at best (i.e., when there is some degree of control such as keeping the pretest-posttest interval short, isolating participants from extraneous environmental stimuli during the interval, frequently checking and calibrating equipment, and using well practiced observers and minimizing their fatigue). Otherwise, this should be considered a *pre-experimental design.*

8. *Time-series design* is where *multiple observations,* spaced at equal time intervals, are made on a single group of participants before and after a treatment. This has *strengths and advantages* over the one-group pretest-posttest design in that a *baseline* or *trend* is established by the pretreatment multiple observations, which reduces *sampling error* and *controls* several extraneous variables: statistical regression toward the mean, testing, instrumentation, and maturation. A *discontinuity* in observations after the treatment — i.e., a shift from a stable baseline or a change in trend — indicates an effect. But this design has a number of weaknesses in that there are *several uncontrolled secondary, extraneous variables:* contemporary history, participant attrition, testing-by-treatment interaction, and participant bias.

It is considered a *quasi-experimental design* primarily due to the extraneous variable of contemporary history, although this major concern might be controlled (e.g., by keeping the pretest-posttest interval short and isolating participants from extraneous environmental stimuli during the interval — whenever practical). Except for participant bias, the other problems are only of *external validity*, not internal validity. *Cross-validation/replication* would be required, of course, for results involving *unpredicted delays* in treatment effects or when treatments are *not specified in advance*, each of which might occur in exploratory studies. This design is most useful for *behaviors that are naturally periodic*, since this would minimize testing effects.

9.   *Equivalent time-samples design* is where two (or more) samples of dependent variable observations are taken, each under different independent variable conditions that are *distributed equivalently over time*, and that are randomly sequenced with unequal temporal intervals. This has *strengths and advantages* over the one-group pretest-posttest design in that *sampling error* is reduced, *contemporary history* is better controlled, and the effects of two (or more) different treatment conditions can be compared. The latter two are also *strengths and advantages* of this design over the time-series design, in addition to the *extraneous factors being controlled* more straightforwardly by the randomized sequence of conditions and averaging.

The *weakness* is that a *testing-by-treatment interaction* is a more likely problem because treatments are repeated, in addition to observations, which might produce *greater sensitivity or resistance* to the treatments. The *particular combination* of treatments might also be important. Thus *external validity* (generalizability) might be limited. This design is most useful when treatment effects are *relatively transient*, since sufficient time must be allowed between presentation of conditions to avoid any possible *carry-over* of the effects of one condition to those that follow. This design is usually considered to be *experimental*, the problems seemingly being limited to only external validity.

10. *Counterbalanced designs* are where *order of presentation* of conditions to participants is *systematically varied* so that it is *balanced*, either within a single set of participants — *intraparticipant counterbalancing*, or across different sets of participants — *interparticipant counterbalancing*, or by combining both approaches. In addition there is *randomized counterbalancing*. These designs attempt to *control order effects* by distributing them *equally* to all the conditions.

The *weaknesses* of these designs is that *carry-over effects*, i.e. *differential/asymmetrical transfer* of order effects across conditions, is possible, and thus the conditions would be *confounded* to some degree with order. Also, a possible *testing-by-treatment interaction* would cause an *external validity problem*. Counterbalancing, however, has an *advantage* over the equivalent time-samples design whenever repetition of conditions is undesirable, i.e., where it might lead to *habituation of effects*. These designs are considered *experimental* if there are no serious carry-over effects.

11. Within-participants (repeated-measures) designs were evaluated in Chapter 7, under "Using Participants as Their own Control with Counterbalancing," and that evaluation is elaborated in the present chapter. The strengths and advantages are (a) reduced statistical error variance associated with individual participant

differences, which is proportional to twice the correlation of measurements from participants taken under the different conditions; (b) economy of participants, since data is gathered for all conditions from each participant; and (c) reduced participant- preparation/set-up time and effort.

12. Within-participants designs have the following general *applications*: (a) they are used whenever their *strengths and advantages* are important; (b) they are necessary for *single-participant and small-N research;* and (c) they are necessary, at least as a component in a mixed design, whenever an *order/sequence effect* (e.g., practice/ learning) is an independent variable.

13. The *weaknesses and disadvantages* of within-participants designs are (a) need and difficulty of *controlling order/sequence effects* and other forms of *repeated-measures experimental contamination* that might exist; and (b) *demand characteristics* leading to *participant bias* are a greater problem.

14. There are several *limitations* of within-participants designs: (a) some types can't be used if treatment effects are *very enduring;* or (b) if *differential- transfer/carry-over effects* are too great; and (c) they can't be used if other forms of *repeated-measures experimental contamination* or *demand characteristics* are too great; or (d) if there are *so many conditions* that participants would become so fatigued and/ or lose so much motivation that they would *quit.*

Review Questions

When answering questions about specific designs, it should be helpful to write out the schematic summary of the designs, which can be checked against those given in Handbook Chapter 9 and Appendix 7.

1. List the five *components,* with their major subcomponents, of *experimental research designs* in terms of what needs to be *specified,* i.e., decided upon.

2. List the three major *determinants* for selecting a particular research design.

3. List the three *criteria for evaluating* a research design, what each is *contingent* upon, and any *relationship(s)* among the criteria.

4. Describe the *one-group posttest-only design,* and explain why it is considered a *pre-experimental design.*

5. Describe the properties of *quasi-experimental designs* relative to experimental and non-experimental designs, and explain when they are used.

6. Describe the difference between *within-participants designs* and *between- participants designs,* and explain a *mixed factorial design.*

7. The *one-group pretest-posttest design,* where a single group of participants is observed both before and after a treatment, has what *advantage* over the *one-group posttest-only design*? Why, however, is it still a very weak design (be as specific as

possible and note the most serious shortcoming)? Under what conditions could it be considered a *quasi-experimental design*, as opposed to *pre-experimental*?

8. The *time-series design*, where *multiple* observations are made on a single group of participants before and after a treatment, has what *strengths and advantages* over the *one-group pretest-posttest design*, and why? What *secondary, extraneous variables* are controlled with this design, and which are not? What is the major reason that this design is considered *quasi-experimental*, and what can be done to minimize this concern? Finally, for *what kind of behaviors* is this design most useful, and why?

9. The *equivalent time-samples design*, where two (or more) samples of dependent variable observations are taken, each under different independent variable conditions that are *distributed equivalently over time*, has what *strengths and advantages* over the one-group pretest-posttest design and over the time-series design? What is the *weakness* of this design? When is this design most useful? Finally, is this design usually considered to be experimental, and why?

10. The *counterbalanced designs* attempt to control what, and in what manner? What is the difference between *intraparticipant counterbalancing* and *interparticipant counterbalancing* (schematically diagram the designs)? What are the *weaknesses* of these designs? What is the *advantage* of these designs over the equivalent time-samples design? Finally, when are these designs considered *experimental*?

11. List three *strengths or advantages* of using *within-participants designs*.

12. List three general *applications* of within-participants designs.

13. List two *weaknesses or disadvantages* of using within-participants designs.

14. List four *limitations* to using within-participants designs.

10

Designs for Experiments — Between Participants and Multiple Treatments

CONTENTS

This is the *second* chapter devoted to research designs, which, as noted earlier, is an extensive and very important topic. *The previous chapter* covered the general concepts, principles, and categories of "Experimental and Quasi-Experimental Designs," followed by coverage of the "Within-Participants Designs." *The present chapter* covers both the "Between-Participants Designs" and the "Multiple Treatment Designs."

Between-Participants, Independent (Randomized), Two-Group Designs

In these designs, dependent variable measures are taken on *different groups of participants/ subjects,* with each group receiving a different value of an independent variable — one of which might be zero (no treatment) or some standard value serving as a control condition. Comparisons are thus made *between* groups of participants.

All the following design examples are for use when there is only *one independent variable,* and just *two levels* (such as presence versus absence of some treatment). The designs are sequenced from the weakest to the strongest, and thus *quasi-experimental designs,* with their deficiencies, are presented first so that the strengths/advantages of *experimental designs* can be better appreciated. (Appendix 7 is a summary of all the designs in this Handbook.)

1. **Nonequivalent Posttest-Only Control-Group Design (Nonequivalent Posttest-Only Design) (Static-Group Comparison Design)**

— Description

- *One group of participants* (top row in the schematic) that has experienced a treatment (X) is observed (O) on the dependent variable, and their scores are compared with those of a *control group* (bottom row) that has *not* experienced the treatment.

    - ∞ The *static group* referred to in the alternative design name is the *comparison* control group (shown below the dashed line), which does not receive the treatment, and which therefore should not change.

- The *dashed line* separating the treatment and control groups is to indicate the probable, and undesirable, *nonequivalence* of the two groups of participants *before the treatment*.

    - ∞ This likely nonequivalence is because in this design the groups are *not randomly formed*, and hence it is quite possible that the two groups are *not comparable* in participant characteristics.

    - ∞ Note that nevertheless this is an *independent-groups design*, since there would be no known relationship (correlation) between the participants in the two groups, even though the participants are *not randomly assigned* to conditions.

- *Descriptive research* with *ex post facto naturalistic observation* would be the typical use of this design (see later for exceptions).

    - ∞ *Independent variables*, in other words, normally are *not purposely, directly manipulated* when using this design.

        - *An event (X) just happens to occur* and is experienced by a group of individuals, some or all of whom are then *selected* for the study *ex post facto* (i.e., after the fact).

        - In addition, a *control group* that is *assumed* to be comparable to the treatment group, except that they have *not* experienced the *naturally occurring event*, is also selected *ex post facto*, and a *dependent variable observation (O)* is then made on them as well as on those who did experience the event.

    - ∞ Example: The *personality characteristics* (such as anxiety and aggressiveness) of individuals who just happen to have experienced some *natural disaster* (such as an earthquake) could be compared against a *control group* of individuals who have *not* experienced that event, but who *should have been similar* to the others *before* the event. The control group could be individuals who happen to have moved away from the area right before the natural disaster.

— Analysis

- Note: *Non-random assignment* of participants to conditions occurs not only in this research design, but in the next two as well, and thus they too involve *nonequivalent groups*.

- These designs, therefore, do not readily lend themselves to *standard statistical analyses*, since for most analyses *random assignment* is a general *assumption* and thus a *requirement*.

  ∞ This is because without random assignment to conditions there are *rival hypotheses* related to *biased selection* (see below), which could account for differences in performance between the experimental and control groups.

  ∞ Hence, the statistical analysis must be *carefully constructed* to fit each specific research situation, and this requires a *considerable amount of knowledge and thought* on the part of the investigator — a detailed discussion of which is beyond the scope of this Handbook. ([See pages 147-205 of Cook, T.D., & Campbell, D.T. (1979). *Quasi-experimentation: Design & analysis issues for field settings.* Chicago: Rand McNally College Publishing Company.]

— Strength and advantage over one-group posttest-only design

- The present design attempts to correct for the deficiency of the one-group posttest-only design (discussed in Chapter 9) through the *addition* of a *control group* in order to provide a *baseline measure* for comparison against the observations made on the treatment group.

  ∞ However, the attempted correction is *less than adequate* (as discussed next) given the probable *nonequivalence* of the control group.

— Weaknesses

There are important uncontrolled secondary, extraneous variables.

1) Selection

  ∞ Because participants are *not randomly assigned* to conditions, and since the independent variable is typically manipulated simply by *measured selection of participants* who *already possess* the desired values, then the two groups of participants would likely *differ in many respects* other than whether or not they experienced the treatment (hence the dashed line separating these probably *nonequivalent* groups).

    - One could not, therefore, be certain that an observed dependent variable difference between the groups was due to the treatment — it could be due to the *confounding extraneous variables* of *individual participant differences*.

    - *Internal validity concerns* would be the consequence.

      Example: Any behavioral or physiological differences observed between *animals found in zoos* and those found in the wild might be due not to the treatment, i.e., the *environmental conditions* of zoos, but rather to individual differences associated with the animals' *ability to escape capture* (and thus be selected), i.e., perception, intelligence, speed, strength, and health.

- Attempts to *ex post facto match* treatment and control participants on the basis of *background characteristics* usually are *ineffective and misleading,* especially if the treatment group sought out the experience.

    Example: Any differences in reading ability between those who have and have not experienced *speed-reading training* could be simply due to *motivational differences* that led one group to seek the training.

    Although it might be possible to *ex post facto* find a control group matched for initial reading ability and IQ, it would be more difficult to do so for motivation.

2) Participant attrition/experimental mortality

    ∞ *Differential drop-out* could also be responsible for any difference in dependent variable values for the two groups.

    - *Internal validity problem* would be the consequence.

        Example: This difficulty is frequently met in studies to determine the intellectual effects of *a college education* by comparing aptitude measures (such as the SAT or GRE) for *freshmen,* who have not yet received the treatment of a college education, with those of *seniors,* who have received the treatment.

        Seniors would hopefully have higher aptitude test scores, but this could be due, at least in part, to the fact that the *less motivated and less intelligent* students drop out before the senior year.

3) Maturation

    ∞ Differences between groups involving *changes internal to individuals* as a function of the *passage of time* is occasionally a possible *confounding* extraneous-variable problem that affects *internal validity* in this design.

    - Example: Consider the case of comparing the *aptitude of first-year versus senior-year college students,* who, on average, would naturally be at rather different levels of cognitive maturation (4 years or so can make a rather big difference).

        Note that, in this particular example, *contemporary history* could also be a confounding extraneous-variable problem, because learning experiences can occur outside of school.

— Applications

- *Exploratory or pilot studies* might employ this design.
- Also it is used when *situations* make this design the *only option:*

    ∞ *Natural settings* (e.g., workplaces, schools, and hospitals) typically do not permit random assignment of participants to conditions due to *pre-established groups/collectives;*

    ∞ *Unplanned events* (e.g., natural disasters and wars) do not permit advance preparation and thus optimal controls;

∞ *Ethical constraints* (e.g., concerns about physical or mental stress or harm) might exclude purposeful, direct manipulation of certain independent variables (such as enforced long-term separation from loved ones).

— Conclusions

- *Pre-experimental* is how this design would usually be classified.

- *Quasi-experimental* status is possible, but only when the following three conditions are met:

    a) The independent variable can be and is *purposely, directly manipulated.*

    - Hence the study is *not ex post facto,* and therefore *more control* is possible over extraneous variables.

    b) The *participants* in the different groups are shown to be *matched before-hand* on several relevant variables (usually by frequency distribution matching, not precision matching).

    - But note that since the matched participants cannot be randomly assigned to groups, the groups are still possibly *nonequivalent* on one or more unmeasured extraneous participant factors.

    c) Which group receives the treatment is *randomly determined.*

    - Example meeting all three of the preceding conditions: *Purposeful, direct manipulation* of the *reinforcement* (independent variable treat-ment) of a desirable behavior in an institutional setting (such as a school), where *pre- established collectives* are found and thus ran-dom assignment to treatment and control groups (such as classes) is not permitted, but where the groups are *matched* on several important extraneous variables, and which group receives the treatment is *determined randomly.*

2. **Nonequivalent Pretest-Posttest Control-Group Design (Nonequivalent Con-trol-Group Design) (Pretest-Posttest Static-Group Comparison Design)**

— Description

- This design is a *combination* of the one-group pretest-posttest design (look-ing just at the top row; see discussion in Chapter 9) and the nonequivalent posttest-only control-group design (which as described earlier lacks the pretests shown on the left).

- *One group of participants* (top row) is observed (O) at some time *prior* to the onset of a treatment (X), and the dependent variable scores are compared with those observed (O) *during or after* the treatment, and any *change* for this first group is evaluated with respect to a possible change for *a control group of participants* (second row) that has *not* experienced the treatment.

- The *dashed line* separating the treatment and control groups is to indicate the probable, and undesirable, *nonequivalence* of the groups of participants in this design *before treatments*.

  - ∞ This is because the groups are *not randomly formed,* and hence, as in the previous design, it is likely that they are *not equivalent* in participant characteristics.

  - ∞ Nevertheless, this is an *independent-groups design,* since there would be no known relationship (correlation) between the participants in the two groups, even though the participants are *not randomly assigned* to conditions.

— Analysis

- See earlier discussion for the preceding design, which notes problems that are applicable to the present design as well.

— Strengths and advantages over nonequivalent posttest-only control group design

1) The *pretest observations* made for both groups provide *baselines* that can be compared to determine the degree of *initial equivalence* on the dependent variable for the two groups.

2) *Nonequivalence can be controlled* — i.e., adjusted for — by analyzing *pretest-posttest changes* for the participants (two approaches are given below), rather than just analyzing their posttest scores (Note: control weaknesses and a limitation are discussed later):

   - ∞ Change/difference/gain scores: (A – B)

     - *Pretest/before (B) score* is subtracted from the *posttest/after (A) score* for each participant.

   - ∞ Change/difference/gain ratio scores: (A – B) ÷ (P – B)

     - This approach measures the *actual change (A – B)* relative to the *potential change (P – B)* by computing the *ratio.*

     - The *optimal or potential (P) score,* however, must be known or determinable in order to do this calculation.

     - *Ceiling and floor truncation effects* are better controlled with this measure, hence the value of the extra effort, since change *ratio* scores (unlike simple change scores) are not restricted by very high or low pretest values — i.e., the *ceiling or floor (P)* is taken into account by the computation of *(P – B),* against which the actual *(A – B)* is compared.

     - Example: Consider, as in Table 10.1, two students among a class that are given equivalent exams *(P* = 100) at the beginning and the end of some course, and where the course grade will be determined not by the last exam, the posttest, but by the *pretest–posttest improvement (gain).*

TABLE 10.1

Example of Pretest–Posttest Change Scores and Change Ratio Scores

Student	A – B = Change Score	(A – B) ÷ (P – B) = Change Ratio Score
S_1	90 – 80 = 10	(90 – 80) ÷ (100 – 80) = 0.50
S_2	60 – 40 = 20	(60 – 40) ÷ (100 – 60) = 0.33

Notice in the example the *reversal* in the comparative amount of change (gain) for the two students that results from the different calculations (10 versus 20 as opposed to 0.50 versus 0.33), and that the *change ratio score* certainly seems to be the more *logical and fair* measure.

∞ Controlling *pretreatment individual differences among participants* — and thus group differences — through the use of a *pretest* to adjust/correct for any differences by computing *change scores* or *change ratio scores* means that:

- *Confounding* would be reduced;
- *Masking* would be reduced as well, due to the resulting decrease in *error variance* associated with individual participant differences, which in turn increases the *power/sensitivity* to detect a treatment effect.

— Strengths and advantages over one-group pretest-posttest design

1) *Contemporary history* (an extraneous *environmental* variable) is better controlled due to the presence of a control group.

2) *Testing and instrumentation* (two forms of *experimental* contamination) are also better controlled due to a control group.

3) *Maturation* (an extraneous *participant* variable) is also usually, but not necessarily, better controlled due to the control group (see "Weaknesses" discussed below).

— Disadvantage

- *Extra time and effort* is required for the additional observations.

— Weaknesses — uncontrolled secondary, extraneous variables

1) Selection

∞ Because participants are *not randomly assigned* to conditions, the *selected groups* could differ in ways other than whether they experienced the treatment (hence the dashed line that separates the groups, indicating assumed *nonequivalence*).

- It is possible that the *pretest* does not *adequately* control the *individual participant differences* between groups.
- Generally this is not likely, but there are several situations where this might occur if the groups *substantially differ* in their selection (see comments below on interactions).

- Thus one could not be certain whether the observed differences between the groups were due to the treatment, or due to *confounding extraneous factors*.

∞ *Internal validity problems* would be the consequence of *interactions* that might occur between *selection/recruitment* differences for the groups and various *extraneous variables*, since this would mean that the *effects* of the extraneous variables *would not be the same* for the different conditions.

- *Maturation* is one such variable that can interact with (be influenced by) *selection differences* that distinguish the experimental/treatment and control groups. This represents *perhaps the most common problem for this design*.

  Example: In a study of the *effectiveness of a psychotropic drug for depression*, the experimental group would naturally consists of depressed patients, but the control group might *inappropriately* consist of patients with a different (nonequivalent) pathology.

  In such a case, any diagnostic test gain unique to the experimental group might not be a treatment effect, but just *spontaneous remission* of the measured psychopathology, which could not occur for the control group not suffering from that pathology — hence the value of the pretest control is undermined.

- *Statistical regression toward the mean* can also interact with *selection,* and this represents a major concern when *just one group* is selected/recruited due to its extreme scores on the pretest or a correlated measure.

  Example: In *remedial-reading programs* in schools, the treatment participants are logically recruited from among students that are the poorest readers, but if the control participants are recruited (inappropriately) from among students that have normal reading skills, then statistical regression could not occur to the same extent for both groups, thus again negating the value of the pretest control.

- Another situation where *statistical regression* is likely to interact with *selection* is when there is an attempt to *match groups* based on the pretest scores or a correlated measure, and when due to the very different distributions (means) of scores within each group, matching requires using participants with the *highest scores from one group* and those with the *lowest scores from the other group.*

  Due to statistical regression toward the mean for each group, the scores of the two groups would tend to change in *opposite directions* from the pretest to the posttest, and because of the *confounding* this regression would be *confused* with a treatment effect.

- *Participant attrition/experimental mortality* can interact with *selection* too; i.e., a particular type of participant might drop out of one but not the other group during a study due to *between-groups participant differences* that are related to *recruitment differences* for the groups.

Example: In a *remedial-reading study* with a nonequivalent control group of normal ability readers, it is primarily in the *treatment group* that those participants with the *lowest motivation* would likely drop out before the posttest.

As a result, the remainder of the treatment group might show a greater gain than the control group on the dependent variable, not because of *the* treatment, but because the low motivation participants are *missing primarily from the treatment group.*

- • *Contemporary history and testing* extraneous variables can similarly interact with selection differences — thereby also causing *internal validity problems.*

2) Testing-by-treatment interaction

- ∞ *External validity problems* (generalizability deficiencies) occur when the pretest produces *sensitization or resistance* to the treatment (note that the internal validity problem of a direct effect of testing is usually controlled in this design by the presence of a control group — but see "Limitation" just below).

— Limitation

- • This design is effective in controlling for extraneous variables only to the extent that the experimental/treatment and control groups are *similar in their selection/recruitment,* and to the extent that this similarity is confirmed by *comparable pretest scores.*

  - ∞ This is because of the just discussed weaknesses involving *interactions* of extraneous variables with *selection.*

  - ∞ Also, a dependent variable might change in a *non-linear* manner over its range, thus reducing the usefulness of difference scores when pretest scores are *very dissimilar.*

  - ∞ Note, however, that even if the groups score *similarly* on the pretest, there still might be *hidden control problems* related to selection differences.

  - ∞ Example: *Belief* in the probable effectiveness of therapy might differ for treatment and placebo-control groups, if *selected from different hospitals,* even when the two groups initially have equally severe problems.

  - ∞ IN SUMMARY: Selection/recruitment equivalence for the groups is extremely important.

— Applications

- • This design is used when the participant population consists of *pre-established groups/collectives,* such as schools with their classes, hospitals with their wards, and businesses with their work shifts.

  - ∞ It is unlikely that a researcher would be *allowed* to precision match or randomly assign participants from such *"naturally assembled" collectives* into treatment and control groups; rather the investigator would have to use the *already existing groups* as is; e.g., entire schools or classes would have to be assigned to the different conditions.

- *More control* would be desired, however, than could be provided by the previously discussed nonequivalent *posttest-only* control-group design.

- Hence, because of its superior control, the nonequivalent *pretest-posttest* control-group design is one of the most *widely used* designs in *educational research.*

    Example: If a school district were considering the addition of expensive *language laboratories* to all its courses teaching Spanish, they might attempt to first evaluate the *effectiveness* of such laboratories by doing a study comparing the language skills gained by just one class or school using the laboratory versus another class or school not using the lab.

    The two *naturally assembled* collectives would be chosen for their likelihood of being *matched* on all important extraneous variables, and this would be *evaluated* by their pretest dependent variable scores; which class or school was assigned to be the experimental treatment group would be *randomly* determined (even though students individually could not be randomly assigned to the different conditions).

- Note: This design, as well as the other between- participants designs, can be *modified* (along with appropriate changes in their names) to include a *second treatment condition* instead of a control condition.

    Example: If a school district or state wanted to evaluate the relative effectiveness of two different instructional strategies for teaching *reading*, such as the *phonics method* (learning to pronounce the phonetic value of letters and syllables) versus the *whole-word method* (the look-say/whole-language approach), they could randomly assign to the two teaching methods *pre-established collectives* that are likely to already be *matched* (such as presumably equivalent classes within a school, or schools within a district), and then compare the difference in *pretest-posttest change scores* for the different *treatment conditions.*

— Conclusion

- *Quasi-experimental* is how this design would be classified.

    ∞ Although the design has important strengths and advantages, it also has serious potential *internal-validity* control problems related to the secondary, extraneous variable of *selection.*

3. Multiple Time-Series Design

$$
\begin{array}{ccccccc}
O & O & O... & X & O & O & O... \\
\hline
O & O & O... & & O & O & O... \\
\end{array}
$$

— Description

- This design represents an *incorporation* of the one-group time series design (top row) into the nonequivalent pretest-posttest control group design (which was just discussed).
  - ∞ In essence, this design can be seen as an *extension* of the nonequivalent pretest-posttest control-group design.
- *Multiple observations* (O) are made before treatment (X) on a group of participants (top row), and the dependent variable scores are compared with those from multiple observations (O) made during or after the treatment, with any *change* being evaluated with respect to that for *a control group of participants* (bottom row) that has *not* experienced the treatment.
  - ∞ *Greater confidence* of interpretation can result from this design due to the *multiple pretest and posttest observations,* and because any evidence of an experimental effect in the treatment group, indicated by a pretest-posttest change, can be tested by *comparison* with the control group observations.

— Analysis

- See earlier discussion for the first design in this chapter, which notes problems that are applicable to the present design as well.

— Strength and advantage over one-group time-series design

- *Contemporary history* (as well as *testing* and *instrumentation*) is better controlled by having the comparison/control group.
  - ∞ The *control is inadequate,* however, to the extent that participants in the experimental and control groups are not equivalent (see "Weaknesses" below).

— Strengths and advantages over nonequivalent pretest-posttest control group design

1) *Sampling error* is reduced by *averaging* multiple measures obtained before versus after treatment, which reduces *error variance* and thus increases *sensitivity/power* to detect an effect.
2) *Trends* can be determined from multiple measures/observations.
   - ∞ This results in better controls for *regression, testing, and instrumentation,* as well as for *interactions of selection* with maturation, etc.

— Disadvantages

- *Extra time and effort* are needed for the multiple pretest and posttest observations.

— Weaknesses

1) Selection
   - ∞ Since participants are *not randomly assigned* to conditions, the two groups could differ in ways other than whether or not they experi-

enced the treatment (hence the dashed line separating the groups, indicating assumed *non-equivalence*).

- *Pretests* might not adequately control the participant differences between groups (not likely, but see below).

- To the extent that there are participant differences between the groups due to *selection/recruitment differences*, they can *interact* with extraneous variables, leading to dependent variable differences that could be *confused* with treatment effects (see next).

∞ Contemporary history-by-selection interaction could occur.

- As noted for other designs, the likelihood of a contemporary history problem is greater when the *pretest-posttest interval* is longer, and when there is less *isolation* of participants from the external environment.

- *Internal validity problems* result from contemporary history-by-selection interactions.

∞ *Treatment-by-selection interactions* are possible too, but this is also true with other designs; e.g., due to using only college students and/or volunteers, or only one species and/or sex.

- *External validity problems* (generalizability deficiencies) would be the consequence for this and other designs.

2) Testing-by-treatment interaction

∞ This occurs when a pretest produces *sensitization or resistance* to the treatment.

- *External validity problems* would again be the consequence.

- This is perhaps a *particular concern* with this design because of the *multiple pretests*, which would be more likely to produce sensitivity or resistance to the treatment than would just a single pretest.

3) Participant attrition/experimental mortality

∞ This too is possibly a *particular concern* with this design due to the *multiple pretest and posttest measures*, as opposed to single measures.

- The probable severity of the problem is dependent on such factors as whether participants have to *return* for each observation, the *duration* of the pretest-posttest observation sequence, and the *difficulty* of the task.

- *Internal validity problems* arises if *dissimilar participants* drop out from the two groups.

- *External validity problems* arise, however, even when *similar participants* drop out from both groups if the dropouts, and as a consequence the individuals that remain, are *not representative samples* of the population.

— Conclusions

- *Quasi-experimental* is how this design would be classified.
    - ∞ But note that this is the *best of the nonequivalent designs*.
        - *Problems* associated with this design might be limited to only those of *external validity.*

4. ***Posttest-Only Control Group Design* (After-Only Randomized Two-Group Design)**

$$R \quad X \quad O$$
$$R \quad\quad\ O$$

— Description

- *Random assignment* (R) of participants (in contrast to the three preceding independent-groups designs) is made to either an *experimental group* (top row) that receives the treatment (X) or to a *control group* (bottom row) that does not, after which an observation (O) on the dependent variable is *compared* for the two groups of participants.
    - ∞ Note: Instead of a control condition, a *second treatment condition* could be used in this and the following designs — with, of course, the appropriate descriptive names, such as the alternatives listed in parentheses.

— Analysis

- ANOVA (analysis of variance) or a *t*-test for independent groups would be used when score data are obtained.

— Strength and advantage over nonequivalent posttest-only control group design

- *Selection* problems are controlled in this design by the use of *random assignment,* which makes it probable that the extraneous variables of *individual participant characteristics* will be *equivalent* between groups.
    - ∞ Hence the *between-conditions/groups variance* (differences) present *before* the administration of treatments is *minimized.*
        - The greater the *number of participants (n)* the better the control is likely to be, i.e., the greater the equivalence of participants in the treatment and control conditions, since random error averages toward zero with increases in *n.*
        - Note that there is *no dashed line* separating the groups (as opposed to the preceding three design schematics), since in this and the following designs it is assumed that the different groups of participants are *equivalent.*
    - ∞ *Confounding* is thus *controlled* in this design, although *partial* confounding is still possible due to chance outcomes associated with random assignment.

∞ To also control *extraneous environmental and experimenter factors,* the two groups should of course be observed in the same time period, in the same surroundings, and with the same measuring instruments (mechanical and human), etc.

∞ *Masking* would remain a possibility, as always, depending on the size of *within-conditions/groups variance* relative to the between-conditions/ groups variance (see Chapters 7 and 8).

— Strength and advantage over one-group pretest-posttest design

• *Experimental contamination* cannot occur from extraneous factors due to *repeated measurements,* since only one measurement is made on each participant; i.e., there can be no effects of *testing (or order/sequence), mortality, regression, maturation, instrumentation, or contemporary history.*

— Weaknesses

1) Experimenter bias/expectancy

∞ This is a potential problem in *all designs* (thus it will not be noted again when discussing any of the remaining designs).

• *Control* procedures involve use of the *single-, double-, or partial-blind techniques, and automation,* in order to keep observers *unaware* of the conditions under which participants are run.

2) Participant bias/expectancy

∞ *Placebo or Hawthorne effects,* etc., which lead to observed differences between conditions regardless of a specific treatment's effectiveness, are possible in *all designs* too (hence this control problem also will not be noted again for any of the remaining designs).

• *Control* procedures involve the use of *deception or concealment,* i.e., *disguised experiments,* or making the control group a *placebo control* rather than a no-treatment control, and/or employing the *double-blind technique.*

— Conclusion

• *Experimental* is how this design would be classified.

∞ Note: This and *all the following designs* are true experimental designs due to degree of *experimental control* that is exerted by the researcher (for extended discussion see Chapter 2 under "Experimental Observation Used in Explanatory/Experimental Research"; and Chapter 7 under "Two Meanings of Experimental Control"):

• *Independent variables and their values* (the conditions) are purposely and directly produced — i.e., controlled.

• *Participant assignment* to groups and which groups receive which conditions are purposely and directly determined through the intentional use of randomization — thus *extraneous participant factors* are controlled.

- *Extraneous variables* of all *other forms* also can be controlled to a high degree in this and following designs.

5. **Pretest-Posttest Control Group Design (Before-After Randomized Two-Group Design)**

$$
\begin{array}{ccccc}
R & O & X & O \\
R & O & & O
\end{array}
$$

— Description

- This design is a *combination* of the One-Group Pretest-Posttest Design (looking just at the top row; see discussion in Chapter 9) and the Posttest-Only Control-Group Design (which as described earlier lacks the pretests shown near the left of the schematic).

- *Random assignment* (R) of participants is made to two groups, and an observation (O) is made on the dependent variable for the *experimental group* (top row) both *before and after* they receive the treatment (X), with any *change* being evaluated with respect to that for the *control group* (bottom row) that does not experience the treatment.

  - ∞ Note: Instead of a control condition, a *second treatment condition* could be used — with, of course, the appropriate descriptive name, such as the alternative listed above in parentheses.

— Analysis

- ANOVA (analysis of variance) or a *t*-test for independent groups would be run on the pretest-posttest change/difference/gain scores, when score data are obtained.

- Preferably, however (again assuming score data) one would run a Two-way ANOVA on the results of a 2×2 Mixed Factorial Design (treatments X tests), with repeated measures on the tests factor: pretest-posttest (see discussion under "Factorial Designs").

  - ∞ In this case, a *treatment effect* would be indicated by a significant *treatments-by-tests interaction*, rather than by a significant main effect for treatment (i.e., conditions).

    - Example of a 2×2 mixed factorial design matrix is shown in the matrix below. Note that *tests* would be a *repeated-measures factor* since both pretest and posttest observations are made for each participant in each condition/group. Mean dependent variable scores would be recorded in the four cells of the matrix.

	TESTS	
	Pretest	Posttest
CONDITIONS		
Treatment		
Control		

Example of a 2x2 Mixed Factorial Design Matrix

∞   This is a *more powerful/sensitive analysis* because the data are more completely used (not just differences of paired scores).

— Strengths and advantages over one-group pretest-posttest design

1) *Contemporary history* and *instrumentation* are controlled if participants in both groups are run during the same time period.

2) *Maturation* and *regression* are controlled by the presence of the control group and random assignment of participants to groups.

3) *Testing* is controlled with respect to any direct effects (internal validity problem) by the presence of the control group.

— Strength and advantage over posttest-only control group design

• *Pretest scores* provide a *check* for whether the random assignment of participants did in fact yield *equivalent groups*.

• *Pretest scores* are used to *adjust/correct* for any *nonequivalence* of participants that remains after randomization (this control is true for both *t*-tests and ANOVA).

∞   Note that the correction and control might not be perfect, which is especially likely *if pretest differences are large* (see the next three designs, which minimize the likelihood of this).

• In addition, note that this correction/control assumes *linearity* of the dependent variable over the range of values involved in the study.

∞   If the *average pretest scores* for the two groups are *equivalent*, then a comparison of just the posttest scores might be sufficient to evaluate the effect of the treatment — assuming there are no *ceiling or floor truncation problems* indicated by the pretest.

• However, this would not be the preferable analysis (see discussion above), since the *error variance* due to *individual participant differences* within-conditions would not be controlled if the pretest scores were not used, and thus the analysis wouldn't be as powerful/sensitive — i.e., *masking* would be more likely to occur.

— Disadvantage

• *Extra time and effort* are needed to make the pretest observation for each participant.

— Weaknesses

1) Participant attrition/experimental mortality

∞   This would be likely to occur to the extent that the *pretest- posttest interval* were not kept short.

∞   *Internal validity problems* arise if *dissimilar participants* drop out from the two groups, leaving different participant types.

∞   *External validity problems,* on the other hand, arise even when *similar participants* drop from both groups, if the dropouts, and as a consequence those remaining, are *not representative samples.*

   2) Testing-by-treatment interaction

      ∞ *External validity problems* (generalizability deficiencies) occur when the pretest produces *sensitization or resistance* to the treatment.

      ∞ Solomon four-group design can be used to control for this.

          • It is a rarely used *combination* of the pretest-posttest control group design and the preceding pretest-only control group design, resulting in a 2 × 2 factorial design having four groups, with the treatment and control conditions occurring both with and without the pretest, thereby allowing both the treatment and testing *main effects* to be evaluated independently, as well as their possible *interaction* (see under "Factorial Designs").

6. Randomized-Blocks Design

BR   X   O

BR   O

— Description of block randomization procedure (BR)

    • This design is an *extension* of the posttest-only control group design (which was discussed earlier).

    • Participants are *assigned to blocks* (B) based on some *shared characteristic,* and then from each block an equal/balanced number of participants is *randomly assigned* (R) to the different conditions (hence the name block-randomization procedure).

      ∞ Example: Participants could be blocked on the basis of *gender,* and then half of each block randomly assigned to each of the conditions.

      ∞ Note: This design and its procedure were *discussed earlier* in Chapter 8 under Randomizing an Extraneous Variable.

— Analysis

    • (See item 2 below regarding *blocks* being analyzed.)

— Strengths and advantages over posttest-only control group design

   1) Individual participant differences *between conditions* are reduced to a greater degree by *systematically blocking and balancing* participants, than by using *random assignment* alone.

      ∞ *Between-groups variance before treatments* is thus minimized as a result of this *quasi-random* assignment procedure.

          • *Confounding* is therefore better controlled.

   2) Blocks can be *analyzed* as an additional variable by expanding to a treatments-by-levels (blocks) factorial design, rather than just *balancing* for the participant extraneous variable blocking factor.

      ∞ *Statistical error variance* would thereby be reduced by an amount equal to the variance accounted for by the *systematized* extraneous variable (the blocking factor), and by its interaction with treatments.

- The amount of variance accounted for would depend on *degree to which the blocking factor is related to the dependent variable,* i.e., the extent to which the blocking factor is an extraneous variable.
- *Masking* would be controlled to this degree.
- *Power/sensitivity* to detect an effect of the independent variable would therefore be greater.

∞ Note: If the between blocks and blocks-by-treatments variance are negligible, then statistically they would not offset the *loss in degrees of freedom* for the error term that are given to those sources of variance.

- In such a situation the randomized blocks design would actually have *less power* than for simple random assignment.
- However, the loss of a few degrees of freedom is less important the *larger the sample size* (and thus the larger the total degrees of freedom); moreover, the blocking factor may be collapsed (ignored) if no effect is found for it, thus saving the degrees of freedom.

— Disadvantage

- *Extra time and effort* might be needed to block the participants.

— Enhancement of pretest-posttest control group design

- *Power/sensitivity* to detect a treatment effect would be greater if participants were *block randomized* on the basis of their *pretest scores* being above versus below the median of all participants, which would result in the following design:

  ∞ Randomized-Blocks Pretest-Posttest Design

BR	O	X	O
BR	O		O

  - Note: Although this design could instead be *diagrammed* so that it reflects the *actual sequence of events,* the resulting schematic (shown just below) would not be consistent with the schematics for other designs.

O	BR	X	O
O	BR		O

  ∞ *Power/sensitivity* would be still greater if, rather than being *blocked* and then *balanced,* participants were instead *matched* on their pretest scores before being randomly assigned to conditions (see "Pretest-Match-Posttest Design" later in this chapter).

  - Therefore, this enhancement, the "randomized-blocks pretest-posttest design," is *not really needed.*

Between-Participants, Related/Matched/Correlated, Two-Group Designs

All design examples are for when there is only *one independent variable,* and just *two levels* (such as presence versus absence of some treatment). (Appendix 7 is a summary of all the designs in this Handbook).

1. **Match by Correlated-Criterion Design**

$$
\begin{array}{ccc}
\text{MR} & \text{X} & \text{O} \\
\text{MR} & & \text{O}
\end{array}
$$

— Description of matching procedure (MR)

- Participants are *matched* (M) in this design on the basis of a *participant criterion* that is *measurable* and *related/correlated* with the dependent variable.

- Then *random assignment* (R) of the matched-participant members is made to either an *experimental group* (top row) that receives the treatment (X) or to a *control group* (bottom row) that does not, after which an observation (O) on the dependent variable is *compared* for the two groups of participants.

  - ∞ Examples: Participants could be matched on IQ or other test scores, years of education, age, degree of pathology, hormone levels, etc. — when such criteria are related/correlated with the dependent variable of the study.

  - ∞ Note: Instead of a control condition, a *second treatment condition* could be used in this and the next design.

— Analysis

- ANOVA (analysis of variance) or a *t*-test for related/correlated measures would be used when score data are obtained; i.e., the same analysis as for the within-participants/groups designs.

— Advantages over between-participants, independent-groups designs

- Matching better controls for *measurable extraneous individual participant differences* than does random assignment by itself; it is an *extended and more precise* form of blocking-and- balancing than the randomized-blocks designs, whenever there are *several values/levels* of the secondary, extraneous participant variable, as opposed to just two or three levels.

  - ∞ Therefore, the *between-groups variance* on the dependent variable *before* any treatment is administered (i.e., the variance *not* related to the treatment) would be reduced.

    - The amount of the reduction would be *contingent on* the extent to which the participant matching criterion/factor is actually *correlated/related* to the dependent variable, i.e., by the degree to which the matching criterion *influences* the dependent variable, and is therefore an *extraneous variable.*

∞ *Statistical error variance* associated with individual participant differences would be partly accounted for, and thus reduced as well.

• The reduction would be proportional to *twice the correlation* of posttest scores under the different conditions for the *matched participants* (as was also noted before for *within-participants* correlated-groups designs, where the correlation should be higher since repeated measures on individual participants represents *extreme matching* — see the *t*-test equation there).

Note: The *correlation of posttest scores* would itself be determined by the earlier mentioned correlation between the *matching-criterion extraneous variable* and the *dependent variable*, which reduces between- groups variance before any treatment is administered.

• *Masking* would be reduced as a result of the decreased statistical error variance.

• *Power/sensitivity* to detect a treatment effect would consequently be increased.

— Disadvantages relative to between-participants, independent-groups

1) *Correlations,* and hence the decrease in error variance, must be sufficiently large to offset the *loss in degrees of freedom* that occurs when *matching* is used (which is similar to the concern discussed earlier regarding analyzing for a blocking factor).

∞ Specifically, for *matched-groups* designs $df = n-1$, where n equals the number of matched pairs, i.e., the number of participants per group; in contrast, for *randomized-groups* designs $df = N-2$, where N equals the total number of participants (naturally $n-1$ is always less than $N-2$, except when N is only 2, which would obviously not be desirable).

∞ If the correlation is not sufficiently large, then matched- groups designs actually would have *less power/sensitivity* than randomized-groups designs.

• Thus matching should be used only when the matching variable is *strongly related* to the dependent variable measure (i.e., correlations greater than 0.5). Otherwise, it is probably best that participants not be matched.

• However, the *greater the number of participants per group,* the less the statistical effect on power of losing degrees of freedom.

But *matching is most likely* to be used when there are *only a few participants* available, or when it is *very expensive* to run many participants, since with simple *random* assignment the *equivalence of groups* in different conditions is *less* likely the *smaller* the number of participants — thus making matching all the more important.

2) Matching procedures also can be *costly and time-consuming,* hence they are *less commonly used* than simple random assignment (see also the "Weaknesses" noted below).

— Advantage over within-participants/groups designs

- *Problems of repeated measures,* especially order/sequence effects, are avoided or reduced (but see "Weaknesses" below).

— Disadvantages relative to within-participants/groups designs

1) *Individual participant differences* are not as well controlled since, due to the many *unmatched variables,* the matched participants would not be as equivalent as when compared to themselves.

2) *More participants* are required to get the same amount of data.

3) *More participant preparation and set-up time* is usually required since more participants would be used.

4) *More time and effort* also are required because of the need to obtain the criterion measurement or measurements on which the participants would be matched.

— Weaknesses

1) Testing-by-treatment interaction

   ∞ *Matching participants* using a *pretreatment measurement,* rather than an already available measure, can lead to a testing-by-treatment interaction.

   ∞ *External validity problems* (generalizability deficiencies) occur when the pretest produces *sensitization or resistance* to the treatment.

   ∞ *Internal validity,* however, is not a problem since the *control group* controls for any *direct effects* of a pretest on the posttest scores.

2) Participant attrition/experimental mortality

   ∞ *Matching participants* using a *pretreatment measurement,* rather than an already available measure, also can lead to the extraneous variable concern of participant attrition.

   ∞ It depends, in part, on whether participants have to *return later* for the posttest, and on the *interval* between measures.

   ∞ *Internal validity problems* arise if dissimilar participants drop from the two groups, which can occur if those in different groups were *not matched* on the factor(s) leading to dropout.

   ∞ *External validity problems* arise even when similar participants drop from both groups, if the dropouts, and therefore those remaining, are *not representative samples.*

2. Pretest-Match-Posttest Design

$$\begin{array}{cccc} MR & O & X & O \\ MR & O & & O \end{array}$$

— Description of matching procedure (MR)

- Participants are *matched* (M) in this design on the basis of the pretest observation (O) on the *dependent variable* itself, rather than on some correlated criterion.

- Then *random assignment* (R) of the matched-participant members is made to either an *experimental group* (top row) that receives the treatment (X) or to a *control group* (bottom row) that does not, after which an observation (O) on the dependent variable is *compared* for the two groups of participants.

  - ∞ Note: For *schematic consistency* the method of assignment is shown in the leftmost column, although pretesting of all participants actually occurs before assignment to conditions.

  - ∞ Also, instead of a control condition, a *second treatment condition* could be used in this design, as for previous designs.

— Analysis

  - ANOVA (analysis of variance) or a *t*-test for related/correlated measures would be used when score data are obtained, with the analysis run on the *posttest data only* since the pretest measure would be used just for matching the participants — i.e., the same analysis as for the within-participants/groups designs.

— Advantage over match by correlated-criterion design

  - Individual participant differences are better controlled.

    - ∞ *Pretest dependent variable scores* should be the participant criterion *most strongly correlated* with posttest dependent variable scores, since the pretest ideally would measure *all* participant factors that could affect the dependent variable.

    - ∞ *Statistical error variance*, therefore, would be more substantially decreased (close to within-participants designs).

      - *Masking* would thereby be more thoroughly reduced.

      - *Power/sensitivity* to detect a treatment effect would consequently be increased to a greater extent.

— Disadvantages relative to match by correlated-criterion design

  1) *More time and effort* might be required to obtain the *pretest dependent variable measurement* on which to match participants, than would be the case to obtain some other correlated criterion for matching, which in fact might already be available.

  2) Testing-by-treatment interaction

    - ∞ *Greater external validity problems* (i.e., generalizability deficiencies due to production of sensitization or resistance to the treatment) exist potentially for this design than for the match by correlated-criterion design, since for the latter design a criterion measure might already be available before the study begins, and thus there would be no pretesting.

    - ∞ Note that there are *no internal validity problems* of testing for either of these matching designs, since the *control group* controls for *direct effects* of a pretest on the posttest scores.

3) Participant attrition/experimental mortality

- ∞ *Greater internal and external validity problems* exist potentially for this design than for the match by correlated-criterion design, since for the latter design, as just noted, a criterion measure might already be available before the study.
  - The degree of these problems depends partly on whether participants have to *come back later* for the posttest, and on the *interval* between pretest and posttest measures.

3. **Yoked Control Group Design**

$$Y(R/MR) \quad X(X') \quad O$$
$$Y(R/MR) \quad (X') \quad O$$

— Description of yoked matching procedure Y(R/MR)

- Participants are *matched/yoked* (Y) in this design for extraneous characteristics of the *experimental situation* (X'), and perhaps also for *environmental factors*, as opposed to being matched for extraneous participant factors.

- However, in addition the participants might *also be matched* (M) on one or more extraneous *participant factors*.

- Then *random assignment* (R) of the participants is made to either an *experimental group* (top row) that receives the treatment (X) or to a *control group* (bottom row) that does not, after which an observation (O) on the dependent variable is *compared* for the two groups of participants.
  - Note: *Yoking participants* was *discussed earlier* in Chapter 7 under "Making Equivalent the Extraneous Variance."

- Control group
  - ∞ Through the *yoking* procedure, each *control group* participant is exposed to the same *quantity*, and also usually the same *temporal distribution*, of extraneous experimental events (X') as their yoked/matched *treatment group* participant.

- Treatment (X)
  - ∞ It is the particular *sequence or temporal relationship of experimental events*, as opposed to just the events themselves, that often constitutes the *treatment*.
    - In this case, the difference between the treatment and control participants would be whether or not there is a *contingent relationship* between the participant's behavior and treatment events.
    - Specifically, the *pattern of treatment* would be *dictated* by the behavior of the participants in the experimental group, and this pattern then would be *imposed* on the corresponding yoked participants in the control group, *irrespective* of their behavior.

- Example: See the example that was given for this control procedure in Chapter 7, regarding a study on whether electrical stimulation of *"pleasure centers"* in the brain is reinforcing — i.e., whether the electrical stimulation needs to be *contingent* on the desired behavior in order to increase its occurrence.

∞ Alternatively, or as well, the *experimental treatment* might consist of some *additional factor* beyond the extraneous characteristics of the experimental situation.

- Example: See the next example under physical yoking.

- Yoking methods

  ∞ Physical yoking

  - *Harnesses* of some sort can be used to physically yoke participants together, as when oxen are *yoked together* to pull a wagon or a plow — hence the design's name.

  - Example: A study was conducted of whether *active/spontaneous movement* or only passive movement is necessary for the *development of visually guided sensory-motor coordination.*

    Kittens served as subjects/participants, and while an *experimental animal moved actively* around a central axis within a vertically striped cylindrical chamber, a *physically yoked control animal* was moved *passively* but identically by being carried in a gondola that was connected to the experimental animal by means of an interconnecting rod and chain-gear system.

    Thus the yoked control kittens were exposed to the *same pattern, quantity, and temporal distribution of visual stimulation,* which, however, was associated only in the experimental animals with the *additional factor* (treatment) of *correlated* motor behavior. Alternatively, one could say that the treatment was change in sensory stimulation being *contingent* upon the motor behavior of the experimental participants.

    The kittens received visual experience only while in the cylindrical chamber. *Results* on later tests of visually guided behavior showed that *active/spontaneous movement was necessary* for visually guided sensory-motor coordination to develop [Held, R. & Hein, A. (1963). Movement produced stimulation in the development of visually guided behavior. *Journal of Comparative and Physiological Psychology, 56,* 872-876.]

  ∞ Electronic yoking

  - Computerized or electromechanical *programming equipment* can be used, instead of physical yoking, to produce *equivalence* in the extraneous characteristics of the experimental situation.

- • Example: See example noted earlier about *reinforcement* and the *contingency of "pleasure center"* stimulation.

— Analysis

- • ANOVA (analysis of variance) or a *t*-test for related/correlated measures would be used when score data are obtained; i.e., the same analysis as for other *matched-participant designs*.

— Strength and advantage

- • *Yoking* is needed when *experimental manipulations unavoidably* (and undesirably) involve the *introduction* of one or more secondary, extraneous variables *along with* the independent variable, *and* the other control techniques are not as satisfactory because different participants in the experimental group — *due to their own performance* — would likely experience *different exposures* to these extraneous variables.

  - ∞ Under these circumstances, *yoked controls* would be necessary in order to match participants on the *amount and/or temporal distribution* of extraneous variables, and thereby *isolate* the independent variable effect.

    - • This is the case for all the examples given for yoking in this chapter as well as in Chapter 7.

  - ∞ However, a *non-yoked control group* would be *sufficient and more appropriate* to control for *experimental contamination* when exposure to the introduced extraneous variable is *not likely to vary* due to the participants' performance.

    - • This is the case, e.g., when controlling for *surgical effects* in brain lesion studies by having a *sham-lesion control group* (covered in Chapter 8 under "Control Groups").

  - ∞ It should be noted that the yoked control group design also incorporates controls for *individual participant differences*.

    - • Participants are randomly assigned to conditions in this design, and also they are sometimes first *matched* for extraneous participant factors (as stated earlier).

— Disadvantages

- • *Time, effort, and cost* of the procedure and apparatus (which might have to be specially built) can be considerable for yoking together the experimental and control participants.

Multiple Treatment Designs

Multiple treatment designs are more complicated than the preceding two- level, one-independent variable designs, but they have distinct advantages. (Appendix 7 is a summary of all the designs in this Handbook.)

1. Characteristics

a. Multiple treatment designs are characterized either by having *more than two levels* of a single independent variable being studied, in which case they are called *multilevel designs;* and/or there is *more than one independent variable* (factor) studied, in which case they are called *multifactor/factorial designs.*

b. *More time, effort, and participants* are usually involved in the use of Multiple Treatment Designs.

c. *Disproportionately more information,* however, is provided by these designs than by the previously discussed one-independent-variable, two-level designs; hence multiple treatment designs are actually *more efficient.*

d. As for the previously discussed designs, multiple treatment designs might also involve *more than one dependent variable,* in which case MANOVA (multivariate analysis of variance), rather than ANOVA (analysis of variance), should be used in statistical analyses.

- Note: The dependent variable(s) measured should be the same for all levels of *all the independent variables,* otherwise there is little advantage over simply studying the different independent-dependent variable relationships in *separate* experiments.

e. *Assignment of participants* to conditions can be by *randomization or matching* for between-participants comparisons, or by using *repeated measurements* for within-participants comparisons — just as for the one-independent-variable, two-level designs covered earlier.

2. Multilevel Designs

$$
\begin{array}{cccc}
\text{R/MR?} & \text{O?} & X_1 & O \\
\text{R/MR?} & \text{O?} & X_2 & O \\
\text{R/MR?} & \text{O?} & X_{3(0)} & O \\
\cdot & \cdot & \cdot & \cdot \\
\cdot & \cdot & \cdot & \cdot \\
\cdot & \cdot & \cdot & \cdot \\
\end{array}
$$

— Description

- *More than two levels* of a *single independent variable* (X) are studied with this design; otherwise, it is just like the two-level designs discussed earlier, i.e., multilevel designs:

  ∞ Can include a *control condition* (X_0), or just *treatments* (X_3).

  ∞ Can involve *within-participants comparisons,* or if *between- participants,* then assignment can be by *randomization* or by *matching* followed by randomization (R/MR?).

  ∞ Can have *posttest-only* or *pretest-posttest* observations (O?).

    - Pretest observations can be used to determine pretest- posttest *change/difference/gain or gain ratio scores.*

- Note: If pretest and posttest observations are both included in the analyses, rather than just their change/difference/gain or gain ratio scores, then there would be an *additional independent variable of "tests,"* and hence the design would be *factorial* (covered later).

— General questions asked and answered by this design

- Consider the *simplest situation* of only three independent variable levels, with one of them being a control condition (X_0), and where there are no pretests, only posttest observations (O):

$$X_1 \quad O$$
$$X_2 \quad O$$
$$X_0 \quad O$$

This can be thought of as a *combination of three two-level designs* within *one* multilevel design that addresses the following issues:

∞ How do the effects of each of the *treatments* (two in this case: X_1 and X_2) compare with that of the *control condition* (X_0)?

∞ How do the effects of the *treatments* (X_1 and X_2) compare with *each other?*

— Analysis

- A one-way ANOVA (analysis of variance), for either independent groups or for related/correlated measures (whichever is appropriate), would be used when score data are obtained.

— Strengths and advantages

1) *More efficiency* is achieved, since fewer participants and less time and effort are required to use *one* multilevel design than to run *several* separate two-level designs in order to make all the possible comparisons among the three or more conditions.

∞ Example: Suppose you wanted to compare the effects of *three levels* (X_1, X_2, and X_0) of some independent variable, with 30 *participants/ subjects studied at each level.*

- If you chose to run *three separate* between-participants two-level designs in order to make the three comparisons required for the three levels taken two at a time, then you would need 60 participants (2×30) in each experiment, for a *total of 180 participants* (3×60).

On the other hand, with a *single* between-participants multilevel design you would need *only 90 participants* to answer the same three questions (3×30).

Thus, in this simplest of cases (i.e., only three levels), a multilevel design would require only *half as many participants* to obtain the same amount of information, and even *more savings* would occur if additional independent variable levels were studied (this is similar to the savings when within-participants designs are used).

Another way to illustrate increased efficiency is when one condition is added to a two-level design, producing a three-level design, then *three questions* can be answered as opposed to just one, and yet *only 50% more participants* would be needed (e.g., 90 versus 60).

2) *Easier control over extraneous variables* is possible in a single multilevel design study, as opposed to several two-level design studies used to answer the same questions.

3) *Enhanced likelihood* occurs in multilevel designs for employing *optimal values of the independent variable* for comparison — since more than two levels are used — thereby maximizing the primary variance and *increasing the power/sensitivity* for finding any independent variable effects that exist.

4) *More accurate determination* can be made of the nature of the *functional relationship* between the independent and dependent variables when using multilevel designs (e.g., linear or curvilinear, monotonic or nonmonotonic relationships) — again as a result of studying more than two independent variable levels (see next).

— Parametric research

- When *multilevel designs* are used to study the relative effects of several levels of a *quantitative* independent variable on some dependent variable, we have what is called *parametric research*.

  ∞ The *purpose* of such research is to determine the *parameters* (values) of the equation that describes the *form/shape* of the functional relationship between those variables (noted in Chapter 7 under "Maximizing Primary Variance").

3. Factorial (Multifactor) Designs

		A		
		a_1	a_2	
	b_1	a_1b_1	a_2b_1	Mb_1
B	b_2	a_1b_2	a_2b_2	Mb_2
		Ma_1	Ma_2	

— Description

- Factorial designs are *different* from all the preceding designs that have been discussed, since these other designs have only a *single* independent variable.

- Factorial designs, in contrast, always have *two or more independent variables/factors* (two factors in in the preceding schematic example: A and B), which are arranged *orthogonally* (perpendicular) to each other; i.e., they are independent of and thus *not confounded* with one another.

- This independence and non-confounding in factorial designs is accomplished by having all levels of each factor *balanced* — i.e., occur *equivalently* — under all levels of the other factors (in the schematic example, levels b_1 and b_2 of Factor B are both balanced under levels a_1 and a_2 of Factor A, and vice versa).

  - ∞ Hence the effects of each variable/factor can be evaluated *separately*, as well as in *combination* (as explained later).

- *Schematics* for factorial designs can be seen to be quite different from those for other designs, i.e., they are a *matrix* with a separate *cell* for each of the *condition/treatment combinations* of the independent variable levels (there are four combinations in the simplest case, as shown in the schematic example: a_1b_1 etc.).

  - ∞ As already indicated, independent variable *factors* are designated in the schematic by different capital letters, whereas the *levels* of those factors are designated by corresponding lowercase letters with subscripts. (Note: M is a *marginal mean*, and represents the *mean dependent variable score* for a given independent variable level. The mean dependent variable scores for particular *combinations* of independent variable levels would be shown in their respective cells. All of this is illustrated in examples later in this chapter.)

- Factorial designs can be either *posttest-only* or *pretest-posttest*.

  - ∞ In the latter case, pretest-posttest *change/difference/gain or gain ratio scores* could be computed; alternatively, and more commonly, the pretest and posttest scores could be analyzed as two levels of an *additional factor* referred to as *tests* or *trials*, which would be a *more powerful/ sensitive analysis* since the data would be more completely used.

— Major forms of factorial designs

- Complete factorial design

  - ∞ *All possible combinations* of the selected values, or levels, of two or more independent variables are studied when using *complete factorial designs* (as in matrix schematic above).

    - This is the *most common form* of factorial designs, and the only type that is discussed here in any detail (examples are given later in this chapter).

- Nested factorial design

  - ∞ When using a nested factorial design, *different levels* of one independent variable — the *nested factor* — are paired with *each* of the different levels of another independent variable, and thus the variables are *not factorially combined* in this design.

- Example: Studying the effectiveness of three different algebra textbooks, with a *different set* of high school classes using each different textbook — thus classes would be a *nested factor* under textbook.

Algebra Book A	Algebra Book B	Algebra Book C
Class 1	Class 3	Class 5
Class 2	Class 4	Class 6

Note that there must be at least *two* classes using each different textbook, otherwise there would be *complete confounding* of the independent variables.

To control for *extraneous variables*, it is also essential that the algebra classes be as *equivalent* as possible in all respects other than the specific textbook that is used.

- Nested factorial designs are *complex*, but if appropriately analyzed, using advanced statistics, the nested design might present no special problems of interpretation.

∞ Hierarchically nested design

- When using a hierarchically nested design, *two or more nested factors* are involved, with one of the nested factors nested under another nested factor, and so on.

  Example: Several different districts of schools might be assigned to each of several different conditions, so that different schools are *nested* within districts, and different districts are *nested* within conditions.

— Distinctions among factorial designs within the major forms

In addition to the preceding major distinctions among factorial designs, they also vary in *four other ways*:

1) How many *independent variables* are studied?

   ∞ There must always be *two or more* independent variables (this is the very definition of factorial designs).

2) How many *levels* are investigated for each independent variable?

   ∞ There must always be *two or more* levels (this is required for *comparison* if one is to determine whether there is an *effect* of an independent variable on some dependent variable).

3) How are *participants/subjects* assigned to conditions?

   ∞ *Between-participants designs* would involve either independent/randomized-groups assignment, or related/matched-groups assignment.

∞ *Within-participants designs* would involve repeated measures on participants (more than one condition per participant).

∞ *Mixed factorial designs* would involve *combinations* of both *independent groups* (i.e., random assignment of participants) for the levels of one or more factors, and *correlated groups* (i.e., matched assignment or within-participants repeated measures) for the remaining factors — thus mixed designs are not necessarily "Between-Within Designs," as some books designate them (i.e., all factors can be *between participants*).

4) How are *conditions/treatments* selected?

∞ Fixed-effects model

• *Systematic choice* of levels for all independent variables defines this model.

Example: Deciding, based on experience and logic, to study the effects of three *systematically chosen* levels of food deprivation (such as 0, 24, and 48 hours), and three *systematically chosen* levels of sleep deprivation (such as 0, 24, and 48 hours) on the rate of maze learning by rats.

• This is the *most common* model of factorial designs.

∞ Random-effects model

• *Random choice* of levels for all independent variables defines this model.

Example: *Randomly* selecting three values from among all possible levels of food deprivation and sleep deprivation (within the limits of causing serious health problems) when studying their effects on the rate of maze learning by rats.

• This is an *uncommon* model of factorial designs.

∞ Mixed-effects model

• *Combinations* of the fixed- and random-effects models define this model; i.e., each approach is used for at least one of the independent variables in the factorial design.

Example: *Systematically choosing* three levels of food deprivation and *randomly choosing* three levels of sleep deprivation in a factorial design study of effects on maze learning by rats.

• This model is *less common* than the fixed-effects model, but *more common* than the random-effects model.

∞ Note: The *choice of models* determines the *appropriateness of generalizations* to other levels of the independent variables, and hence *confidence* in the generalizations.

• Generalizations to *non-sampled levels* would be inappropriate for a fixed-effects model if an *arbitrary, non-representative criterion* were used to systematically select levels.

This common approach is used when an experimenter wants to see how participants perform under certain *specific conditions,* and only wants to generalize to *other individuals,* not to other conditions/levels (e.g., *non-studied* dosages of a drug that is investigated).

- The *greatest confidence* possible when generalizing to other independent variable levels would result from a *quasi-random approach* to the random-effects model, since properly chosen *restrictions* on randomization — i.e., some *systematization* — should yield the *most representative* levels (see Chapter 11 on "Sampling and Generalization").

    Example: Randomly choosing one level each from the lowest, middle, and highest third of the *population* of feasibly studied *independent variable levels/values,* such as the magnitude or frequency of rewards.

- ∞ Note: The choice of models also determines the proper *statistical error term* in analyses, and thus the *power/sensitivity* to detect independent variable effects.

— Rules for shorthand designation of factorial designs

- *Matrix* that was illustrated earlier in the schematic diagram is a 2×2 factorial design, using shorthand designation, and it is the simplest factorial design possible — consider instead the more complex example of a 3×4 factorial design:

    - ∞ *Number of numbers* in the designation indicates the number of *independent variables* studied (two in this example, as there are two numbers in the designation, i.e., 3 and 4);

    - ∞ *Magnitude of the numbers* indicates the number of *levels* investigated for each independent variable (3 levels of one variable and 4 levels of the other variable in this example);

    - ∞ *Product of the numbers* (which is never shown, but is implied in the designation) indicates the number of *combinations* of levels of the independent variables investigated, i.e., the number of condition/ treatment combinations, and hence the number of cells in the factorial matrix (12 in this example, which is the product 3×4).

- Incorporating into the shorthand designation also the *method of participant assignment* to conditions/levels, as well as the names of the independent variables and their levels, we might have, e.g.: a $3 \times 2 \times 2$ *mixed factorial design* (Reward: high, medium, and low; × Gender: male and female; × Tests: pretest and posttest) with *repeated measures* on one factor (Tests).

— Analysis

- When score data are obtained, a two-way ANOVA (analysis of variance) would be used if there are two independent variables, a three-way ANOVA if there are three independent variables, etc.

∞ The ANOVA would be specified, depending on which is correct, as either randomized/independent between-groups, or matched/related between-groups, or repeated-measures/within-groups, or mixed-groups (i.e., independent-correlated; with additional descriptive information — see example).

- Example: Three-way mixed-groups ANOVA, with repeated measures on one factor (note how the designation differs somewhat from that commonly used for factorial designs).

— Strengths and advantages

1) *More efficient design:* Fewer participants and less time and effort are required to gain information about the effects of two or more independent variables when they are studied using a *single* factorial design, rather than using two one-factor designs.

∞ With factorial designs, from *each participant* data are *simultaneously* gathered for a level of *every* independent variable (e.g., a_1 and b_1 or a_2 and b_1 in the matrix schematic example illustrated earlier).

∞ Thus *multiple hypotheses* can be tested *simultaneously,* i.e., hypotheses regarding *separate (independent) effects* of *different* independent variables.

- As noted earlier, under each level of every independent variable the levels of all other independent variables are *balanced*, thus there is *no confounding*, and the *average* effects (*main effects —* see below) can be determined (e.g., $Ma_1 - Ma_2$ and $Mb_1 - Mb_2$ in the matrix schematic shown earlier, where the two levels of factor B occur equally often under the two levels of factor A, and vice versa).

- Example of efficiency: Suppose you wanted to know for *rats* whether variation in *hours of food deprivation* (independent variable A) affects the *rate of bar pressing* for food (the dependent variable); in addition, you might also want to know whether variation in the *magnitude or type of food reward* (independent variable B) also affects the same dependent variable, i.e, the rate of bar pressing.

  If you wanted to investigate each of these two problems/questions separately, with 30 participants randomly assigned to each of, say, two levels of each independent variable, then 60 rats would be required for each of the two studies, for a *total of 120 participants.*

  However, if a factorial design were used to answer both questions in a single experiment, then a *total of only 60 participants* would be required: i.e., 15 rats under each of the four treatment combinations of independent variable levels, which would yield 30 animals under each of the two levels for each of the independent variables (see Table 10.3).

Thus, in this simplest of cases, only *half as many participants* would be needed to obtain the same amount of information when using a factorial design — and even greater efficiencies would occur in more complex cases.

TABLE 10.3

Example of Efficiency of a Factorial Design
Testing Two Independent Variables
Simultaneously

		Factor A		
Factor B		a_1	a_2	**Total**
b_1		15	15	30
b_2		15	15	30
Total		30	30	

Note: Values shown are the number of participants.

2) *Interaction effects* of independent variables can also be evaluated/measured when using a factorial design — in addition to the already noted *main effects* (i.e., average effects).

∞ These are determined from information about *combined/joint effects* of independent variables (i.e., *simple main effects* — as described below).

3) *More power/sensitivity* is often possible for the detection of independent variable effects when using a factorial design.

∞ *Systematization of secondary, extraneous variables* that are not eliminated or held constant, but which instead are *balanced* using the randomized-blocks (treatments-by-levels) factorial design, permits the *analysis* of extraneous variable effects — thereby *reducing the statistical error variance*, which is the *unaccounted* for variance of a dependent variable.

• Note too the *additional benefit* that valuable information is gained about the *effects of extraneous variables* controlled in this manner (see earlier discussion of this in Chapter 7, under "Treatment Groups and Systematizing Extraneous Variables").

— Main effects

• The *overall*, or *average/mean*, effect of an independent variable on a dependent variable is called the *main effect* of that variable.

∞ The *main* effect is thus the *difference in effects* of the different levels of an independent variable when *averaged* over *all levels* of all the other independent variables (also called the *differential main effect*, or the *independent effect*).

∞ Main effects are *determined* by a *comparison* of the *marginal matrix means* for *columns* or for *rows*, depending on the particular arrangement of the factors in the matrix.

∞ In a 2×2 Factorial Design, the main effects of A and B are:

$$A = Ma_1 - Ma_2 = (a_1b_1 + a_1b_2)/2 \quad - \quad (a_2b_1 + a_2b_2)/2$$
$$B = Mb_1 - Mb_2 = (a_1b_1 + a_2b_1)/2 \quad - \quad (a_1b_2 + a_2b_2)/2$$

Note: Here a_1b_1 represents the mean dependent variable score for participants receiving this particular treatment combination, and similarly for a_1b_2 and a_2b_1 and a_2b_2 — see first factorial design schematic matrix.

— Simple main effects

- The *selective*, or *unaveraged*, effects of an independent variable on a dependent variable is called the *simple main effect* of that independent variable.

  ∞ The *simple* main effect is thus the *difference in effects* of the different levels of an independent variable when considered at *specific, selected levels* of the other independent variables (rather than averaged over all levels of all the other independent variables, which would be the main effect).

  ∞ In a 2×2 factorial design, simple main effects of A and B are

$$A = a_1b_1 - a_2b_1 \quad \text{as well as} \quad a_1b_2 - a_2b_2$$
$$B = a_1b_1 - a_1b_2 \quad \text{as well as} \quad a_2b_1 - a_2b_2$$

— Interaction effects

- The *unique combined/joint* effect of two or more independent variables on a dependent variable is called the *interaction effect* of those independent variables

  ∞ An *interaction* effect is indicated by an *inequality* of the *difference in effects* that the different levels of an independent variable have at the *various levels* of the other independent variable(s).

    - Hence interactions are *differences among simple main effects* — e.g., differences in relative/differential effects of the two levels of Factor A at the two levels of Factor B, and vice versa (this is illustrated later in the chapter):

$$a_1b_1 - a_2b_1 \quad \text{versus} \quad a_1b_2 - a_2b_2$$

      and similarly for Factor B

$$a_1b_1 - a_1b_2 \quad \text{versus} \quad a_2b_1 - a_2b_2$$

    - Example: See the first hypothetical, data matrix computation example for a factorial design presented a few pages later in this chapter, and ignore for now the detailed discussion below the matrix.

  ∞ When there is an interaction, then the effects of one independent variable are *dependent* on the *level of the other independent variable(s)* at which its effects are measured (Note: *ordinal versus disordinal interactions* are compared in Chapter 11, under "Generalization/External Validity Concerns").

- Example: The relative/differential effects of different forms/levels of *psychotherapy* might not be the same (i.e., they might be greater or less) when combined with different types of *psychotropic drugs*, and vice versa.

• Interactions signify *mutual influence* among the independent variables of the study with regard to dependent variable effects.

∞ Effects of Factor A would be influenced by (and thus dependent on) Factor B, and vice versa — an *interdependence*, at least statistically.

- This explains the *uniqueness* of the combined effects.

• Interactions also signify the *nonadditivity of main effects*, which is a reflection of the *unique* combined/joint effects resulting from the *interdependence* of the independent variables.

∞ Thus, when there is an interaction, the effect of a *condition/treatment combination* would *not* be predictable from just the *sum* of the corresponding condition/treatment *main effects*, for example:

$$a_1b_1 \neq a_1 + b_1$$

∞ Interactions therefore represent something *other* than a simple summation of effects of independent variables: they indicate that the *whole is greater or less than the sum of the parts*, i.e., that the total is *different* than the sum.

- Example: Consider water, where the fire-extinguishing properties of the molecule are not predictable from the sum of the properties of the atoms, i.e., $H_2O \neq 2H + O$ (liquid $H_2O \neq$ the sum of two flammable, explosive gases).

∞ In a 2×2 factorial design, the *interaction* between A and B (which is *symbolized* as AB or $A \times B$) is

$$AB = (a_1b_1 + a_2b_2)/2 \quad - \quad (a_2b_1 + a_1b_2)/2$$

- The interactions are thus *determined* by *comparison* of the means for the *matrix diagonals* (as opposed to columns or rows) — however, this is only true in these simple 2×2 designs, since there are no complete diagonals in larger, more complex factorial designs.

- Note that mathematically if there were *no interaction effect*, and hence the effect of treatment combinations (e.g., a_1b_1) were simply equal to the *sum* of their corresponding main effects (e.g., $a_1 + b_1$), then the difference between the means for the diagonals would equal zero — thus appropriately indicating no interaction:

$$AB = (a_1 + b_1 + a_2 + b_2)/2 \quad - \quad (a_2 + b_1 + a_1 + b_2)/2 = 0$$

- Interactions are actually the *non-additive sources of variation* (sums of squares) seen after subtracting/removing the additive effects of conditions — and sometimes they are referred to as the *multiplicative effects* of conditions.

  - ∞ Note: The following explanation is rather technical, but is clarified by two computational examples. (Explanation and examples are provided for those who appreciate mathematical confirmations. Others can skip over to "Graphic Illustration and Interpretation of Interactions.")

  - ∞ If the *grand mean* of a set of data is computed, by taking the *average of the mean dependent variable scores* under all the condition combinations (e.g., a_1b_1 etc.), then the *effect of each condition* (e.g., a_1) can be computed simply by calculating the *difference/deviation* of the mean under each condition (the matrix margin means) from the grand mean (see hypothetical examples below for clarification).

  - ∞ If mean dependent variable score under given *condition combination* (e.g., a_1b_1) is not equal to *additive effect* of conditions, i.e., to grand mean plus sum of effects/deviations under individual conditions making up that combination ($a_1 + b_1$ in this case), there must be a *non-additive source of variation*, i.e., an interaction/multiplicative effect.

  - ∞ *Interaction effects* can be computed by *subtracting the additive effect* from the condition combination mean, as the sum of additive and multiplicative (interaction) effects should equal condition combination mean.

  - ∞ The following *hypothetical data-matrix computation example* illustrates these principles for an *AB interaction* — note the differences present between simple main effects ($a_1b_1 - a_2b_1 \neq a_1b_2 - a_2b_2$ and similarly $a_1b_1 - a_1b_2 \neq a_2b_1 - a_2b_2$); also note differences between diagonal means:

Data-Matrix Computation with AB Interaction

		A		
		a_1	a_2	
B	b_1	10	20	15
	b_2	14	56	35
		17	12 38	33

Grand mean = $(10 + 20 + 14 + 56)/4 = $ **25**
Deviation/effect of $a_1 = 12 - $ **25** $= -13$
Deviation/effect of $b_2 = 35 - $ **25** $= +10$
Additive effects of $a_1 + b_2 = $ **25** $+ (-13) + 10 = 22$
Condition combination $a_1b_2 = 14 \neq a_1 + b_2 = 22$
So, there is a *non-additive* source of variation:
Interaction effect $a_1 \times b_2 = a_1b_2 - (a_1 + b_2) = 14 - 22 = -8$
Note that $a_1b_2 = (a_1 + b_2) + (a_1 \times b_2) = 22 + (-8) = 14$

∞ The following *hypothetical, data-matrix computation example* illustrates *no AB interaction* — note the lack of differences between simple main effects ($a_1b_1 - a_2b_1 = a_1b_2 - a_2b_2$ and similarly $a_1b_1 - a_1b_2 = a_2b_1 - a_2b_2$), and also note that the diagonal means are equal:

Data-Matrix Computation with no AB Interaction

		A		
		a_1	a_2	
B	b_1	10	20	15
	b_2	30	40	35
		25	20	30	25

Grand mean = $(10 + 20 + 30 + 40)/4 = $ **25**
Deviation/effect of $a_1 = 20 - $ **25** $= -5$
Deviation/effect of $b_2 = 35 - $ **25** $= +10$
Additive effects of $a_1 + b_2 = $ **25** $+ (-5) + 10 = 30$
Condition combination $a_1b_2 = 30 = a_1 + b_2 = 30$
So, there is *not* a non-additive source of variation:
Interaction effect $a_1 \times b_2 = a_1b_2 - (a_1 + b_2) = 30 - 30 = 0$
Note that $a_1b_2 = (a_1 + b_2) + (a_1 \times b_2) = 30 + 0 = 30$.

— Graphic illustration and interpretation of interactions

— Interactions are *most clearly comprehended* when the data for condition/treatment combination mean effects are presented in graphic form (although tables are usually best for main effects).

Graphic Presentation of Condition/Treatment Combination Mean Effects

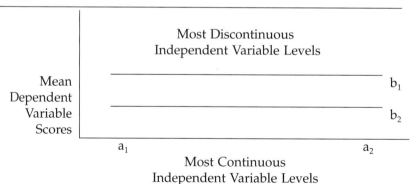

∞ Mean *dependent variable scores* would usually be plotted on the vertical axis, or *Y-axis* (the ordinate).

∞ Levels of the *independent variable* that is most *continuous* would be plotted on the horizontal axis, or *X-axis* (abscissa).

• Example: *Hours of food deprivation* is a continuous variable, since its values lie along a continuum that can be represented by both whole and fractional units, i.e., the values do not have to vary in discrete steps. (For further discussion see Chapter 4 under "Discrete Versus Continuous Variables.")

∞ Levels of the other, more *discrete/discontinuous independent variable*(s) would be plotted as *discrete/separate lines.*

• Example: *Gender* is a discrete/discontinuous variable, since we normally think of gender in terms of only whole-unit values, or categories: male versus female.

∞ *Interactions* between the independent variables would be indicated by *nonparallel lines;* e.g., one might rise while another drops along the Y-axis for different values along the X-axis (a *disordinal,* and major, interaction), or one might just rise or drop more rapidly than another (an *ordinal* interaction; see Chapter 10).

• Note: *Graphic illustrations* of all the following *hypothetical data-matrix examples* are provided in a single figure placed at the end of these examples as a summary. (This allows readers to plot their own graphs to see if *they* can determine whether there is any *evidence* of interactions.)

— Interpretation of 2 × 2 factorial design data

• In factorial designs with two independent variables, any of eight different *combinations of findings* can occur for the two possible main effects and the interaction effect, since each of these three potential effects can either be present or not, and 2 × 2 × 2 = 8. (All eight possibilities are discussed in the five hypothetical data matrix examples shown below.)

• Moreover, within the possible combinations of outcomes for factorial designs, the precise *form/nature* of main effects and interactions, along with their *magnitude,* can vary considerably.

• Note that any differences between the dependent variable means in the *hypothetical examples* that follow are only *suggestive* of independent variable effects — the actual *significance* of effects, if any, would have to be determined using appropriate *inferential statistical analyses.*

EXAMPLE 10.1

Factorial Design Data Suggesting Main Effect for
Only One Factor and no Interaction

		A		
		a_1	a_2	
	b_1	10	10	10
B				
	b_2	20	20	20
		15	15	15

∞ A *main effect* for only *one* factor is suggested in Example 10.1, and there is *no* evidence of an *interaction*.

- Evidence for a main effect for *Factor B* is indicated by the difference between the *row means* for levels b_1 and b_2 ($10 \neq 20$).

- Evidence against a main effect for *Factor A* is indicated by the lack of difference between the *column means* for levels a_1 and a_2 ($15 = 15$).

- Evidence against an *interaction* between the independent variables A and B is provided by the fact that there is *no difference in the relative effects* of the two levels of factor A at the two levels of factor B:

$$a_1b_1 - a_2b_1 = a_1b_2 - a_2b_2$$

$$10 - 10 = 20 - 20$$

- It follows that there also would be no difference in the relative effects of the two levels of Factor B at the two levels of Factor A, since interactions represent *interdependence*, at least statistically:

$$a_1b_1 - a_1b_2 = a_2b_1 - a_2b_2$$

$$10 - 20 = 10 - 20$$

- Confirmation of the lack of interaction is also provided by the fact that there is no difference between the *diagonal means* of this 2×2 factorial design:

$$(a_1b_1 + a_2b_2)/2 \quad - \quad (a_2b_1 + a_1b_2)/2 \quad = \quad 0$$

$$(10 + 20)/2 \quad - \quad (10 + 20)/2 \quad = \quad 15 - 15 \quad = \quad 0$$

- *Graphing* the means for condition/treatment combination effects (see later) would produce *parallel lines,* indicating once again that there is no interaction between the independent variables (note that the lines would be parallel since *equal differences* between table values must yield *equal distances* between graph values).

- Note: *It is also possible* to have only a main effect for *Factor A,* rather than for *Factor B,* in which case the means for columns not rows would differ; moreover, it is *possible* to have *no main effects or an interaction effect,* i.e., nothing (the clearest case being when the means in all the cells are equal, e.g., 10).

∞ *Main effects* for *both* Factor A and Factor B are suggested in Example 10.2, but there is *no* evidence of an *interaction*. (The description of how all this is determined was given in Example 10.1.)

EXAMPLE 10.2

Factorial Design Data Suggesting Main Effects for
Both Factors with no Evidence
of Interaction

A

	a_1	a_2	
b_1	10	16	13
b_2	20	26	23
	15	21	18

B

18

- Note: *Not only can there be main effects for all factors without any interaction, but there also can be interactions without any main effects* (see Example 10.5).

- *Graphing* the means for condition combination effects would produce *parallel lines*, indicating *no interaction*.

EXAMPLE 10.3

Factorial Design Data Suggesting Main Effects
for Both Factors with Evidence
of Interaction

A

	a_1	a_2	
b_1	10	14	12
b_2	10	20	15
	10	17	15

B

12

∞ *Main effects* for *both* factors are suggested in Example 10.3, and there is evidence of an *interaction*.

- Evidence for an *interaction* between the independent variables A and B is provided by the fact that there is a difference in the relative effects of the two levels of factor A at the two levels of factor B:

$$a_1b_1 - a_2b_{1\mu} \neq a_1b_2 - a_2b_2$$
$$10 - 14 \neq 10 - 20$$

- It follows, as noted earlier, that there would also be a difference in the relative effects of the two levels of Factor B at the two levels of Factor A, since interactions represent *interdependence*, at least statistically:

$$a_1b_1 - a_1b_2 \neq a_2b_1 - a_2b_2$$

$$10 - 10 \neq 14 - 20$$

- Confirmation of the interaction evidence is also provided by the presence of a difference between the *diagonal means* of this 2×2 factorial design:

$$(a_1b_1 + a_2b_2)/2 \neq (a_2b_1 + a_1b_2)/2$$

$$15 \neq 12$$

- Note: *Interpretation of main effects must be qualified, and are of less interest, whenever a significant interaction occurs.*

  Observe in Example 10.3 that a *simple main effect* of Factor B is seen at only one level of Factor A: a_2.

  Interactions thus reflect on the *generalizability of main effects*, and it is for this reason that they, and hence factorial designs, are very important.

  Example: Consider that the data matrix in Example 10.3 might represent possible differences that would be found for the average number of *errors* that rats make over several trials of running to the goal box in a complex maze, when measuring the effects of a high versus low magnitude of food reward (b_1 versus b_2) under a high versus low level of food deprivation (a_1 versus a_2).

- *Statistical analyses* should be run for *simple main effects* when a significant interaction is found.

  Post hoc tests (also called *a posteriori tests*, in contrast to *a priori tests*) are appropriate when effects have *not been predicted*, but instead are suggested *after inspection/analysis* of the data that's been obtained (see statistics books for specific tests).

- *Graphing* the means for condition combination effects given in the data matrix would produce *non-parallel lines*, representing evidence for an *interaction*.

∞ A *main effect* for only *one* factor is suggested, and there is evidence of an *interaction* in Example 10.4.

EXAMPLE 10.4

Factorial Design Data Suggesting
Main Effect for Only One Factor
With Interaction

		A		
		a_1	a_2	
	b_1	10	24	17
B				
	b_2	20	6	13
		22	15	15

22   15   15   8

- Note (expanding on points made in the earlier example): *Interpretation of main effects* (generalizability) must be *qualified* whenever there is a significant interaction — *regardless* of whether main effects are found.

  Although, in this example, there is no evidence of a main effect for *Factor A*; observe that there are apparent *simple main effects* for Factor A, but they are *opposite/reversed* at the two levels of Factor B (b_1 and b_2), and they are *equivalent in absolute magnitude* (±14) — hence they *cancel* each other out when they are averaged to determine the main effect.

  Moreover, although there is evidence of a *main effect* for *Factor B*, note similarly that the relative effects of the two levels of B are *influenced by (dependent on)* the level of Factor A at which they are measured, and indeed the *simple main effects* of Factor B are also *reversed*, but they are *not equivalent* — thus they don't cancel out the main/overall effect when averaged.

  Example: Consider that the data matrix in Example 10.4 might be representative of what would be found in a study of *cerebral lateralization of specialization* in humans for the number of correct identifications out of 30 of very brief flashed visual stimuli presented to the left versus right halves of the retinas, and thus the left versus right brain hemispheres (a_1 versus a_2), when the stimuli are geometric patterns versus words (b_1 versus b_2). It should be pointed out that the left hemisphere of the human brain is typically superior at language tasks, and the right hemisphere at visuospatial tasks.

- *Graphing* the means for condition combination effects given in the data matrix would produce *intersecting lines going in opposite directions*, indicating that there is a *strong interaction*.

  Non-parallel lines that do not intersect can be *hypothetically extrapolated* (extended) so that the lines then *do intersect* for independent variable values that weren't measured; however, it is not certain that the intersection would occur in *reality* if the dependent variable actually were measured for these other values — it's only a *prediction/estimation*.

- Note: *It is alternatively possible* to have an interaction along with a main effect for only *Factor A*, rather than an interaction along with a main effect for only *Factor B* (as in the present example), in which case the means for columns rather than rows would differ.

EXAMPLE 10.5

Factorial Design Data Suggesting Main Effects for Neither Factor but with Interaction

		A		
		a_1	a_2	
	b_1	10	20	15
B				
	b_2	20	10	15
		20 15	15 10	

∞ *Main effects for neither factor* are suggested in Example 10.5, but, nevertheless, there is evidence of an *interaction*.

- Note: It is possible to have interactions without any main effects when (as in this example) *there are simple main effects for both factors that are opposite and symmetrical/equivalent*.

  Indeed, it is because of this *special form of interaction* that there are no main effects.

- *Graphing* the means for condition combination effects would produce *symmetrically intersecting lines going in opposite directions*, indicating that there is a *strong interaction*.

— Limitations on number of independent variables studied

- In theory

  ∞ *No absolute limits exist* for the number of independent variables that can be studied in factorial designs.

  - The *same principles* discussed for 2×2 designs apply regardless of how many independent variables (or levels of independent variables) there might be in a design.

- In practice

  ∞ *Difficulties become greater* in several ways when actually conducting research as the number of independent variables in a design increases:

  - When there are additional *between-groups* factors in the design, there would be an *increase* in the number of *participants* needed.

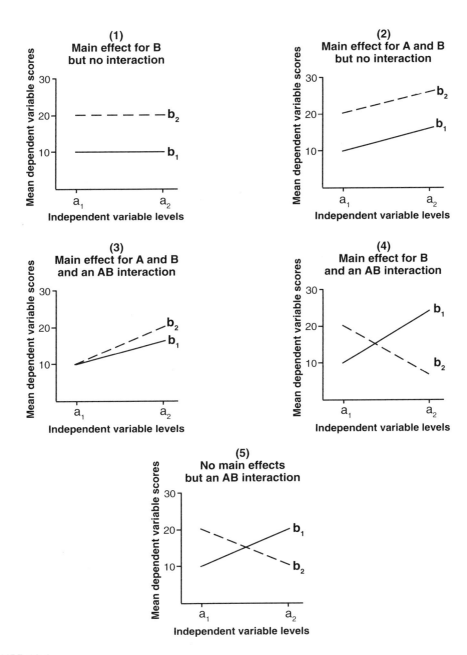

FIGURE 10.1
Graphs of the hypothetical data for the five preceding examples of a 2 × 2 factorial design. The graphs illustrate evidence for various combinations of a main effect for independent variable A, a main effect for independent variable B, and/or an AB interaction (see text).

- When there are additional *within-groups* factors in the design, then the *condition combinations* that the participants would be *run under* would *increase*.

- When there are additional *within-* or *between-groups* factors, *simultaneous manipulations* of independent variable conditions would become *more challenging*.

- When there are additional factors, *data interpretation complications* would *increase* since *interactions* would potentially increase in both number and complexity.

  Note that when there are *two independent variables* (A and B), there is only one interaction possible (AB). However, when there are *three independent variables,* (A, B, and C), then there are four interactions possible: three two-way interactions (AB, AC, and BC) and one three-way interaction (ABC).

  Three-way interactions occur when the *effect* on a dependent variable is a *unique joint/combined function* of *three* independent variables — the effect of each one being influenced by the other two in a manner that must be *described and explained* (which sometimes can be rather challenging).

  To help interpret complex interactions, *simple interaction effects* (like simple main effects) can be analyzed — i.e., the interactions of two or more independent variables at *selected levels,* or combinations of levels, of the other variables.

  Examples: the AB simple interaction at level c_1 of Factor C; or the AC simple interaction at level b_3 of Factor B; or the BC simple interaction at level a_2 of Factor A.

— Application and importance of factorial designs

- *Behaviors and cognitions* commonly have *multiple determinants,* and these determinants might influence the effects of each other, i.e., they might interact.

  ∞ *Searching for interactions* is therefore *important* to the advancement of understanding, prediction, control, and theory construction (i.e., the formulation of a systematic body of knowledge).

  - As noted earlier, interactions *reflect on the generality of main effects,* and thus they affect the *interpretation* of results found for independent variable manipulations.

  - More to the point, interactions — the influence/dependence among independent variables — *can only be studied using factorial designs.*

  ∞ *Interactions* might, in fact, be the *principal interest* of a study.

  - A simple, common example would be a 2×2 *mixed factorial design* (Tests: pretest and posttest; × Treatment: present and absent) with repeated measures on one factor (Tests).

    The *interest* in the study would *not* be the *main effects* for the tests or even for the treatment independent variable, but rather the *interaction* between the variables, which would indicate whether there was a difference in the pretest-posttest change for the treatment condition relative to the control condition, and hence an effect of the treatment.

Note: Using a factorial design and a two-way analysis of variance would be a *more powerful research approach* than just analyzing difference/change scores with a *t*-test or one-way analysis of variance (discussed earlier under the "Pretest-Posttest Control Group Design").

Review Summary

When studying the following material, it will be useful to refer to Appendix 6, showing "Experimental and Quasi-Experimental Designs: Category-Subcategory Relationships," and Appendix 7, showing a "Schematic Summary of Designs." The preceding chapter covered Within-Participants Designs, the present chapter began with between-participants designs.

1. *Nonequivalent posttest-only control-group design* is where one group of participants that has experienced a treatment is observed on the dependent variable and compared with *another group* that has not experienced the treatment. The two groups, however, are *not randomly formed,* and thus most likely are *not equivalent* in terms of participant characteristics. This between-participants, *independent, two-group design* is typically an *ex post facto naturalistic study,* in that the independent variable is usually not purposely, directly manipulated — the event of interest just happens to occur for some group of individuals.

   The *strength* of this design over the one-group posttest-only design is the addition of a *control group* to provide a *baseline measure.* However, it is less than adequate, and this is a weak design due to *uncontrolled secondary, extraneous variables* that could be *confounding:* selection, participant attrition, and maturation. Thus, this is usually a *pre-experimental design.* It's *quasi-experimental* only if the independent variable can be and is *purposely, directly manipulated,* and the participants in the different groups are shown to be *matched beforehand* on several relevant variables — although still likely to be *nonequivalent* on other factors, since they are not randomly assigned.

2. *Nonequivalent pretest-posttest control-group design* is a combination of the one-group pretest-posttest design and the nonequivalent posttest-only control-group design. Its *strengths* and *advantage* over the nonequivalent posttest-only control group design are that *pretest observations* for the two groups can be compared to determine if there is *initial equivalence* on the dependent variable, and if there is not, then a control/adjustment for *individual participant differences* can be made by analyzing *pretest-posttest changes.* Change *ratio scores* are best since they measure *actual change* relative to *potential change,* and thus better control for *ceiling and floor truncation effects.* This design's advantage over the one-group pretest-posttest design is that *contemporary history, testing, and instrumentation* are better controlled due to the control group, as is usually *maturation.*

The weaknesses are several possibly uncontrolled *secondary, extraneous variables*: *selection* (individual participant differences might not *all* be adequately controlled by the pretest) and *interactions* between selection and maturation (perhaps the most common problem with this design), statistical regression toward the mean, participant attrition, contemporary history, and testing. Hence, this is a *quasi-experimental design*. However, because of its relative strengths, it is one of the most widely used designs in educational research, where the participant population consists of *naturally assembled groups*.

3. *multiple time-series design* is a combination of the one-group time series design and the nonequivalent pretest-posttest control group design. Its *strength and advantage* over the former is that *contemporary history* (as well as testing and instrumentation) is better controlled by having the comparison control group. The strengths and advantages over the nonequivalent pretest-posttest control group design are that *sampling error* is reduced by averaging multiple measures, and *trends* are determinable, which controls for several secondary, extraneous variables.

    The *weaknesses* are several uncontrolled variables: selection (as for the preceding design) along with possible contemporary history-by-selection and treatment-by-selection interactions, as well as a possible testing-by-treatment interaction and participant attrition (which could be a particular problem due to the multiple tests). Although this is a *quasi-experimental design*, it is the *best of the nonequivalent designs*.

4. *Posttest-only control group design* involves *random assignment* of participants to a treatment and a control group (in contrast to the preceding nonequivalent designs) with only a *posttest* measure on the dependent variable. The *strength and advantage* over the nonequivalent posttest-only control group design and that *selection* problems are controlled by *randomization,* which makes it probable that the extraneous variables of *individual participant characteristics* will be *equivalent* between groups. The strength and advantage over the one-group pretest-posttest design is the avoidance of *experimental contamination* from extraneous factors due to *repeated measures*.

    The *weaknesses* are possible experimenter bias and participant bias, but these are common to *all* designs and *can be controlled,* e.g., by the double- blind technique and use of a placebo control group. Thus this is a *true experimental design,* as are all the following designs, because of the degree of *experimental control* over both independent and extraneous variables.

5. *Pretest-posttest control group design* is a combination of the one-group pretest-posttest design and the posttest-only control-group design just discussed. The *strengths and advantages* over the former are that contemporary history and instrumentation can be controlled, as well as maturation, statistical regression toward the mean, and the direct effects of testing. The strength and advantage over the posttest-only control group design is that the *pretest* provides a *check* for whether random assignment of participants yields *equivalent groups,* and if not, the pretest can be used to *adjust/correct* for nonequivalence. The *weaknesses*

are possible participant attrition and a testing-by-treatment interaction (external validity issue).

6. *Randomized-blocks design* is where participants are *assigned to blocks* (groupings) based on some *shared characteristic,* e.g., ethnicity, and then from each block an equal number of participants is *randomly assigned* to the different conditions. This is an extension of the posttest-only control group design. The *strengths and advantages* are that *individual participant differences between conditions* would be reduced to a greater degree, and also *blocks* could be *analyzed* as an additional variable, rather than simply *balancing* for the extraneous variable blocking factor. This should reduce *statistical error variance,* and thus *masking,* in addition to *confounding* being controlled.

7. *Match by Correlated-Criterion Design* is where participants are *matched* on the basis of a *participant criterion* that is *measurable* and *correlated* with the dependent variable. Then they are *randomly assigned* from their matched groupings to the different conditions, and a posttest measure is taken.

   The *advantages* over between-participants, *independent-groups* designs is that *matching* better controls for *measurable extraneous individual participant differences* than just random assignment. It is an *extended and more precise* form of blocking-and-balancing than the preceding design, whenever there are several levels of the secondary, extraneous participant variable. *Statistical error variance* associated with individual participant differences would thus be reduced (in proportion to 2×, the correlation of posttest scores for matched participants), which would lead to reduced *masking* and hence increased *power,* if the correlation is sufficiently large to offset the loss in degrees of freedom that occurs when matching is used.

   The *advantage* over within-participants/groups designs is that the problems of *repeated measures* are avoided. The *disadvantages* are, e.g., that individual participant differences are not as well controlled, and more participants are required for the same amount of data. The *weaknesses* are a possible testing-by-treatment interaction, and participant attrition.

8. *Pretest-match-posttest design* is where participants are *matched* on the basis of the *pretest observation* on the dependent variable, and then they are *randomly assigned* from the matched groupings to the conditions, and a posttest observation is taken. The *advantage* over the match by correlated- criterion design is that *individual participant differences* are better controlled, since *pretest scores* should be the participant criterion most strongly correlated with posttest dependent variables scores. The *disadvantages* are due to the required *pretest,* and involve possibly greater time and effort being required, testing-by-treatment interaction, and participant attrition.

9. *Yoked control group design* is where participants are *matched/yoked* for extraneous characteristics of the *experimental situation,* and perhaps also for *environmental factors,* as opposed to extraneous participant factors. The control group is exposed, through *yoking,* to the same *quantity* and also usually to the same *temporal distribution* of extraneous experimental events received by the *treatment group.* The *treatment* would consist of a particular *order or relationship of experimental events* or

an *additional factor*. Yoking may be either *physical or electronic*. The *strength and advantage* of this design is the control for extraneous factors introduced due to the participants' *participation* in the experiment, i.e., *experimental contamination*. Specifically, the experimental situation might involve the undesired *simultaneous introduction* of more than one influential variable, in which case the extraneous variable(s) would need to be controlled.

10. *Multiple treatment designs* involve either *more than two levels* of a single independent variable being studied, and thus are called *multilevel designs,* or there is *more than one independent variable* studied, and thus are called *factorial designs*. Although *more time, effort, and participants* are usually involved, they provide *disproportionately more information* than the non- multiple treatment designs — hence they are *more efficient. Assignment of participants* to conditions can take all the varieties discussed for the previous designs: *randomization* or *matching* for between-participants comparisons, or taking *repeated measures* for within-participants comparisons.

11. *Multilevel designs* can ask two general questions: (a) How do the effects of each of the treatments compare with that of a control condition? (b) How do the effects of the treatments compare with each other? The *strengths* and *advantages* of these designs are (a) they are more efficient, (b) it's easier to control extraneous variables (than if using several two-level designs), (c) the likelihood of finding any effects is increased, and (d) functional relationships can be more accurately determined. They are used in *parametric research* involving *quantitative* independent variables.

12. *Factorial designs* are the only ones that allow the effects of *two or more independent variables* that are arranged *orthogonally*, i.e. *balanced* with respect to each other, to be evaluated *separately* as well as in *combination* (multiple determinants). *Complete factorial designs,* the most common form, are where *all possible combinations* of the selected levels of two or more independent variables are studied. *Nested factorial designs* are where *different levels* of one independent variable, the *nested factor,* are paired with the different levels of another variable — thus not factorially combined.

Factorial designs are *further distinguished* in terms of (a) how many independent variables are studied (two or more), (b) how many levels are investigated for each independent variable (two or more), (c) how participants are assigned to conditions (between, within, or mixed, i.e., between-within), and (d) how conditions are selected (fixed/systematic, random, or mixed).

13. In the *shorthand designation* of factorial designs (e.g., $2 \times 3 \times 4$); the *number of numbers* indicates the number of independent variables studied (three in this example); the *magnitude of the numbers* indicates the number of levels investigated for each independent variable (2, 3, and 4 in this example); and the *product of the numbers* indicates the number of combinations of levels of independent variables investigated (24 in this example).

14. The *strengths and advantages* of factorial designs are (a) they are *more efficient*, (b) *interaction effects* can be evaluated as well as *main effects*, (c) *more power/sensitivity*

is often possible for detecting effects. *Main effects* are the *overall/average/mean* effects. *Simple main effects* are the *selective/unaveraged* effects of independent variables at *specific, selected levels* of other independent variables. *Interaction effects* are the *unique combined/joint* effects, i.e., the *difference in relative effects* of the levels of an independent variable at *different levels* of other independent variables. Interactions indicate *influence/dependence* among independent variables, and thus *nonadditivity* of main effects — whole is different than the sum of parts.

15. Main effects are usually best seen from *tables,* while interactions are most clearly *comprehended* when data are presented *graphically.* Interactions between independent variables are suggested by *nonparallel lines* when graphed. But the *significance* of any apparent effects depends on *inferential statistical analyses.* Note that it is possible to have main effects for all factors *without* any interaction, and it is also possible to have interactions *without* any main effects. The *interpretation of main effects* must be *qualified,* and are of *less interest,* whenever a significant interaction occurs. This is because *interactions,* which are very important, reflect on the *generalizability of main effects* regardless of whether or not they are found.

Review Questions

When answering questions about specific designs it should be helpful to write out the schematic summary of the designs, which can be checked against those given in Chapter 10 and Appendix 7. The preceding chapter covered within-participants designs, the present chapter began with between-participants designs.

1. Describe the *nonequivalent posttest-only control-group design,* and explain why it is typically a *pre-experimental design.* List the two *conditions* that must be met for this design to be considered *quasi-experimental.*

2. Describe the *strengths/advantages* of the *nonequivalent pretest-posttest control-group design* over the preceding design, and explain why. Also, discuss the *advantages* of using *change ratio scores* rather than simply *change scores.* Finally, describe this design's advantage over the one-group pretest-posttest design.

3. Write out the schematic summary for the *multiple time-series design* and explain why it is the *best* of the nonequivalent designs.

4. What is the major *advantage* of the *posttest-only control group design* over the *nonequivalent* posttest-only control group design, and what specifically is responsible for this? Also, is this a true *experimental design,* and why?

5. Discuss one *advantage* and one *weakness* of the *pretest-posttest control group design* relative to the posttest-only control group design.

6. Discuss two *strengths* of the *randomized-blocks design* over the posttest- only control group design. Be sure to be complete.

7. Discuss, in as much detail as you can, the *advantage* and possible major *disadvantage* of the *match by correlated-criterion design* relative to the posttest-only control group design, and what the important implication of this is for when matched-participants designs should be used. Also, state one *advantage* and a *major disadvantage* of this design relative to within- participants designs.

8. State and explain the *advantage* of the *pretest-match-posttest design* over the match by correlated-criterion design, and list the three possible *disadvantages*.

9. What is the *yoked control group design* specifically used to *control,* and what are two *alternatives* for the general nature of the *treatment?*

10. Name and describe the two forms of *multiple treatment designs,* and list the *disadvantages* and explain the *advantage* of these designs.

11. State the two types of questions addressed by *multilevel designs,* and list the four *strengths and advantages* over two-level designs.

12. Describe what *factorial designs* allow one to do. Then name and describe the two *major categories* of factorial designs, and also describe the four characteristics that *further distinguish* among factorial designs. Finally, describe a *mixed factorial design.*

13. Explain the *rules for shorthand designation* of factorial designs. Then state for a *3 × 4 × 5 factorial design,* how many *independent variables* are studied, how many *levels* of each independent variable are investigated, and how many *combinations* of levels of independent variables there are.

14. Discuss the three *strengths and advantages* of factorial designs. Then define *main effects, simple main effects,* and *interaction effects.* Finally, state as thoroughly as possible what interactions *indicate.*

15. Draw a *graph* of the following *data* for treatment combination means, and discuss whether there is evidence for a *main effect* of independent variables A and B, and/or for an *interaction effect*: $a_1b_1 = 20$, $a_1b_2 = 10$, $a_2b_1 = 6$, and $a_2b_2 = 24$. Discuss a couple of reasons why the *interpretation of main effects* must be *qualified* whenever there is a *significant interaction* — regardless of whether or not main effects are found.

11

Sampling and Generalization

CONTENTS

Before discussing *sampling and generalization* in some detail, this chapter begins by providing *definitions of the major concepts* and elucidating the *fundamental decisions* involved in sampling with respect to the implications for generalization. This is followed by coverage of the various *statistics and parameters* that can be computed for samples and populations, respectively.

Definitions (See Appendix 8)

What are *populations and samples*, and what are the processes of *sampling and generalization*?

1. **Populations**
   — Populations are the *entire collections* of individuals, objects, or events of *interest* to which one wants to *generalize* the results of research — in Boolean terminology, populations are referred to as *universal sets*.
     • Examples: All *people* who live in the United States, all *primates* in the world, all possible *values* of some *independent variable*, all possible *dependent-variable measures* for some independent variable

2. **Samples**
   — Samples are *subsets* of individuals, objects, or events that are *selected* for study from populations — which involves the process of *sampling*.
     • Examples: A random subset of *people* from the city where the scientist is conducting research, a convenient subset of *monkeys* from a university's

vivarium, a systematically selected subset of the *independent-variable values* that are of research interest, and a logical subset of *dependent variables* that are appropriate measures of the effects of an independent variable to be studied.

3. **Sampling**

— Sampling is the *process of selecting* (obtaining/drawing) a sample subset from a population (see later in chapter for techniques).

4. **Generalization**

— Generalization is the *inductive inferential process* of *extending the findings* from a sample to a population, which might involve the use of *statistical estimation* from sample statistics to population parameters (discussed later in this chapter).

Fundamental Decisions When Selecting a Sample of Research Participants

The choices that are made when selecting samples have important *ramifications* for the process generalization.

1. **Population from Which the Participants Will Be Selected**

— *Population that is sampled* determines the *larger group* (population) to which the sample results can be *appropriately generalized.*

2. **Number of Participants Who Will Be Selected**

— *Size of the sample* influences mainly the *reliability* of the sample and hence the reliability of generalizations to the population sampled.

3. **Technique by Which Participants Will Be Selected**

— *Method of sampling* influences mainly the *validity* of the sample and hence the validity of generalizations to the population sampled.

Note: These issues are also applicable to *samples other than participants,* i.e., to other variables of the study (as discussed later in this chapter).

— Example: When studying the possible effects (the *dependent-variable values*) of a new drug for the treatment of obsessive-compulsive disorders, (1) *what population/range* of drug dosages (the *independent-variable values*) should be sampled for study, (2) *what number* of different dosages should be sampled, and (3) *what sampling technique* should be used to select the dosages for study?

Statistics and Parameters (See Appendix 8)

There are a variety of *descriptive statistic measures* that can be computed with *sample data,* several of which are also used with *inferential statistics* for estimation and hypothesis testing of *population parameters*. These are described below. Statistics were also discussed earlier in Chapters 3 and 8.

1. **Parameters**

    — Descriptions (numerical indices) of *population characteristics* are called *parameters.*

      • Examples: Mean, variance, and correlation — just as for statistics.

2. **Statistics**

    — Descriptions (numerical indices) of *sample characteristics* are called *statistics —* more specifically, they are *descriptive statistics.*

      • Statistics can also be used to *infer/estimate* the values of corresponding *population parameters* and to *test hypotheses* about those values — in which case we would additionally use *inferential statistics* (both types of statistics are discussed next).

    — Descriptive Statistics

      • There are *three categories of descriptive statistics,* all of which are used to *organize and summarize* (i.e., describe) the data obtained from *samples*: they are measures of *central tendency, dispersion/variability,* and *association/relationship.*

      1) Measures of the central tendency of distributions

        ∞ Mean

          • *Average* of all the scores (values) in a distribution is called the *mean:* $M = \Sigma X / n$ (where M is the mean, Σ indicates the sum of, X represents the individual scores, and n equals the number of scores).

            Example: The mean of 2, 4, 4, 5, and 10 equals 5.

          • The mean is the *most commonly used* measure of central tendency.

        ∞ Median

          • *Midpoint* of scores in a distribution, above and below which lie an equal number of values, is called the *median;* it is the middle value of a set of scores, or the mean of the two middle values if there is no one middle score.

            Example: The median of 2, 4, 4, 5, and 10 equals 4 (actually this is only roughly the midpoint in this case, because there are two 4s at the middle of the sequence, and one 4 cannot be less than another).

          • The median is useful because it is *insensitive to extreme/deviant scores,* and because it is applicable to *ordinal data* — whereas neither is the case for the mean.

            Note: Chapter 4 covered "Scales of measurement for variables and their data types."

        ∞ Mode

          • *Most frequent/common* score (value) in a distribution is called the *mode.*

            Example: The mode of 2, 4, 4, 5, and 10 equals 4.

- The mode is rarely used except for *nominal scale data* (frequencies) — for which neither the mean nor median is really applicable.

2) Measures of the dispersion/variability of distributions

∞ Range

- *Largest value* minus *smallest value* in a distribution is called the *range*.

  Example: The range of 2, 4, 4, 5, and 10 equals 8.

- This is the *simplest measure* of dispersion, but it is *rarely used* because it reflects only the extreme end-scores of a distribution.

∞ Mean deviation

- *Average* (mean) of the *absolute deviations* of all the values from the mean of the distribution is called the *mean deviation*:

$$MD = \Sigma \,|\, X - M \,|\, /n.$$

  Note: The term *"absolute"* refers to the value of a number without regard to its sign (i.e., ignoring whether it is positive or negative), and it is symbolized by | |.

  Example: The mean deviation of 2, 4, 4, 5, and 10 equals 2.

- The mean deviation is not useful for statistical inferences about population parameters, but sometimes it is preferred for *descriptive purposes* because it is *less sensitive to large deviations* than the range or standard deviation.

∞ Variance

- *Average* (mean) of the *squared deviations* of all values from the mean of the distribution is called the *variance*: $s^2 = \Sigma(X - M)^2/n$.

  Example: The variance of 2, 4, 4, 5, and 10 equals 7.2.

- The variance is commonly used in *inferential statistics* (see below).

∞ Standard deviation

- *Square root* of the variance is called the *standard deviation*: $s = \sqrt{s^2}$.

  Example: The standard deviation of 2, 4, 4, 5, and 10 equals the square root of 7.2, which is 2.68.

- The standard deviation is useful as a *descriptive statistic*.

3) Measures of the association/relationship of variables

- Correlation coefficient

- Depending on type of data involved, there are different *computational formulas* for *correlation coefficients* (consult statistics books for details; for further discussions of correlations, with examples, see Chapter 1 under "Description," and Chapter 2 under "Correlational Study").

∞ *Values* of correlation coefficients can range from +1 through 0 to –1 for score and ordinal data, and these values indicate the *degree and type of association* found between the measured variables.

+1 represents a perfect *positive/direct* relationship, in which *higher* scores on one variable are associated with *higher* scores on the other variable.

−1 represents a perfect *negative/inverse* relationship, in which *higher* scores on one variable are associated with *lower* scores on the other variable.

0 indicates that there is *no* relationship/association.

Note: There are *tests of statistical significance*, i.e., measures of the reliability/chance likelihood for non-zero values in the *sample data*, which are used to estimate the value in the *population*.

- *Square of the correlation coefficient* (r^2) is known as the *coefficient of determination* and indicates the proportion of variance of one variable that can be accounted for (predicted and perhaps explained) by variation of another variable — this is a very useful *strength-of-association measure* (discussed further in Chapter 12, under "Statistical Significance" versus "Practical Significance").

    Example: r^2 between grades and IQ or studying

— Inferential statistics

- There are several types of *inferential statistics*, which are used for *estimation and hypothesis testing*, i.e., for making inferences (general statements) about *population parameters* based on the reliability of *sample data-descriptive statistics* (discussed in both Chapter 8 under "Principles of Minimizing Error Variance," and throughout Chapter 12; also see Appendix 15 for a table of statistical tests for different designs and levels of measurement).

    ∞ Examples: Chi-square, *F*-ratio, and *t* statistic

3. **Accuracy of Parameter Estimation**

— The accuracy of estimating a *population parameter* from a *sample statistic* depends on *how* the sample is selected and also on the sample *size* (see earlier coverage in this chapter).

- These mainly relate, respectively, to the *validity and reliability of sampling*, which are discussed next and which are central to *hypothesis testing and statistical significance* (the topics of Chapter 12).

Sampling Reliability

Both *validity and reliability* were discussed previously in Chapter 4 under "Accuracy and Consistency of Variables," and in Chapter 8 under Minimizing Error Variance. These are very important *general concepts* that are relevant to sampling. *Sampling validity* is covered after the discussion of *sampling reliability.*

1. **Definition**

   — *Consistency/stability* of a sample statistic used to estimate a population parameter is what is meant by *sampling reliability*.

   - This can be evaluated by *repeatedly* obtaining sample statistics from a population, and comparing the values for *similarity*.

2. **Determinant (See Appendix 2)**

   — The relative size of the *variable/inconsistent error component* of a measure, in comparison to the true score component, determines sampling reliability (see Chapter 4, under "Components of Measures").

   - Note: Variable error, which is random, *averages toward zero* as the *sample size increases* — a principle noted before.

     ∞ Recall that σ_M, the *standard error* (i.e., the standard deviation) of a distribution of *sample means*, which indicates the expected magnitude of the *sampling error of the mean, decreases* in proportion to the *square root* of *sample size* (n).

       - $\sigma_M = \sigma/\sqrt{n}$ where σ is the *population* standard deviation (*standard error of the mean* was covered earlier in Chapter 8, under "Techniques for minimizing error variance," and it is further discussed below).

     ∞ Hence, conversely, *sampling reliability increases* with (is directly proportional to) the *square root* of the *sample size*: $R \propto \sqrt{n}$.

   - Law of diminishing returns

     ∞ *Less increase* in reliability occurs for each additional *constant increment* in sample size, because the relationship between reliability (R) and sample size (n) is exponential ($\sqrt{}$); hence there are *diminishing returns* for successive increments in sample size, which limit the sample sizes researchers use.

       - Example: Going from an n of 20 to an n of 80 (fourfold larger and an increase of 60 in sample size) would increase R by 100% or 2× (i.e., double the reliability), because $\sqrt{(80/20)} = \sqrt{4} = 2$; but then going from an n of 80 to 140 (again an increase of 60 in sample size) would increase R by only 32% because $\sqrt{(140/80)} = 1.32$.

3. **Sampling Distributions**

   — Normal distributions (See Appendix 9)

   - Most dependent-variable data in the behavioral sciences have a *frequency/probability distribution* that has the appearance of a *bell-shaped curve* and thus at least approximates what is referred to as the *normal curve*, or *normal distribution*.

     ∞ Specifically, scores/values are most numerous in the *middle* of their range and decline in frequency (fairly symmetrically) with increasing distance toward the *tail ends* of the curve.

- These distributions are *characterized/specified* by two statistical measures: their *mean (M)* and their *standard deviation (s)*.

- *Central tendencies* of mean, median, and mode are the *same* for any given normal distribution, but they can *differ* among distributions.

- *Variability* too can differ from one distribution to another, with the normal curve appearing *tall and narrow* when the standard deviation (or variance) of scores is small and being *low and wide* when the standard deviation (or variance) is large.

- *Specific proportions* of the distribution of scores are always contained within *specific areas of the curve*:

  68.26% fall within ± 1 standard deviation of the mean.

  95.44% fall within ± 2 standard deviations of the mean.

  99.74% fall within ± 3 standard deviations of the mean.

- Hence, if we know an individual's score as well as the mean and standard deviation of the normal distribution of scores, then we also know the individual's *relative rank in percentiles (centiles)*.

  ∞ *Percentiles* indicate the *percentage of scores in a distribution* that fall below any given score — a very useful measure.

    - Example: For an IQ test with a mean of 100 and a standard deviation of 15, an individual with an IQ of 115 would have scored 1 standard deviation above the mean, and thus higher than 84.13% of all the others — i.e., higher than the 50% below the mean, plus the 34.13% (or 1/2 of 68.26%) within 1 standard deviation above the mean — hence this person is at about the 84th percentile.

- Standard scores (z scores)

  ∞ *Standard scores* are *converted/transformed scores* that indicate how many standard deviations each score is from the mean of the distribution of scores; i.e., their *distance from the mean* stated *in standard deviation units*: $z = (X - M)/s$.

  ∞ This *linear, z transformation* can be used to convert a set of scores into a set of *standard scores (z scores)* that would always have a *mean of zero* and a *standard deviation of one*.

  ∞ *z scores* not only permit the computation of *percentile ranks* (as discussed above), but they are also useful because they allow the *comparison of scores across normal distributions* that have *different* means and variances and even very different units of measurement.

    - Example: It would be possible to show that a person is either more or less intelligent than emotionally stable based on his/her *z scores* (say $z = 0.75$ versus $z = -1.5$) and thus to show *relative percentile ranks* on tests of these two different characteristics, even though the *raw score measures* would be on entirely different test scales, and therefore could not themselves be directly compared.

— Distribution of sample means

- Technically, a *distribution of sample means* is referred to as a *sampling distribution of means* (they signify the same thing).

- Because a dependent variable's values/scores within a *population* will almost always vary, they will also vary for *samples*, and as a result of chance factors the *sample means* of values will vary too.

- When the *population distribution* of dependent-variable values is *normal*, as it typically is, then the *distribution of sample means* will also be normal (and, as discussed in Chapter 12, so would the distribution of *differences between pairs of sample means* — e.g., those found for different independent-variable conditions).

- Standard error of the mean (σ_M)
  - ∞ The *standard deviation* of a distribution of sample means (computed when samples are repeatedly taken) is referred to as the *standard error of the mean* (σ_M), and it serves as a *measure of sampling reliability* (see earlier in this chapter).
    - An *unbiased estimate* (s_M) can be obtained, and typically is, from the standard deviation (s) of a *single sample*: $s_M = s / \sqrt{(n-1)}$ (this was also discussed in Chapter 8 under Techniques for Minimizing Error Variance).
    - s_M indicates the *expected variability* in a *distribution of sample means*, such as would be obtained by chance when drawing samples of participants/subjects for the *groups* assigned to the different conditions in a study. Thus it indicates the *expected magnitude of the error* for any given *sample mean* with respect to the actual *population mean* (see next).

4. Sampling Error

— *Deviation*, by some amount, of a *sample statistic* from the true value of the *population parameter* that it is used to estimate is called *sampling error*.

- Deviations are due to *chance variation* when drawing a sample's few cases from the population's many cases.
  - Note that when *differences* are found between condition groups in research, a possible cause is just *sampling error* (chance variation), rather than effects of the conditions.

— *Expected magnitude* of the sampling error is provided by the *standard error* of the sampling distribution for the *statistic* that corresponds to the population *parameter* being estimated — a common example being the standard error of the *mean* (as just discussed).

— *Larger samples* result in the following:

- *Smaller sampling error* (recall n in *denominator* of the standard error equation)

- *Greater sampling reliability* (due to the *smaller* sampling error)

- *Increased power/sensitivity* to detect any treatment effects using inferential statistical tests (due to the *greater* sampling reliability)

  - ∞ Example: Consider the difference *in expected sampling error* and thus *reliability* for the extreme cases of a sample size equal to one versus a size of nearly the entire population.

  - ∞ Hence, given larger sample sizes and thus greater degrees of freedom, *statistical significance* is reached for *smaller F and t values* in statistical test tables and therefore with *smaller treatment effects* — which means that there is *greater power.*

— Determining sample size for research

  - The *larger the sample size the better,* because that *increases the power/sensitivity* to detect treatment effects in research; but it is also associated with *greater costs* in terms of time, energy, and money — and in addition there is the *law of diminishing returns.*

  - *Precedent* is the primary guide used by most researchers, but that does not necessarily lead to the best decision, which is complex.

  - The number of participants needed to *adequately test a research hypothesis* depends on the probable strength of relationship/effect between the independent and dependent variables (primary variance); the expected variability of the data (error variance); the research design that will used, e.g., between- versus within-participants designs; the power of the statistical analysis chosen, e.g., parametric versus nonparametric tests; whether the tests will be one-tailed or two-tailed; and the statistical significance level that will be used (see earlier chapters and Chapter 12 for discussion of these issues; for more details see (1. Cohen, J. (1988). *Statistical power analysis for the behavioral sciences.* Hillsdale, NJ: Lawrence Erlbaum Associates. 2. Kraemer, H. C. & Thiemann, S. (1987). *How many subjects? Statistical power analysis in research.* Newburry Park: Sage Pub.)

5. Implication of Reliability with Regard to Accuracy

— High reliability, i.e., *consistency/stability,* implies *accuracy* when sample statistics are used to *estimate* population parameters, which is central to hypothesis testing and statistical significance (Chapter 12).

  - However, reliability is a *necessary but not sufficient condition* for accuracy, because in addition to *variable error,* there is also the possibility of *constant error* when making measurements (as discussed earlier in Chapter 4, under "Consistency versus Accuracy: Reliability versus Validity").

Sampling Validity

Validity of sampling is an even *greater concern* than reliability of sampling. After all, what is the good of being *consistently wrong?*

1. **Definition**

    — *Accuracy/correctness* of a sample statistic with respect to the corresponding population parameter is what is meant by *sampling validity.*

2. ***Determinant* (See Appendix 2)**

    — The relative size of the *constant/systematic error component* of a measure, in comparison to the true score component, determines sampling validity (see Chapter 4, under "Components of Measures").

    • Constant/systematic error depends on the degree to which the sample is *unrepresentative* and thereby the extent to which the sample statistic/ estimate is *biased.*

    ∞ *Unrepresentative/biased* samples are *inaccurate* and thus *invalid;* i.e., they yield data and therefore statistics that are *altered systematically* from the population parameter values.

    ∞ *Representative/unbiased* samples, in contrast, are *valid.*

    • *Accuracy/validity* is determined more by *how* a sample is selected than by the *size* of a sample (note: the various specific sampling techniques are discussed at the end of this chapter).

3. **Causes of Biased/Invalid Sample Statistics (See Figure A8.2)**

    There are *two basic causes* for samples being biased and thus invalid.

    a. *Including* in a sample some *individuals/elements* that are *not members* of the population of interest is one cause of biased samples (illustrated in Figure A8.2 by the *sample* at the left, which projects *outside* the target population in this Venn diagram).

    • Example: Including in a study on *birth control practices* of the *married population* some *unmarried couples* who happen to be living together — certainly this *unrepresentative sample* would likely yield data that are *biased*, thus resulting in statistics that are *inaccurate* and hence *invalid* for married individuals.

    b. *Excluding* from a sample some *elements* of the population of interest is the other cause of biased/invalid samples (illustrated in Figure A8.2 by the sample at the right in the Venn diagram).

    • Example: Sampling only from an *experimentally accessible population*, consisting of, say, *college students*, in a study of AIDS awareness and "safe-sex" practices throughout America.

    Most likely the results would be *unrepresentative* of the awareness and behavior of the *target population*, i.e., *all Americans*, given the differences in distribution of age, education, marital status, etc. between those included in the study and those that were excluded (note: target versus experimentally accessible populations are formally defined below under "Generalization/External Validity Concerns").

4. Requirements for Ensuring a Valid/Representative Sample

— *Precise specification* for the *population(s)* to be sampled of all the *relevant characteristics* and their *distributions* is clearly essential for selecting representative and thus valid samples.

- Only if the important properties of the population(s) are know is it possible for researchers to ensure that those properties are *equivalent* in the *sample(s) selected* so that they are *not biased* (illustrated ascentral sample in the botom panel of Appendix 8).

- Examples: Important characteristics for *populations of participants* and hence for samples are sex, race, age, education, and any relevant disorders — all of which should be noted with their distributions, e.g., 55% female and 45% male.

5. Importance of Unbiased/Representative Sampling for Accuracy

- To have *accuracy* and thus *validity* when sample statistics are used to estimate *population parameters,* it obviously essential that unbiased/representative samples be obtained.

- *Accuracy,* however, is dependent not only on *sampling validity,* but also on *sampling reliability* — although not to the same degree.

  - ∞ This is because the *lower* the sampling reliability, the *greater* the probable *sampling error* (see earlier discussion).

Generalization/External Validity Concerns[7]

There are *three categories of validity concerns* when making generalizations: *participant* (or subject), *ecological* (or environment), and *temporal* (or time). All of these involve the issues just covered regarding *sampling validity.*

1. Participant (Subject) Validity

— *Participant validity* is the ability to *correctly generalize* from a *sample of participants* to the *population of individuals that is of interest.*

— Types of populations (see Figure A8.2)

Two types of populations must always be considered, *target* populations and *experimentally accessible* populations.

1) Target population

  - ∞ A *target population* is the *actual population* to which the findings of research are to be *generalized.*

    - Example: All the vertebrate animal species in the *world*

  - ∞ Such populations are *not typically used* to obtain samples.

- It is usually *too difficult* to fully access such a population to get a *representative/random sample,* given that these populations are normally *quite large* and *very dispersed.*

2) Experimentally accessible population

∞ An *experimentally accessible population* is a *convenient population,* i.e., one that is *readily available* to the researcher.

- Example: A *school's colony* of laboratory-bred white rats

∞ Such populations are the type *typically used* for obtaining study samples.

- However, such a population is *rarely representative* of the *target population,* of which it is a *sample* (elaborated below).

— Two-step inferential process normally involved in generalization

- The process of *generalization* (conceptually, if not in reality) usually involves *two steps* that correspond to the just noted *two types of populations* that can be sampled; but in most situations, unfortunately, *neither* of these two steps is likely to be *valid.*

1) From the sample to the experimentally accessible population

- *Good confidence in the validity* of this first step of generalization is *possible* if *random selection/sampling* is carried out to ensure that the sample is *representative* of the readily available population. But this is *not typically the case,* as when a sample is *restricted* to only those humans who will *volunteer* or to only those animals that can be *readily caught.*

2) From the experimentally accessible population to the *target population*

- *Poor confidence in the validity* of this second step of generalization is *ordinarily* the case, because the experimentally accessible population (and thus also the sample) is *rarely representative* of the target population.

  Values of the various characteristics in the target population (e.g., education and age distribution for humans) usually do not occur with the same *relative frequencies (distribution)* as in the experimentally accessible population, and some elements/values of the target population might even be *excluded.*

— Selection-by-treatment interactions

- When a sample is *not representative* of the population due to the *selection/sampling procedure,* the concern is that it's quite possible there would be interactions of *experimental treatment conditions* with one or more *participant-characteristic differences* that exist between the *sample* and the *target population.*

∞ Such interactions would mean that treatment effects are *influenced* by the participant-characteristic differences.

- Hence there would be poor confidence in the *accuracy/validity of generalizations* from an unrepresentative sample to the target population.

- At a minimum, *different* unrepresentative samples would yield *inconsistent results*, leading to a *reliability* problem.

- Note, however, that *interactions* of treatments with participant characteristics are usually *less serious* and thus less of a generalization problem for behavioral phenomena near the *reflex level* of organisms (e.g., studies involving simple reaction times to different stimuli), than for phenomena at the *cognitive level* (e.g., studies of concept formation).

∞ *Degree of generalization/external validity problem* that results, if a selection-by-treatment interaction actually does occur, depends on the *type/form of the interaction*.

- Forms of interactions[21,23]

  There are *two types* of interactions: *ordinal and disordinal.*

  1) Ordinal interactions

     Ordinal interactions are those in which the *rank order* of the dependent-variable means, which indicate the relative effects of the different levels of an independent variable, *remains the same* across the different levels of second independent variable (Figure 11.1).

     In the *context of generalization* discussed here, these would be interactions in which the rank order of the dependent variable means is the *same for the population and sample participants* (and this would also apply to variations in environments or times).)

ORDINAL INTERACTIONS

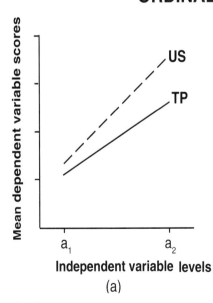 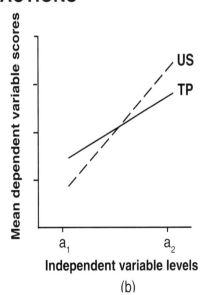

FIGURE 11.1
Hypothetical ordinal interactions of treatments (a_1 versus a_2) with selection (target population (TP) versus an unrepresentative sample (US)). In ordinal interactions, the rank order of treatment effects remains the *same*, and whether the lines plotting these treatment effects intersect (b) or not (a) is irrelevant (see also Figure 11.2). What is relevant is that the lines go in the *same direction* when there are ordinal interactions.

So, although there would be a difference in *relative effects* of treatments/conditions for the target population versus sample participants, there would be no reversals in the rank order of effects.

Generalizations are therefore only *slightly limited*, because the problem for ordinal interactions is *quantitative* rather than qualitative (in contrast to the case for disordinal interactions, as noted below).

- *Graph of the data* for the target population versus the sample would have non-parallel lines, but the lines would go in the *same general direction* — not opposite directions (e.g., *both* rising and thus showing increasing dependent variable scores for higher independent-variable levels, as in Figure 11.1a and b).

2) Disordinal interactions

- *Disordinal interactions* are those in which the *rank order* of the dependent variable means, which as already noted indicate the relative effects of the different levels of an independent variable, *changes* across the different levels of a second independent variable (see Figure 11.2).

  In the *context of generalization* discussed here, these would be interactions in which the rank order of the dependent-variable means is *different for the population versus sample participants* (and this would also apply to variations in environments or times).

DISORDINAL INTERACTIONS

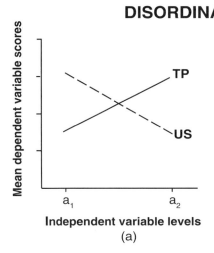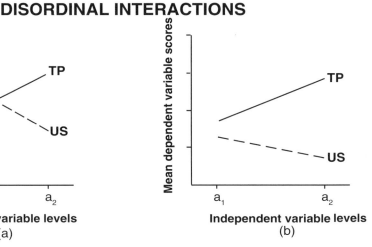

FIGURE 11.2
Hypothetical disordinal interactions of treatments (a_1 versus a_2) with selection (target population (TP) versus an unrepresentative sample (US)). In disordinal interactions, the rank order of treatment effects *changes*, but, as for ordinal interactions (see Figure 11.1), whether the lines plotting these treatment effects intersect (a) or not (b) is irrelevant. What is relevant is that the lines go in the *opposite directions* when there are disordinal interactions.

Thus, not only would there be a difference in *relative effects* of treatments/conditions for the target population versus the sample participants, there would actually be *reversals* in the rank order of effects (or *no differences* in effects for one of them).

Generalizations would therefore be *invalid*, unlike the case for ordinal interactions, because the relative effects of different levels of the independent variable on the dependent variable would be *qualitatively* different, i.e., typically *flip-flopped*.

- *Graph of the data* for the target population versus the sample would have non-parallel lines that, in fact, go in *different directions* (one rising while the other falls, as in Figure 11.2a and b).

  Note: As illustrated in the figures, it is *not necessary* that the lines *cross* for interactions to be *disordinal*, just that they *don't* go in the same direction; furthermore, *ordinal* interactions can themselves have *crossing lines*, so long as the lines *do* go in the same general direction.

2. Ecological (Environmental) Validity

— *Ecological validity* is the ability to *correctly generalize* from a *study's environmental conditions* — which represent a sample that includes *experimenter factors (such as independent variables and their levels, and dependent variable measures)* — to the *real-world* environmental conditions that are of interest.

- To be *completely* generalizable, treatment effects in a study must be *independent of*, i.e., not interact with (be influenced by), any *differing* environmental conditions — which is unlikely.

  ∞ *Degree of any generalization/external validity problem* again depends on the *type of interaction*, if there is any (see above regarding ordinal interactions versus disordinal interactions).

— Forms of environmental threats to external, or internal, validity

  There are *six categories of environmental threats*, most of which are elaborated elsewhere in this Handbook as *extraneous variables*, that must be controlled.

  1) Physical-setting effects

     ∞ Physical-setting effects result from any unrepresentativeness of the *experimental environment* (e.g., the apparatus and characteristics of the research room or other study environment) with respect to the *real world*.

  2) Experimenter characteristic effects

     ∞ Experimenter characteristic effects result from any influences on the study of the researcher's *bias, performance, appearance, or personality.*

  3) Hawthorne effects

     ∞ Hawthorne effects are caused by participants reacting to *knowing that they are being observed/studied* and thus behaving differently than they otherwise would.

4) Novelty or disruption effects

∞ Novelty or disruption effects result from participants reacting to the *newness or unusualness of treatments or tasks,* in addition to simply reacting to the treatments and tasks themselves.

5) Multiple treatment effects

∞ Multiple-treatment effects are *practice-fatigue* influences, associated with *order/sequence,* when individuals are observed under more than one condition.

6) Testing effects

∞ Testing effects occur when *pretest measures* cause *sensitization or resistance* to independent-variable treatment(s) and/or any *direct influences* a pretest might have on posttest measures.

3. Temporal (Time) Validity

— *Temporal validity* is the ability to *correctly generalize* the experimental results *across time,* i.e., from the time of the study to other times.

• To be completely generalizable, treatment effects in a study must be *independent of,* i.e., not interact with (be influenced by), differences in the *time* of the research and its application.

∞ *Degree of any generalization/external validity problem* again depends on the *type of interaction,* if there is any (see earlier regarding ordinal interactions versus disordinal interactions).

∞ *Magnitude* of an interaction also is important for temporal, as well as for participant and ecological, generalization validity.

— Forms of time-variation threats to external validity

There are *three primary categories* of *time-variation threats.*

1) Seasonal variation

∞ Fixed-time variation

• Fixed-time variation represents change that occurs for many people at specific *predictable times,* such as during certain seasons or holidays or on certain weekdays or weekends.

Examples: *Spring and Christmas seasons/holidays* can affect people's behavior; hence the results of research done in one of these seasons might not be generalizable to other times, i.e., there would not be external validity.

∞ Variable-time variation

• Variable-time variation represents change that is *unpredictable,* except from individuals' *personal events* that are not usually known to the investigator.

Example: Annual recurrence of grief or depression on the *anniversary* of a loved one's death

Note: This form of temporal validity problem applies particularly to *small-N research*, in which there are only very few participants and perhaps just one.

2) Biorhythmic variation

∞ Biorhythmic variation represents cyclical changes due to the *biological clocks* within individuals.

   • Examples: *Menstrual/estrous cycles* (about a month in humans); *circadian rhythms* (about a day); and the *Basic Rest–Activity Cycle* (about 90 min in humans, running throughout the day and night and playing a role in dream cycles, daydreaming, and subtle performance variations in such things as perceptual thresholds and problem-solving ability).

   • Note that rhythms in behavioral activity, hormone levels, etc., can be fairly *synchronized for groups of individuals,* and thus research findings obtained, e.g., at one time of day or month, might not be generalizable to other times.

3) Noncyclic variation

∞ Noncyclic variation represents changes in characteristics/traits of individuals due to *maturation* or *history,* and thus the changes are ordinarily not repetitive.

   • Note: noncyclic changes can be fairly *widespread in groups* as a result of a *shared history,* such as a political movement, war, or economic event.

      Examples: The *Feminist Movement,* the *Viet Nam War,* and the *Great Depression* of 1929.

      Therefore, research done with individuals of a fairly narrow *age range* might not be generalizable to people of other ages — not only because of differences in *physical age,* but due to differences in the *historical events* that they have experienced.

Sampling Techniques

There are *four general sampling techniques,* most of which have alternative names: *haphazard/catch-as-catch-can/convenient sampling, strict random sampling, quota/stratified random sampling,* and *bellwether/area/block sampling.* As noted earlier, the sampling technique that is chosen influences the *validity/accuracy of generalizations* because of differences in how *representative* the resulting samples are likely to be.

The following discussion of sampling techniques is in the context of populations of *participants/subjects.* However, the principles involved (although not necessarily the difficulties) are also true for selecting samples from among the possible *independent-variable levels* that might be studied, the *extraneous-variable levels* that might be involved (such as environmental conditions and times of day), and the alternative *dependent variables* that might be measured.

1. **Haphazard/Catch-as-Catch-Can/Convenient Sampling**
   — *Selection* in this type of sampling is from a *limited, readily available portion* of the population.
   - *Experimentally accessible populations* are sampled with this technique, rather than the true *target population(s)* of interest.
     - ∞ Even within this limited population, sampling is *often not random,* but rather *catch-as-catch-can;* hence it is difficult to specify *what population* is actually represented by a sample.
   — *Samples* resulting from this technique are likely to be *unrepresentative/biased* with respect to the population(s) of interest.
   - *Generalizations* must therefore be made with *great caution.*
     - ∞ Note: *Contradictory findings* that occur in the research literature might be partially due to the use of this catch-as-catch-can sampling technique.
   — Haphazard sampling is the *most common technique* used for participant populations.
   - This is because such sampling is *convenient and inexpensive,* i.e., it doesn't require much time, effort, or money.
   - Example: Most psychological research on *humans* has been conducted just on the limited population known as *college students* and, for any given study, usually only those at a *specific university* and often only individuals especially *readily available* to the experimenter and free for the taking (such as introductory psychology course students), and then only if they *volunteer.*
     - ∞ Clearly it is *questionable* just how representative these *convenient samples* of participants would be of all humans.
   — Haphazard sampling is the *opposite* of strict random sampling.

2. **Strict Random Sampling**
   — *Selection* in this type of sampling is by a *completely random process,* i.e., each element in the population would have an *equal probability* of being included (see Chapter 8 under "Randomizing").
   — *Samples* resulting from this technique are likely to be *representative/unbiased* with respect to the population(s) of interest.
   — Strict random sampling, however, is a *rarely employed technique* for participant populations.
   - This is because it is a *difficult, time-consuming, and costly* method of sampling — unless the target population is small.
   - Example: Sampling from a target population of *all people on Earth* versus a population of just the *people in some small town.*

3. Quota/Stratified Random Sampling

— *Selection* in this type of sampling is by a *random process* but with *restrictions/ constraints*.

   • Hence this technique incorporates some *systematization*.

   ∞ *Quasi-random sampling* is what this is called (but it is also sometimes referred to as *limited-/pseudo-random sampling*).

— *Samples* resulting from this technique are likely to be *more representative* of the population(s) of interest than those resulting from strict random sampling.

   • This is because *quota restrictions* are used specifically to make certain that the *proportions* of key characteristics in a sample are the *same* as in the target population.

   ∞ To accomplish this, a population of interest is first *stratified* (divided/ blocked) into *subsets* of potential participants based on one or more important characteristics that they share, such as their sex, ethnic background, and whether they went to college; then from each of the different blocks/strata of the population, random samples are selected in the *same proportions (hence quotas)* as the blocking characteristics occur in the population.

   • Example: Suppose the human population were exactly *one-half male* and *one-half female*; then the population would be *stratified* (divided/blocked) *by sex*, and a *quota* of half the sample would be taken randomly from the male subset of the population, and the other half of the sample would be taken randomly from the female subset/block.

— Quota/stratified random sampling is *often used* for *small* populations, because in those situations strict randomization would be less likely to produce representative samples; however it is a *rarely employed technique* for *very large* target populations of participants, in which strict random sampling would likely produce representative samples.

   • Note: Quota sampling is sometimes also used to enhance the representativeness of *nonrandom (e.g., haphazard) sampling*.

4. Bellwether/Area/Block Sampling

— *Selection* in this type of sampling is *nonrandom* but instead involves a *systematic* choice of a *representative* subset of the population.

   • The *sample subset* would be one that is *known from prior experience* to be *unusually accurate* in reflecting the behaviors or characteristics of interest in the target population.

— Bellwether sampling is *occasionally employed* by social psychologists and pollsters, who use as samples *selected geographical areas or neighborhood blocks* that are *particularly representative* of the population of interest.

   • Example: *Bellwether precincts* are sometimes used to predict the *voting behavior* of the entire population of the United States.

- When bellwether samples are available, this technique is usually *easier* than random forms of sampling for *very large populations*.

Review Summary

1. A *population* refers to an *entire collection* of people, objects, or events of interest to which one wants to generalize (the *universal set*). A *sample* refers to a *subset* of people, objects, or events that are *selected* for study from a population. *Sampling* is the *process of selecting* a sample from a population. *Generalization,* in turn, completes the cycle, and is an *inductive inferential process* of extending the findings from a sample to a population, which might involve *statistical estimation* from *sample statistics* to *population parameters.*

2. The *fundamental decisions* when selecting a *sample* of participants (similarly for other samples) are (a) The *population* from which participants will be selected, which determines the population to which one can generalize; (b) the *number* of participants that will be selected, which influences the reliability of generalizations; and (c) The *technique* with which subjects will be selected, which influences the validity of generalizations.

3. *Parameters* are descriptions of *population* characteristics. *Statistics,* on the other hand, are descriptions of *sample* characteristics, which are used to *estimate* the values of corresponding population parameters and to *test hypotheses* about those values. *Descriptive statistics* are those used to *organize and summarize* data obtained from samples (e.g., mean, standard deviation, and correlation coefficient). *Inferential statistics,* on the other hand, are used for *estimation and hypothesis testing,* i.e., for making inferences about *population parameters* based on the reliability of *sample- data descriptive statistics* (e.g., F-ratio of the analysis of variance).

4. *Sampling reliability* is the *consistency/stability* of a sample statistic used to estimate a population parameter. It is determined by the size of the *variable/inconsistent error component* of a measure. Variable error averages toward zero as the sample size increases. Hence sampling reliability is *directly proportional to the square root of the sample size.*

5. Most dependent-variable data in the behavioral sciences have a *frequency/ probability distribution* that has the appearance of a *bell-shaped curve* and approximates what is referred to as the *normal curve* or *normal distribution. Specific proportions* of the distribution of scores are always contained within specific areas of the curve. Hence, if we know an individual's score as well as the mean and standard deviation of the normal distribution of scores, then we also know the individual's *relative rank in percentiles (centiles). Standard scores (z scores)* are scores expressed in terms of how many standard deviations they are from the mean of the distribution of scores. They not only permit the computation of percentile ranks but

allow the *comparison of scores across normal distributions* that have different means and variances and even very different units of measurement.

6. A *sampling distribution of means* is a *distribution of sample means* that would result when repeated samples are taken and the means calculated. This is usually a *normal distribution*. The *standard error of the mean* is the *standard deviation* of the distribution of sample means, and hence a *measure of sampling reliability*. *Sampling error* is the *deviation* of a sample statistic from the true value of the population parameter due to chance variation. Its *expected magnitude* for a mean is given by the *standard error of the mean*. *Larger samples* have smaller sampling error, thus greater sampling reliability and hence increased power to detect any treatment effects using inferential statistics. High reliability implies *accuracy*; however it is a *necessary but not sufficient condition* for accuracy because *constant error* is possible.

7. *Sampling validity* is the *accuracy/correctness* of a sample statistic with respect to the corresponding population parameter. It is determined by the size of the *constant/systematic-error component* of a measure, which depends on the degree to which the sample is *unrepresentative* and thus the extent to which the statistic/ estimate is *biased*. The *causes* of biased/invalid sample statistics are (a) the sample *includes individuals/elements* that are *not members* of the population of interest; or (b) the sample *excludes elements* of the population of interest. To ensure a *valid/ representative/unbiased sample* it is required that there be precise specification of all the *relevant characteristics* and the *distributions* of those characteristics in the *population(s)* to be sampled.

8. There are three *categories of external validity concerns* when making generalizations: *population* (participant), *ecological* (environmental), and *temporal* (time). These are all dependent on *sampling* validity. *Population validity* is the ability to *correctly generalize* from the sample of participants to the population of individuals that are of interest. (The other forms of external-validity concerns are discussed later.)

9. There are *two types of populations*: (a) the *experimentally accessible population*, i.e., the convenient population that's readily available to the experimenter; and (b) the *target population*, i.e., the larger population to which the findings are to be generalized. Thus *generalization* normally involves a *two-step inferential process*: (a) from the sample to the experimentally accessible population and (b) from the experimentally accessible population to the target population. Although there is usually *good confidence* in the first step's validity, there is normally *poor confidence* in the second step's validity because the experimentally accessible population (and hence the sample) is *rarely representative* of the target population.

This *nonrepresentativeness* can lead to *selection-by-treatment interactions* and thus *externally invalid experiments*. The generalization problem is greatest if the interaction is *disordinal*, i.e., when there are *reversals* in the differential effectiveness of treatments for the sample versus the target population. If, on the other hand, the interaction is *ordinal*, i.e., the rank order of treatment effects is the *same* (no reversals), then the generalization problem is only *quantitative*, not qualitative, and hence is less serious.

10. *Ecological validity* is the ability to *correctly generalize* from the study's *environmental conditions*, including *experimenter factors*, to the real-world environmental conditions of interest. There are several forms of *threats* to this validity: (a) physical-setting effects (representativeness), (b) experimenter characteristic effects (bias, performance, etc.), (c) Hawthorne effects, (d) novelty or disruption effects, (e) multiple treatment effects (practice and fatigue), and (f) testing effects (sensitization or resistance, as well as possible direct effects).

11. *Temporal validity* is the ability to *correctly generalize* the experimental results *across/ over time*. There are several forms of *threats* to this category of external validity: (a) *seasonal variation*, which can include *fixed-time variations* occurring at specific predictable times, e.g., holidays and weekends; as well as *variable-time variations* that are unpredictable, except from individual personal events; (b) *biorhythmic variation* due to biological clocks within individuals, e.g., circadian rhythms; and (c) *noncyclic variation* due to maturation or history.

12. *There are four basic sampling techniques:*

    a. *Available/catch-as-catch-can/haphazard sampling,* in which selection is from a limited, experimentally accessible portion of the population and is thus likely to be *unrepresentative*

    b. *Strict random sampling,* in which selection is by a random process and is thus likely to be *representative*

    c. *Quota/stratified random sampling,* in which selection is *quasi-random* using quotas from the various strata (e.g., males vs. females) to ensure that the proportions of key characteristics in the sample are the same as those of the population, thus *better ensuring representativeness*

    d. *Bellwether/area/block sampling,* in which selection involves a *systematic choice* of a *representative subset* of the population, based on previous experience

Review Questions

1. Define and contrast *populations, samples, sampling, and generalization.*

2. What are the three *fundamental decisions* that must be made when *selecting a sample of participants*?

3. Explain the difference between *parameters* and *statistics,* and differentiate between *descriptive and inferential statistics,* giving an example of each.

4. Define *sampling reliability,* and explain what it is determined by (there are three aspects to this determination).

5. Describe the properties and usefulness of *normal distributions* and the relationship to *standard scores (z scores).*

6. What is a *sampling distribution of means*? What is the *standard error of the mean*, and how does this relate to *reliability* and *sampling error*? Finally, what are three qualities associated with *larger samples*?

7. Define *sampling validity*, and explain what it is determined by and why high reliability is a necessary but not sufficient condition for it. State the two reasons why a sample would *not* be *representative* of a population and therefore why it would be *biased*. Finally, state the *requirements* for ensuring a *valid/representative sample*.

8. Define *population validity*, and name the other two categories of *external-validity* concerns.

9. Describe the two different *types of populations* that usually exist, and in terms of these state the *two-step inferential process* normally involved in *generalization*. Then explain the typical difference in *confidence* regarding the validity of the two steps. Finally, with respect to *external validity* discuss the two forms of *interactions* that can occur and their implications.

10. Define *ecological validity*, and then list and describe as many forms of *threats* to this category of external validity as possible.

11. Define *temporal validity*, and then describe three forms and subtypes of *threats* to this category of external validity.

12. Define and contrast four different *sampling techniques*. If *random selection* were not possible, which of these four would be the best solution, and why?

12

Hypothesis Testing and Statistical Significance

CONTENTS

The material covered in this last chapter of the Handbook is rather technical, but it builds on information that was presented in the preceding chapter on "Sampling and Generalization," as well as on information in Chapter 8, regarding "Minimizing Error Variance" (the reader might want to review those pages now). Much of the material covered in the present chapter is also usually discussed in statistics courses, so the information should not be entirely new to everyone. An important function of this chapter, in fact, is to demonstrate the interrelatedness of *hypothesis testing* in research and the *statistical analyses* of data.

After the initial coverage of a number of important concepts and principles, they are then elaborated and clarified using several figures placed in the Appendices for ease of reference. This chapter concludes with a discussion comparing the concepts of *statistical significance* and *practical significance*.

Types of Hypotheses

There are two general categories of hypotheses in scientific investigations and statistical analyses: *research hypotheses* and *null hypotheses*.

1. **Research (Scientific/Working/Alternate) Hypothesis**
   — The *predicted relationship* between independent and dependent variables, i.e., what the investigator *expects* to find, is called the *research hypothesis*.

- Often this hypothesis is derived from a *theory,* and thus it involves *deductive inference*; otherwise it is usually based on past observations or possibly just a hunch.

- Frequently the research hypothesis includes a statement about the expected *directional nature* of the relationship.

  ∞ Example: "If the magnitude of punishment (the independent variable) is *increased* in value, then the frequency of criminal behavior (the dependent variable) will *decrease* in value."

- Sometimes, however, the research hypothesis is a *nondirectional* statement, i.e., simply that there is *some* relationship/influence.

  ∞ Example: "If college students join fraternities or sororities (the independent variable), then their grades (the dependent variable) will be *affected.*"

— Research hypotheses hold that the samples obtained under different independent-variable conditions will represent *different populations,* i.e., different *dependent-variable distributions* (e.g., mean scores).

— A research hypothesis corresponds to the *alternate hypothesis* (H_a or H_1) in statistical analysis.

- Note: Research hypotheses *cannot be statistically tested directly.*

  ∞ The *reason* for this is that the research hypothesis *doesn't specify* the *exact amount* of effect/relationship expected.

  ∞ Thus support must be obtained *indirectly* by gathering evidence that enables *rejection of the null hypothesis,* which is the opposite of the research hypothesis.

2. Null Hypothesis

— Generally the *null hypothesis* is that *no relationship* exists between the independent and dependent variables (this hypothesis is broader, however, when the research hypothesis is directional — see below).

- *Null* literally means "of no consequence, effect, or value"; thus in research and statistics it signifies *zero* effect/relationship (H_0).

— This hypothesis holds that *after treatments* the samples obtained under the different independent-variable conditions will still represent the *same underlying population,* i.e., the same *dependent-variable distribution* (e.g., the same mean score or μ).

- The null hypothesis is the *convention* of hypothesizing that *any differences* that might be found for the *dependent-variable scores* under various independent-variable levels are simply due to *sampling error,* i.e., *chance variation* when drawing a sample's few cases from the population's many cases, as well as to other forms of *random error* (see Chapter 11 under "Sampling Reliability," and also Chapter 8 under "Minimizing the Error Variance").

— The null hypothesis (H_0) is the one that is actually *tested statistically:*

H_0 is that $\mu_1 = \mu_2$ when H_1 is that $\mu_1 \neq \mu_2$ (nondirectional hypothesis)

H_0 is that $\mu_1 \leq \mu_2$ when H_1 is that $\mu_1 > \mu_2$ (a directional hypothesis)

H_0 is that $\mu_1 \geq \mu_2$ when H_1 is that $\mu_1 < \mu_2$ (also a directional hypothesis)

- Note the *broader null hypotheses* that occur whenever the research hypothesis is *directional*.
- Null hypotheses are *set up for possible rejection* so as to provide support for the *research/alternate hypotheses* (as already stated, and elaborated below).
  - ∞ This requires that a *probability* be determined statistically for the likelihood that obtained differences occurred only by *chance* — and thus, *conversely*, for the probability that there is a *real effect* (of some sort — see under "Proof versus Disproof" below).

Testing the Null Hypothesis: Two Possible Outcomes (See Appendix 10)

The *outcome* of statistically evaluating a null hypothesis is just the *opposite* for the corresponding research hypothesis with regard to their *rejection*.

1. **Rejecting (Failing to Accept) the Null Hypothesis**
   - *Rejection* of the *null hypothesis* occurs when the results of analyses are found to be *statistically significant*, i.e, when it is determined that there is only a *low probability* (e.g., <.05) that *chance alone* is the cause of observed dependent-variable differences (see Chapter 8 under "Minimizing the Error Variance" for an explanation of statistical tests of significance).
   - Thus, it is concluded that the *sample differences* (actually, those obtained as well as those that would be even greater) are *unlikely* when *only chance* is operating; hence the differences are *probably* due, at least in part, to the effect of some *systematic factor*.
   - The *research hypothesis* is therefore *not rejected* (in other words, we "accept" the research hypothesis).
     - When this research decision is *correct*, it is sometimes referred to as a *"hit"* (see later in chapter for discussion of possible *errors*).
   - Note: *Statistically significant differences* between *sample statistics* are those that are large enough to *reject the null hypothesis* that the corresponding *population parameters* (e.g., means) are actually *equal* (i.e., there is only a *very low probability* that they are the same).
2. **Failing to Reject (Accepting?) the Null Hypothesis**
   - *Failure* to reject the *null hypothesis* occurs when results of analyses are found to be statistically *insignificant*.

— Thus it is concluded that the *sample differences* are *not very unlikely* when *only chance* is operating; hence the differences are taken as being in fact due just to chance, such as *sampling or other forms of random error*.

— The *research hypothesis* is therefore *rejected* (in other words, we "fail to accept" the research hypothesis).

- When this research decision is *correct*, it is sometimes referred to as a *"correct rejection"* (as noted, errors are discussed later).

— Note: It is more accurate and thus preferable to speak in terms of *rejection* rather than *acceptance* of hypotheses; hence *"failing to reject"* the null hypothesis, or *any hypothesis*, is more appropriate terminology than is *"accepting"* the hypothesis (as explained next).

Proof versus Disproof: Certainty versus Support

The following discussion may seem a bit like splitting hairs, but the *distinction* that is made is actually a very important one.

1. **Scientific Acceptance is Tentative**

— Experimental results usually *do not prove* a hypothesis or theory.

- Thus scientists typically speak in terms of *probabilities* and *significance/confidence levels*.

— *Successful* research hypotheses or theories, in fact, are those that are tested and *escape being disproved* — which is the result of finding data that lead to the rejection of the corresponding null hypothesis.

- Consider the important fact that in research we usually only study *samples*, not entire *populations*, and what is found for samples might not be true for entire populations; it is only an *estimate*.

- Thus the research hypothesis is *supported/sustained* but *not proven beyond a doubt*, i.e., not verified with *absolute certainty*.

  ∞ Hypotheses and theories, therefore, can generally only be *rejected* or *fail to be rejected*; they cannot be truly accepted.

  ∞ Hence there is the *proscription* (sometimes ignored) against stating that a hypothesis is *"accepted"* and even against using the phrase *"fail to accept"* (preferred terminology being "reject").

- An analogy is our *judicial system*, in which you are assumed innocent until "proven" guilty *beyond a reasonable doubt*; similarly, we assume that the *null hypothesis* is true unless it can be rejected *beyond* some *probability level* (e.g., .05).

  ∞ Note: if you are found *"not guilty"* as a result of a trial, it isn't necessarily the case that you are *"innocent,"* only that there is *reasonable doubt* about your guilt — similarly, if the null hypothesis is rejected, it is *not proven* that the research hypothesis is correct; it is just that there is reasonable doubt about the null hypothesis.

∞ Hence, as for many things in life, there are *more than two alternatives:* is not everything just black or white (see the following discussion).

2. Plausible Rival Hypotheses

— *Rejecting the null hypothesis* means that differences in the data found under different conditions cannot *reasonably* be attributed to *chance events;* therefore, causal factors of the *research hypothesis,* i.e., the independent variable(s), *might* be responsible.

— *Extraneous variables,* however, might *also* account for the results if they were *inadequately controlled* and at least *partially confounded* with one of the independent variables.

• Extraneous variables are therefore *plausible rival hypotheses* to a research hypothesis.

∞ Hence, and this is very important, when the null hypothesis is rejected, the *fewer the plausible rival hypotheses* (i.e., the fewer the alternative explanations), the *greater the degree of confirmation* of a research hypothesis.

• Due to the difficulty (if not impossibility) of controlling all extraneous variables *perfectly,* it follows that a research hypothesis typically *cannot be proven beyond any doubt.*

— Note: *Null hypotheses* also cannot be proven, since in general it is *not possible to prove a negative* (e.g., that there are *no* differences between males and females or that there are *no* angels).

• You can only *fail to reject* a null hypothesis; you can't accept it, because *independent-variable effects* can be *masked* by extraneous variables and by having too small a sample.

∞ Hence the difficulty that researchers experience when they attempt to publish studies with *negative results* if the study does not also contain some positive findings.

3. Principle of Parsimony

— This principle asserts that the *most general and simplest explanation* or the one involving the *fewest assumptions* is to be preferred — i.e., we should favor economy/frugality in the expenditure of resources to achieve some end (this principle was also discussed earlier in Chapter 3 under "Hypothesis Formulation").

• The Principle of Parsimony is appealed to when *summarizing a body of research literature* involving studies that *lack optimal controls,* such as when *quasi-experimental designs* are used.

• It suggests that if *several sets* of related experimental results can be explained by a *single research hypothesis,* as opposed to *differing* uncontrolled extraneous variables in each of the several studies, then an effect of the *independent variable* — the research hypothesis — would be the *more tenable/logical conclusion.*

∞ The *more numerous and independent/different* the *tests of disconfirmation* that a research hypothesis can survive, the *stronger is its support/confirmation* because then the less numerous and plausible any rival, invalidating hypotheses become.

• Hence the value of *"replication"* with some variation.

Potential Errors During Statistical Decision Making (See Appendix 10)

There are *two types* of possible errors that can occur when deciding whether the differences in mean-dependent-variable scores under different conditions are caused by the *chance effects of extraneous variables* or to some *systematic factor,* such as the *manipulations of an independent variable.* These errors were noted earlier in Chapter 7 under "Two Meanings of Experimental Control," and again in Chapter 8 under "Minimizing the Error Variance." The following will expand on the earlier coverage and relate these errors to decision making based on the *statistical analyses* of data.

1. **Type I or Alpha Error**

— *Rejecting the null hypothesis* when, in actuality, it is *correct/true* is called a Type I or alpha error.

• This leads to the *erroneous conclusion* that the *independent variable* of a research hypothesis *did* produce an experimental effect, when in fact it *did not* — i.e., the differences were produced only by *chance.*

• Hence this error is also sometimes referred to as a *"false alarm."*

— *Probability of this error* equals not the *achieved level of statistical significance* (e.g., .02) but rather the *chosen required level* (e.g., .05) for claiming that an effect or relationship exists, and this level is called *alpha* (α) — which is why *Type I error* is also called *alpha error.*

• This provides a *very useful way to remember* the nature of the error: *alpha* (α) is the first letter of the Greek alphabet, hence *Type I error*; and alpha is the level of statistical significance that is chosen for rejecting the null hypothesis, and thus it equals the *probability* that results actually due to *chance* would be *considered significant* — which in turn is the probability of the *erroneous rejection of the null hypothesis,* i.e., Type I error.

∞ There is an *easier way to remember* what this error signifies, although it is much less instructional: Type *One* error is when you claim that you *won* — i.e., that you found an effect of an independent variable — when in fact you *didn't* win (note the mnemonic/memory aid is that "won" sounds like "one").

— *Conservative/high levels* of statistical significance/confidence (i.e., *low* α probability levels that the results would be due only to chance) are chosen to *minimize the likelihood* of making this embarrassing Type I/alpha error.

2. Type II or Beta Error

— *Failing to reject the null hypothesis* when, in actuality, the hypothesis is *incorrect/false* is a Type II or beta error.

- This would lead to the possibly *erroneous conclusion* that the *independent variable* of a research hypothesis *did not* affect the dependent variable, when in fact it *might have* — i.e., *chance alone* wasn't responsible, so something *systematic* was responsible (however, as noted earlier, the systematic factor could be an *extraneous variable* rather than the independent variable).

- Hence this error is also sometimes referred to as a *"miss."*

  ∞ Note that Type II error is just the *opposite* of Type I error.

— *Probability of this error,* which cannot be determined in advance, is referred to as *beta* (β) — which is why *Type II error* is also referred to as *beta error.*

- Determination of this error's probability would require knowing *before* the study both the *primary/experimental variance* and also the *error/random variance* (an illustrated explanation is given later in this chapter); but note that if the magnitude of the independent variable effect — the primary variance — were known *in advance,* then in fact there would be *no need* to run the study!

  ∞ Although the *absolute probability* of a *Type II error* cannot be known in advance, the *relative probability* can (this is discussed latter).

— *Liberal/low levels* of statistical significance/confidence are chosen to minimize the likelihood of this error — when it is important to do so.

- *Conventional levels of statistical significance* for rejecting the null hypothesis are α =.05 or α =.01, both of which are rather *conservative/high levels*; i.e., low probabilities that result might be due just to chance and thus *low probabilities for Type I error.*

  ∞ However, these conventional levels are actually *arbitrary values* for statistical significance, and *to reduce the probability of a Type II error,* a higher probability value could be used, e.g., α = .2 (see just below for implications of this).

_____

Researcher's Dilemma and the Chosen Level of Significance/Confidence

Whether to run the *risk* of a Type I or Type II error is a matter of *utility.*

— *Probabilities* of the two types of error tend to be *inversely related*; i.e., the probability of one usually increases as the other decreases (illustrated explanation is given later) — thus a *choice has to be made*:

1) *More conservative, high levels of statistical significance/confidence* can be chosen (e.g., α =.01, rather than α =.05).

   ∞ Probability of a *Type I error* would *decrease*, but then the probability of a *Type II error* would *increase*.

   • This is *best done when* the consequences of mistaking a *chance difference* for a *genuine difference* would be too costly.

      Example: When studying whether *psychosurgery* (removing parts of the brain) is an effective treatment for pathological behaviors, such as frequent and unexplainable outbursts of violence, we would want to be *very conservative* about claiming useful effects; after all, the surgical removal of parts of the brain is irreversible, and deleterious side effects are common (such as epilepsy or loss of motivation or emotion).

2) *More liberal, low levels of statistical significance/confidence* could instead be chosen (e.g., α =.05 or perhaps even .10 or higher, rather than α =.01).

   ∞ Probability of a *Type I error* would *increase*, but then the probability of a *Type II error* would *decrease*.

   • This is *best done when* the consequences of mistaking a *true difference* for just a *chance difference* would be too costly; i.e., when there is a *search for all promising leads*, and no genuine prospect should be overlooked.

      Example: When conducting studies to find cures for *terminal diseases*, such as cancer or AIDS, or for such disorders as schizophrenia or alcoholism, we would want to be *fairly liberal* when evaluating evidence for a possibly effective treatment — especially in the *early stages of research* — so as not to throw away any leads.

   • Note: *Preliminary investigations* or *exploratory studies* in new areas of research might involve *very liberal levels* of significance in the range of .1 to .3 in order to further increase *statistical power* to find effects when they exist (see below). *Replication*, of course, would be very important when using such liberal levels of statistical significance.

Power/Sensitivity of Statistical Tests and Designs

An *extremely important concept* in research is the *power/sensitivity* of statistical tests and designs. Indeed, this has been referred to throughout the Handbook. Now it is time to discuss this concept in more detail.

1. Definition

— Power/sensitivity is the *probability of rejecting the null hypothesis* when indeed it is *false*, i.e., of making a *correct decision* — it thus represents the *probability of detecting an effect* of the independent variable when, *in truth*, there is an effect.

- Power equals *one minus probability of a Type II error* (which is *failing* to reject the null hypothesis when it is *false*): Power = 1 – β.

- *Relative power*, however, is all that can be known, not the absolute power, because the *value of* β cannot be known in advance of conducting the study (as already noted).

2. Determinants of Relative Power

There are *five major determinants* of the relative power/sensitivity of research designs, all having to do, in one way or another, with *statistical issues.*

a. Statistical tests and their different underlying properties

An important determinant of power/sensitivity is whether one uses *nonparametric* or *parametric statistical tests.*

- Nonparametric statistical tests

  ∞ Examples: chi-square test, Mann-Whitney U test, sign test, and Wilcoxon matched-pairs signed-ranks test.

  ∞ *No assumptions* are made, when using nonparametric statistical analyses, about the underlying *population parameters* (see below under parametric statistical tests) — hence nonparametric statistical tests are often referred to as *distribution-free statistics.*

  ∞ Nonparametric tests are *more broadly applicable* because of their fewer assumptions, and they are also usually *relatively easy* to calculate.

- Parametric statistical tests

  ∞ Examples: analysis of variance (ANOVA) and *t*-test.

  ∞ *Assumptions are made*, when using parametric statistical analyses, about the underlying *population parameters* of the samples on which the tests are performed: i.e., that the distributions are *normal*, that the variances under the different conditions are *equivalent*, and that the treatment effects and error effects are *additive.*

    - It has been determined, however, that fairly sizable *departures* from these assumptions still yield valid statistical analyses and thus often can be *tolerated*; hence the statistical tests are said to be *robust.*

    - But the meeting of a fourth assumption of statistical tests is considered essential — that the dependent-variable values are *independent* of each other, i.e., that any given score is *not influenced by or related to* any other score.

      This is a reasonable assumption if participants are selected at *random* and if only *one* value of each dependent variable is used for each participant in the statistical analysis, e.g., a representative or mean value (see statistics books for other situations).

∞ Note: *Ratio- or interval-scale score data* (as opposed to ordinal-scale ranking data or nominal-scale frequency data) are required for the use of parametric statistical tests.

- *Parametric tests have more power* than nonparametric tests.

    ∞ Knowledge is power: *knowing the underlying distributions* reduces the probability of making Type II/beta errors.

    ∞ Parametric statistical tests also are more powerful because they *use more of the potential information* in the data (e.g., intervals rather than just ranks of dependent-variable scores).

    - Example: A *t*-test, which uses the *amounts/intervals* by which scores differ, is more powerful than a sign test, which just uses information about whether scores are *greater or lesser* under one condition than another.

      Note: there are differences even among *parametric tests*: an analysis of variance for both *pretest and posttest scores* is more powerful than a *t*-test for just the *difference scores* between the pretest and posttest.

b. Directionality of the tests of statistical significance

   Another determinant of power/sensitivity is whether the statistical test used is *two-tailed or one-tailed*.

   - Two-tailed tests

     ∞ *Nondirectional hypotheses* are involved in such tests.

     ∞ They are used when predicting that a *difference* in mean dependent-variable scores will be found for different independent-variable levels but *not* also predicting the *direction* of effect.

     ∞ These tests are called two-tailed because the difference can fall in *either tail/end* of the *chance sampling distribution* of differences between pairs of means (as noted in Chapter 11, the distributions are usually normal/bell-shaped frequency/probability curves; sampling distributions of differences between pairs of means are extensively covered later in this chapter and in the appendices).

   - One-tailed tests

     ∞ *Directional hypotheses* are involved in such tests.

     ∞ They are used when actually predicting the *direction* of the difference in mean dependent-variable scores that will be found for different independent-variable levels, rather than just predicting that there will be an effect of *some sort*.

     ∞ These tests are called one-tailed because the difference between means must fall in a *specified tail/end* of the *chance sampling distribution* of differences between pairs of means.

   - *One-tailed tests have more power*, due to lower β, but *only* if the independent-variable effect is actually *in the predicted direction*.

∞ Knowledge is power — again due to *reduced likelihood of error.*

- If the *direction of effect* can be *predicted,* then the probability is *cut in half* that *chance* or sampling error would produce a difference (equal to or greater than observed) which could be mistaken for an independent variable effect.

- In contrast, when you are *less certain* about the outcome of an experiment, i.e., the research hypothesis is *nondirectional,* then a *larger difference* between conditions (relative to statistical error variance) must be found before deciding that the difference isn't caused by chance.

- *Two-tailed tests* are thus *more conservative* and hence *less powerful* than one-tailed tests; i.e., it is harder to reject the null hypothesis with a non-directional hypothesis than when the direction of an effect can be correctly predicted.

∞ There is *controversy,* however, about the appropriateness of using *one-tailed tests,* because researchers might change their hypotheses after gathering and looking at the data.

- *In practice,* most investigators now prefer using the more conservative two-tailed tests, with *sufficient power* ensured by other determinants such as larger sample sizes and better control over extraneous variables (discussed in earlier chapters and covered below).

- In any case, *decisions* regarding the use of one-tailed tests, as well as *predicting* the direction of an effect, *must* be made at the time an hypothesis is formulated, i.e., *before* the data are collected.

  If a decision is made *after an initial test* indicates that the two-tailed statistic is not significant, but a one-tailed test would be; or even more egregious, if the *hypothesis is changed* to correspond to the direction of an observed apparent effect, then the chosen α level *underestimates* the true *Type I error probability.*

c. Significance/confidence level (α) chosen/required

- *More liberal significance levels* (e.g., .05) yield *lower* β and hence *more power* than do higher significance levels (e.g., .01).

  ∞ *Type I error,* however, would be *more probable* (discussed earlier as well as later in this chapter).

d. Ratio of primary variance to statistical error variance

- *Greater ratios* of independent-variable treatment effects versus masking extraneous variable effects result in *greater power* (illustrated, along with other determinants, later in this chapter).

  ∞ Hence the value of *maximizing* the primary/experimental variance and *minimizing* the error/random variance, while also *controlling* the secondary/extraneous variance (discussed in Chapters 7 and 8).

e. Sample size (*n*)

- *Greater sample sizes yield greater power.*

    ∞ *Sampling error is decreased* (i.e., standard error of the mean is reduced) as sample size gets larger, which in turn leads to *increased reliability* and hence increased power/sensitivity (discussed in Chapter 11 under "Sampling Reliability").

Sampling Distributions of Differences Between Pairs of Means (See Appendix 11)

Consideration of these *distributions* will *clarify* many points already made.

1. **General Description**

   — Sampling distributions of differences between pairs of means are:

   1) *Theoretical* probability distributions;

   2) Show the *frequency* of values for *differences* between pairs of sample means;

   3) *Would* be produced if someone were to repeatedly take random samples for two conditions/populations and do the computations.

   - Example: In contrast to a *theoretical* distribution, we could produce an *actual* sampling distribution by measuring *research skills* year after year for *samples* of students who had versus had not taken a *course in research design* (these would be the two independent variable *conditions*) and then plotting the *frequencies* of the observed *differences* between the *pairs of sample means.*

   - Just as the *distributions* of sample means for *each* of the two individual conditions/populations would tend to be *normal,* i.e., bell-shaped curves, so would a *plot* of the *differences between the pairs* of sample means (this was noted earlier in Chapter 11).

   - The *expected mean/average* of the *sampling distribution of differences* between the pairs of sample means would equal the *population difference* between the means under the two conditions.

   - Moreover, the *variance* of the *distribution of differences* would equal the sum of the variances for the two separate conditions and could be *estimated from* the size and variance of a *single pair of samples,* as is done in *tests of statistical significance.*

2. **Overlapping Distribution Curves**

   — Two contrasting *hypothetical situations* — one in which the *research hypothesis* (*H₁*) *is true,* and the other in which the *null hypothesis* (*H₀*) *is true* — would have overlapping sampling distributions to a certain degree (as shown in Appendix 11) when they are graphed on the same continuum of values (abscissa/*x*-axis) for the *differences between pairs of sample means.*

— Note: *Only one* of the two hypothetical distributions *could actually be true* in a given situation, but we *wouldn't know* which one — that is, what must be determined by gathering *empirical research data.*

- After obtaining research data on some dependent variable under different independent-variable conditions, the *statistical task* is to determine the *probability* that a difference as great as (or greater than) that obtained between a pair of *sample means* represents a value from the distribution that would result when the *null hypothesis is true.*

  ∞ If that probability is low enough, say <.05, then it is concluded that the obtained difference probably came from the distribution for when the *research hypothesis is true.*

- Distribution when the research hypothesis is true

  ∞ The *mean/average of the differences* between pairs of *sample means* would have an *expected value,* when the research hypothesis is true, equal to the difference between the means for the *two populations* — i.e., equal to the difference between the means for *all* individuals of interest under the two separate independent-variable conditions.

    - *Expected variance* of the distribution would be the same as for when the null hypothesis is true and could be *estimated from a single pair of samples* (as noted earlier).

- Distribution when the null hypothesis is true

  ∞ The *mean/average of the differences* between pairs of *sample means* when the null hypothesis is true would typically equal *zero,* because in reality the sample means would come from the *same population* in that there would be no effect of independent-variable conditions.

    - Under such circumstances, *any differences* found between mean dependent-variable scores for pairs of samples would simply be due to *chance/sampling error.*

3. Critical Values/Cutoffs and Areas Under Overlapping Distribution Curves

— Probability of a Type I error (α)

- This probability is equivalent to the *rejection area(s)/critical region(s)* for the *null hypothesis* (H_0) theoretical sampling distribution curve.

  ∞ Rejection area(s) are established by the *statistical significance level* (α) that is chosen and by whether there is a *one-tailed or two-tailed test* of the research hypothesis.

    - Together these set the *critical value(s)/cutoff(s),* on the theoretical sampling distribution of differences between pairs of means, for rejecting the null hypothesis — i.e., they set the *decision rules.*

    - The *percentage of the area* under the *null hypothesis* sampling distribution curve that falls *beyond the critical value(s)* represents _, which is the probability of a Type I error: Claiming that the independent variable of a research hypothesis *did* produce an experimental effect, when in fact the difference was produced only by *chance* (i.e., it's part of the null hypothesis sampling distribution).

- Note: For a *two-tailed test*, in which the research hypothesis is non-direction, there would be a hypothetical H_1 distribution both to the right *and* to the left of the hypothetical H_0 distribution, and the *critical values* on the H_0 distribution would correspond to $1/2\ \alpha$ (although the second hypothetical H_1 distribution is not diagrammed (see Appendix 14, which is discussed later)).

— Probability of a Type II error (β)

- This probability is equivalent to the *percentage of the area* under the *research hypothesis* (H_1) sampling distribution curve that falls *below the upper critical value* and/or (depending on the directionality of the research hypothesis) *above the lower critical value* for rejecting the *null hypothesis*.

  ∞ In other words, it is the percentage of the area under the *research hypothesis* (H_1) sampling distribution curve that falls *within the nonrejection/acceptance region* for the *null hypothesis* — thus leading to the *erroneous* rejection of the research hypothesis.

— Power of the Test ($1 - \beta$)

- Power is equivalent to the *percentage of the area* under the *research hypothesis* (H_1) sampling distribution curve that falls *beyond the critical value(s)* for rejecting the null hypothesis, i.e., *outside the nonrejection/acceptance region* for the null hypothesis.

  ∞ This is a very important point; it is the *probability* — when in fact there really is a *true* effect — that the *difference* between a *single pair* of sample means, obtained under different conditions, will be considered to be *statistically significant*; i.e., *not likely to be caused by chance*, but rather by an *effect of the independent variable* of the research hypothesis.

— Inverse relationship between type I and II errors (See Appendix 12)

- If the statistical significance level (α) were *raised* (e.g., from .05 to .01), then as a result of the *change in critical value(s)* the probability of a *Type I error* (α) would be decreased (as shown in Appendix 12), but the probability of a *Type II error* (β) would be increased — and thus the *Power* ($1 - \beta$) would be decreased.

- In contrast, if the statistical significance level (α) were *lowered*, (e.g., from .01 to .05), then as a result of the *change in critical value(s)* the probability of a *Type I error* (α) would be increased, but the probability of a *Type II error* (β) would be decreased — and thus the *Power* ($1 - \beta$) would be increased.

∞ Adjusting the *statistical significance level* (*α*) is clearly not the most desirable means for increasing *power/sensitivity*, because it requires also increasing the probability of *Type I Error.*

— Increasing power by reducing Type II error without increasing the Type I error (See Appendices 13 and 14)

- *Probability* of a Type II error (β) and the Power (1 − β) cannot be known *exactly* (as indicated earlier) unless one knows *in advance* both the *magnitude of the independent variable effect* (i.e., the average difference) and the *sampling distribution variance.*

  ∞ This is because taken together these determine the *degree of overlap* between the two theoretical sampling distributions.

- *Reducing the overlap of the two sampling distributions* (as can be seen in Appendix 13) would lead to *a decrease* in the probability of a Type II error (b) and thus an *increase* in Power (1 − b), *without affecting* the probability of a Type I error (a).

- *Two strategies* can accomplish this (as suggested above):

  a) *Increase the magnitude of the independent variable effect* (the primary/experimental variance), thus moving the distribution for the research hypothesis further away from the distribution for the null hypothesis.

  b) *Decrease the variance of the sampling distribution* (the statistical error variance) — for which there are *two tactics:*

     - *Increase the sample size (n).*

       Recall that the *standard error of a sampling distribution of means,* which indicates the expected magnitude of the *sampling error of the mean,* is decreased by increasing the sample size: $\sigma_M = \sigma / \sqrt{n}$ (discussed earlier in Chapter 11, and Chapter 8 under "Techniques for Minimizing Error Variance").

     - *Reduce the masking of extraneous variable effects.*

       This involves the use of various control techniques (discussed in Chapters 7 and 8).

  ∞ Hence the *value* of maximizing the primary/experimental variance and minimizing the error/random variance, while also controlling the secondary/extraneous variance, i.e., *maximizing the signal to noise ratio* (as often stated before).

- *More powerful statistical tests* represent an additional strategy for increasing power/sensitivity (as also noted earlier).

  ∞ *Parametric tests* are more powerful than nonparametric tests, but there are certain assumptions and requirements that should be met.

  ∞ *One-tailed tests,* moreover, are more powerful than two-tailed tests when the prediction is in the correct direction (this is illustrated in Appendix 14).

Statistical Significance versus Practical Significance

It is often automatically assumed that if results are *statistically significant* then they are also of *practical significance*, i.e. that there is a *strong, rather than weak, relationship* between the independent and dependent variables. This is *not* necessarily the case, however, as will now be explained. But first a review and elaboration are necessary of some points already covered. At the end of this chapter two other issues related to the pragmatics of research are discussed.

1. **Likelihood/Reliability versus Magnitude/Strength of Effects**[22]
    — In the researcher's quest to obtain a *statistically significant* result, i.e., a *reliable effect*, a more relevant question is often overlooked:
       • Is the *magnitude of effect* — the degree to which a dependent variable is affected by an independent variable — *meaningfully significant*; i.e., are the observed dependent-variable differences among conditions great enough to be of *practical importance*?
    — Statistical significance (See Appendix 15)
       • This form of significance addresses the *likelihood/reliability of effects*, i.e., the *probability* that a *true relationship* exists in the *population* between specific independent and dependent variables, rather than just chance effects.
          ∞ *Tests of statistical significance* are used to evaluate the *likelihood* that a research hypothesis is correct, based on the computed *reliability* of measured dependent-variable differences among the means of *samples* that are obtained under the different independent-variable conditions.
          ∞ More specifically (as already noted), the *level of statistical significance achieved* in an experiment indicates the *probability* that the dependent-variable differences (equal to or greater than those observed) are due simply to *chance effects of extraneous variables*.
          ∞ Conversely, *one minus the level of statistical significance* equals the *probability* that the dependent-variable differences (equal to or greater than those obtained under the different conditions) are due to some *systematic factor(s)*.
             • In *well-designed experiments*, with optimal control of extraneous variables, the *systematic factor* should be a purposely and directly manipulated *independent-variable*, rather than *confounding extraneous variables*.
       • *Statistically significant differences* among sample statistics, such as the mean dependent-variable values under different independent-variable conditions, are those large enough to *reject the null hypothesis* that the corresponding population parameters are actually *equal*.
          ∞ This, however, is *not a real measure* of the *magnitude* of independent-variable effects — *the primary variance* — since both statistical error

variance and sample size (degrees of freedom) are also involved in the computation; e.g., F-ratio $= MS_{\text{treatments}}/MS_{\text{error}}$ with $MS = SS/d_f$ (see Part Two, Chapter 8, under "Minimizing Error Variance").

- *With sufficiently* large participant sample sizes or sufficiently *small* statistical error variance, even *trivial* dependent-variable differences in means can be found to be statistically significant, i.e., *reliably different*. This is clearly an *extremely important point.*

  Example: *Performance differences between males and females* on tasks of verbal ability, quantitative ability, and visual-spatial ability, even when *statistically significant*, are generally quite small — often accounting for only 1–5% of the *population variance.*

— Practical significance

- This form of significance addresses the *magnitude of effects* or *strength of association/relationship* between independent and dependent variables (rather than likelihood/reliability of effects).

  ∞ Example: In a study conducted between 1985 and 1989 in which about 6,500 students were randomly assigned to *small versus large class sizes* in kindergarten through third grade, test scores later sampled in the fourth, sixth, and eighth grades showed that students assigned to smaller classes early on were from 6 to 13 months ahead in reading, math, and science in the later grades.

- Magnitude-of-effect measures

  ∞ The *simplest* indices of what is commonly called *effect size* are referred to as *magnitude-of-effect measures.*

  ∞ These are based on the *difference between means* for an experimental condition versus either the control condition or another experimental condition; or alternatively they're based on the difference between the mean for one of the conditions versus the mean for all of the conditions (the grand mean).

  - The *larger the differences* the *greater the importance* — or practical significance — of a study's findings.

- Strength-of-association measures

  ∞ *More sophisticated* indices of *effect size* are referred to as *strength-of-association measures.*

  ∞ These are based on some type of *correlation* between the independent and dependent variables or on the *proportion of variability* in the dependent-variable scores that may be attributed to the effects of the independent variable.

  - The *proportion* would be the *ratio of the primary variance to the total variance*, which is usually expressed in the form of a *squared correlation coefficient.*

- Coefficient of determination (r^2)

    This is an example of a strength-of-association measure for correlational studies, as opposed to measures used for experimental research (see below), and it is computed from the Pearson product-moment correlation coefficient r (noted earlier in Chapter 11, under "Statistics and Parameters").

    r^2 indicates the proportion of variance in one variable that is associated with, accounted for, or predicted by variation in another variable.

∞ The *higher the strength of association* (range of values is .0 to 1.0), the *greater the importance* (practical significance) of the findings — i.e., the *stronger the relationship* between variables, such as the *effect* of an experimentally manipulated independent variable on a dependent variable.

- If *less than 5%* (.05) of the variance in the dependent variable can be accounted for by an independent variable, then this would be considered a *weak* association/effect.

- If, on the other hand, *more than 10%* (.10) of the variance in the dependent variable can be accounted for by an independent variable, then this would actually be a *stronger* relationship than is found for the vast majority of research findings in the behavioral sciences.

∞ *Several measures* of strength of association are available for use in *experimental research*, depending on the type of data and the design.

- Eta squared (η^2)

    This is a *rough estimate* for the *fixed-effects model* analysis of variance (see Chapter 10).

    Sometimes it is referred to as a *correlation ratio*.

    It is the *ratio* of the *between-groups* sum of squares to the *total sum of squares*: $\eta^2 = SS_{treatment}/SS_{total}$

    When there is more than one independent variable, any $SS_{treatment}$ or $SS_{interaction}$ can be in the numerator of the ratio.

    Contrast this ratio to the *F*-ratio test noted earlier, which is a comparison (in its simplest application) of the *between-conditions* variance to just the *within-conditions* variance and which is affected by the *degrees of freedom* (df).

    Major disadvantage of η^2 is that it gives an estimate of the strength of association only in the *sample*, rather than for the *population*.

- Omega squared (ω^2)

    This is a *more precise estimate* of the strength of association in a *population*.

It is another of several measures used to determine the *relative magnitude* of a treatment effect, i.e, the *proportion* of the *total variance* of a dependent variable that is accounted for by an independent variable or by interaction of independent variables.

Mathematically, ω^2 is the *ratio* of the *expected mean square (EMS)* for the treatment effect of interest, versus the *sum* of expected mean squares for all the effects (note: these are expected *population* mean squares, not the actual mean squares obtained).

Different specific formulas exist for *different designs* and can be rather complex (see statistics books).

∞ *Major point*: In addition to statistical tests of significance, it is *strongly recommended* that strength-of-association measures *always* be computed and reported for experimental studies, since such measures tend to inspire *programmatic research* — i.e., a sequence of interrelated research studies.

• *When the null hypothesis is rejected,* then together the measures of statistical *and* practical significance provide a *more complete evaluation* of the research hypothesis.

If the *proportion of variance* accounted for by one or more independent variables is small (i.e., there is not a substantial *effect size*) and thus a statistically significant (i.e., reliable) effect is *weak in magnitude/practical significance,* then *further research* should be conducted in which the effects of *extraneous variables* would be better controlled and in which one or more *additional independent variables* would be studied in an attempt to account (through their main effects and interactions) for a greater percentage of the variance of the dependent variable.

Hence, strength-of-association measures provide *feedback and guidance* that lead to *further and more sophisticated research* (i.e., programmatic research) to better explain the phenomena of interest.

Note: Even if a *strong* strength-of-association measure is obtained, this does not guarantee that the discovered relationship can be *generalized* to other situations (i.e., extraneous variable conditions) — it is important, therefore, that such possibilities be *confirmed* through additional/programmatic research.

• *When the null hypothesis is not rejected,* in contrast to the preceding situation, it is possible that this is because the *sample size* was too small and/or the *statistical error variance* was too large for the independent variable effects to achieve *statistical significance* (i.e., a reliable effect).

If, however, the *strength-of-association* measure is at least *moderate in magnitude*, then this suggests that it would be advisable to *repeat* the study with a *larger* sample size as well as *better controlled* extraneous variables in order to increase the *statistical power* to find a reliable effect of the independent variable(s), assuming there is one.

Note: *Failure to replicate* the *results* of a study that had found a significant effect could be due primarily to a *smaller sample size* in the replicate and hence *less statistical power* — the likelihood of which would be indicated by a magnitude of the *effect size* that is about as large or larger than in the original study.

2. Other Issues Related to the Pragmatics of Research

— It seems fitting to conclude this Handbook by restating *two important concerns* about research that were raised much earlier but which certainly deserve repeating.

- First, are the expected results of a study *substantial enough* to be worth the probable expenditure of *time, effort and money?*

    ∞ The issue here is that there should be some *balance* between the potential *benefits and costs* of research.

- Second, is the expected outcome of a study *important enough* to possibly warrant the use of *questionable methods?*

    ∞ The issue here is that if there are any *ethical problems* with the way in which research participants would be obtained, informed, or treated, *do the ends truly justify the means?*

— As scientists, we should never lose sight of these serious *practical and moral issues* while grappling with the many *technical challenges* of conducting research that is excellent both in design and execution.

Review Summary

1. There are two general *categories of hypotheses* in scientific investigations and statistical analyses: (a) The *research (scientific/working/alternate) hypothesis*, which is the *predicted relationship* between independent and dependent variables (can be either a directional or nondirectional statement); and (b) The *null hypothesis*, which is that *no relationship* exists (zero effect).

2. It is the *null hypothesis* that is *statistically tested*. It is set up for possible *rejection* in order to provide *support for the research hypothesis*. This requires that a *probability* be statistically determined for the likelihood that differences occurred only by *chance* — and thus, conversely, for the probability that there is a *real effect*. The *null hypothesis is rejected* if results are found to be *statistically significant*, i.e., when

it is determined that there is only a *low probability* (e.g., < .05) that chance alone is the cause of observed dependent-variable differences. Therefore, the *research hypothesis is not rejected*. In contrast, one would *fail to reject the null hypothesis* if results are found to be statistically *insignificant*, and thus the *research hypothesis would be rejected*.

3. *Scientific acceptance is tentative* in that experimental results typically *do not prove* a hypothesis or theory. Thus we speak of *probabilities* and *significance/confidence levels*. Hypotheses and theories, therefore, can generally only be *rejected* or *fail to be rejected*, they can't be truly accepted. *Successful* research hypotheses or theories, in fact, are those that are tested and *escape being disproved*. In addition to *chance effects*, it is possible that *extraneous variables* might also account for the results if they were not adequately *controlled* — they represent *plausible rival hypotheses* to the research hypothesis.

4. There are *two types of possible errors*, which are opposites, that can occur when deciding whether the differences in mean dependent-variable scores under different conditions are due to *chance effects of extraneous variables* or to the effect of some *systematic factor*:

   a. *Type I or alpha error* occurs when you *reject* the null hypothesis, but it is actually *correct/true*. The *probability* of this error equals the chosen level of required statistical significance, which is referred to as *alpha* (α) — hence the alternative name for this type of error.

   b. *Type II or beta error* is when you *fail to reject* the null hypothesis, but it is actually *incorrect/false*. The *probability* of this error is referred to as *beta* (β). Although its *absolute value* cannot be determined in advance — since one would have to know both the primary and error variance — the *relative value* can be determined (covered later).

5. The *researcher's dilemma* is whether to run the risk of a Type I error or a Type II error. The *probabilities* of the two errors tend to be *inversely related*. When a *more conservative*, high level of significance is chosen (e.g., $\alpha = .01$ vs. .05 or more), the probability of a Type I error would *decrease*, but then the probability of a Type II error would *increase*. The opposite would be true when a *more liberal*, low level of significance is chosen. A more conservative level is best when the consequences of *mistaking a chance difference for a genuine difference* would be too costly. A more liberal level is best when the consequences of *mistaking a true difference for just a chance difference* would be too costly.

6. *Power/sensitivity* of statistical tests and designs is defined as the probability of *rejecting the null hypothesis when it is false* — hence a *correct decision*. This corresponds to the probability of *detecting an effect* of the independent variable when *in truth* there is an effect. It is equal to one minus the probability of a Type II error (*Power = 1 – α*). There are several *determinants* of power: (a) *Parametric statistical tests* have more power than nonparametric tests; (b) *One-tailed tests*, which involve directional research hypotheses, are more powerful (when correctly predicting the direction) than two-tailed tests, which involve nondirectional research hypotheses; (c) *Lower, more liberal required significance levels* (e.g.,

.05) yield lower β and hence more power than higher significance levels (e.g., .01); (d) *Ratios of primary variance to statistical error variance* that are greater yield greater power; and (e) *Greater the sample size*, the greater the power.

7. *Sampling distributions of differences between pairs of means* are theoretical probability distributions that show the *frequency* of values for *differences* between *pairs of sample means* that could result from repeated random samples for two conditions/populations. These tend to be *normal distributions*, and the *mean* of the distribution of differences would equal the *population difference* between the means for the two conditions, and the *variance* of the distribution of differences would equal the *sum* of the separate variances for the two conditions and could be estimated from the size and variance of a single pair of samples.

8. *Overlapping sampling distributions* can be drawn for differences between pairs of sample means that would be found under *two hypothetical situations*, only one of which could actually be *true* in any given instance: (a) the distribution when the *null hypothesis* is true, in which case the average of the differences between pairs of sample means would usually equal zero, i.e., no effect; and (b) the distribution when the *research hypothesis* is true, in which case the average of the differences between pairs of sample means would have an expected value equal to the difference between the means for the two condition populations.

9. A *number of concepts* can be illustrated using the overlapping theoretical probability distributions (see Handbook Appendices 11–13):

   a. *Probability of a Type I error* (α) is equivalent to the *rejection area(s)/critical region(s)* for the *null hypothesis* sampling distribution curve and is the *percentage of area* under the curve that falls *beyond* the *critical value(s)/cutoff(s)* set by the *significance level* (α) that is chosen and by whether there is a *one-tailed* or *two-tailed test* of the research hypothesis.

   b. *Probability of a Type II error* (α) is equivalent to the *percentage of area* under the *research hypothesis* sampling distribution curve that falls *within the nonrejection/ acceptance region for the null hypothesis.*

   c. *Power of the test* ($1 - \beta$) is equivalent to the *percentage of area* under the *research hypothesis* sampling distribution curve that falls *outside the rejection/acceptance region* for the null hypothesis, i.e., *beyond the critical value(s)* for rejecting the null hypothesis.

   d. *Power can be increased* by reducing Type II error *without* increasing Type I error, through *reduction of the overlap* of the two sampling distributions — which requires *increasing the magnitude of the independent variable effect* (primary variance) and/or *decreasing the variance of the sampling distribution* (statistical error variance).

10. *Statistical significance* addresses the *likelihood/reliability* of effect(s), i.e., the *probability* that a true relationship exists in the *population* between an independent variable and a dependent variable. *Practical significance*, on the other hand, addresses the *magnitude* of effects, i.e., the *strength of relationship* between the independent and dependent-variables. This is a very important

issue that is often overlooked. Statistical significance is *not* a pure measure of the magnitude of independent variable effects — the primary variance — *because statistical error variance and sample size* are also involved in the computation of statistical significance.

11. There are different types of measures of practical significance. *Magnitude- of-effect measures* are the simplest and are based on the *difference* between means. The *larger* the differences, the *greater* the importance of the findings. *Strength-of-association measures* are more sophisticated indicants of effect size. They are based on some type of *correlation* between the independent and dependent-variables or on the *proportion of variability* in the dependent variable scores that may be attributed to the effects of the independent variable (squared correlation coefficient). Two examples are *eta squared* (η^2) and *omega squared* (ω^2). The *higher* the strength of association (values range from .0 to 1.0), the *greater* the importance of the findings, i.e., the more powerful the relationship.

12. It is *strongly recommended* that in addition to statistical tests of significance, strength-of-association measures be computed and reported. *When the null hypothesis is rejected,* then together the measures of statistical and practical significance provide a *more complete evaluation* of the research hypothesis. *When the null hypothesis is not rejected,* it could be because the *sample size* is too small and/ or the *error variance* is too large to achieve statistical significance. If the strength-of-association measure is at least *moderate,* then logically it would be advisable to *repeat the study* with a *larger* sample size and *better controlled* extraneous variables in order to increase the statistical power.

13. There are two other *important issues/concerns* related to the *pragmatics of research* that were discussed earlier in the Handbook: (a) are the expected results of running a study *substantial enough* to be worth the probable expenditure of *time, effort and money?* (b) is the expected outcome *important enough* to warrant the use of *ethically questionable methods?* While grappling with the many *technical challenges* of conducting research that is excellent in both design and execution, scientists should never lose sight of these serious *practical and moral issues.*

Review Questions

1. Define and contrast a *research hypothesis* versus a *null hypothesis.*

2. Explain which hypothesis is statistically tested, and state the relationship between *statistical significance* and *rejecting* or *failing to reject* the null hypothesis versus the corresponding research hypothesis. Also, explain what it means to obtain *statistical significance* at say $P < .01$ (be precise and thorough — this concept is covered in more than one place in Chapter 12.)

3. Why is it that you can *never prove* that your research hypothesis is correct, even when statistical significance is achieved? Explain in terms of *tentative acceptance and plausible rival hypotheses.*

4. Define and contrast *Type I (alpha) error* and *Type II (beta) error*. Also discuss the *probability* of each type of error as thoroughly as possible.

5. Describe the *researcher's dilemma* with regard to running the risk of a Type I error or a Type II error. Explain the basis for making *decisions* regarding the two types of errors.

6. Define *power/sensitivity* of statistical tests and designs in terms of probabilities, including the *relationship* to either Type I error or Type II error. Discuss five ways to *increase* the statistical power/sensitivity of a research design. Be sure to define and contrast *one-tailed tests* versus *two-tailed tests* of statistical significance, and explain the relevance of this distinction in terms of *power/sensitivity* (covered at more than one point in Chapter 12.)

7. Define and describe (as thoroughly as possible) *sampling distributions of differences between pairs of means.*

8. Draw *overlapping sampling distributions* for when hypothetically the *null hypothesis* is true versus when the *research hypothesis* is true. State for each case what the *expected value of the average* of the differences between pairs of sample means would be.

9. Use the above drawing and additional ones to illustrate what determines the *probability* for *Type I error and Type II error*, as well as to illustrate the determinants of *power/sensitivity* and how it can be increased.

10. Define and contrast *statistical significance* versus *practical significance.*

11. Name and describe the two general *types of measures of practical significance*. State the relationship of their *obtained values* to the importance of findings.

12. Discuss at least two specific reasons/advantages for calculating *strength-of-association measures* between the independent and dependent variables?

13. Name two other *important issues/concerns* related to the *pragmatics of research* that were discussed earlier in the Handbook, and discuss their relevance with respect to the *technical challenges* of conducting research.

Appendices

Appendices

Appendix 1
Variables and Relationships

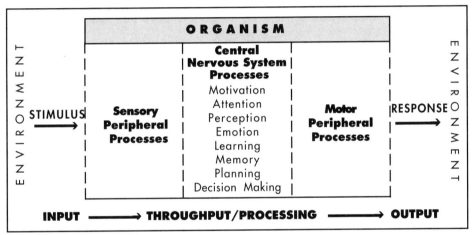

FIGURE A1.1
Relationship of the organism to its environment. *Note:* The organism includes its anatomy and physiology, as well as psychological processes.

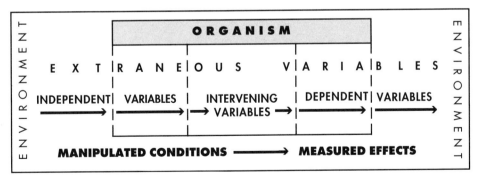

FIGURE A1.2
Relationships among experimental variables. *Note:* Independent, dependent, and extraneous variables can be in the external environment or within the organism, central or peripheral, sensory or motor. Central intervening variables are optional.

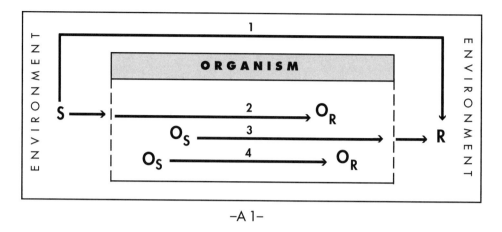

-A 1-

FIGURE A1.3
Relationships investigated experimentally. *Note:* Effects of stimulus (S) independent variables on response (R) dependent variables are studied with, each being either in the environment or the organism (O). Intervening-variable effects are inferred.

Appendix 2
Reliability and Validity

Applications and Relationships to the Theoretical Components of Any Measure

Experiments	Dependent Variable	=	f (Independent Variable)	+	(Extraneous Variables)		
Communications Model	Reception	=	f (Signal)	+	(Noise)		
Components of any Single Measurement (such as the above)	Measure	=	True Score	+	Error Score		
					Constant Error	+	Variable Error
Set of Measurements	Total Variance	=	Primary/ Experimental Variance	+	Secondary/ Extraneous Variance	+	Error/Random Variance
Statistics	SS_{total}	=	$SS_{treatment}$	+	$SS_?$	+	SS_{error}

FIGURE A2.1

Measurement of dependent variables in *experiments* and the reception of information in *communication systems* (mechanical and living) are parallel applications illustrating principles of the theoretical components of any measure and their relationship to reliability and validity. *All measures,* in fact, can be considered to consist of *components* associated with the important concepts of reliability and validity. With respect to the *true score component* of a measure and thus the *primary/experimental variance* of a set of measures: reliability (consistency/ stability) relates to the relative size of the *variable error* and thus the *error (random extraneous) variance*, and validity (accuracy/correctness) relates to the relative size of the *constant error* and thus *secondary (systematic extraneous) variance*. $SS_?$ represents the possible separate *sum of squares* for secondary (consistent effect), extraneous variables. This can be computed when the variables are *systematized,* as opposed to when their extraneous variance is *confounded* with the primary variance of an independent variable or when it is combined with the error/random variance of extraneous variables that can cause *masking* of the primary variance.

Appendix 3
Proposals and Research Reports

Components of Research Proposals

1. **Study Title:** Provide a *concise, self-explanatory, summary label* that indicates the problem area, usually by stating the independent-dependent variable relationships to be studied; that gives some clue to the research method; and that specifies the subject population, if not humans in general.

2. **Name(s) of Investigator(s):** Include *all* those who will play a substantial role.

3. **Research Problem:** State the *purpose* of the research succinctly and unambiguously in a single sentence, possibly in the form of a *question*. Indicate why the proposed research is important; i.e., give the theoretical and/or practical significance of the study.

4. **Literature Review:** Summarize the *relevant* theoretical foundation and previous research that are related to the problem(s) of the proposed study.

5. **Research Hypothesis:** State hypotheses *explicitly* in an "if ... then ..." or mathematical form, i.e., as predicted relationships between independent and dependent variables (general/conceptual hypotheses may also be provided). Relate the hypotheses to previous research and theories; i.e., give rationale.

6. **Participants:** Indicate the relevant *descriptive characteristics* (e.g., sex, age, and education), manner of *selection/sampling*, any *compensation*, manner of *assignment* to conditions, and the *number* per condition and in total.

7. **Apparatus/Materials:** Describe the paper and pencil tests, stimuli, and equipment to be used, and give the publisher or manufacturer and model number whenever possible. Also describe, when pertinent, the features of the research facility/laboratory, e.g., the room dimensions, lighting, climate control, sound attenuation, windows, and furnishings (tables, chairs, etc.); or the characteristics of the outdoor environmental setting of the study.

8. **Procedure:** Describe *step-by-step and in detail* exactly what will be done by the experimenter and subjects — including instructions, tasks, administration of stimuli, observation/recording of responses, and ethical considerations. (Additional aspects of the procedure are covered in components 9–12 directly below.)

9. **Operational Definitions of Variables:** Define the independent and dependent variables *clearly and unambiguously* in terms of how they will be produced, manipulated, or measured. Describe the *levels* of all independent variables.

10. **Extraneous Variables and Controls:** Specify the *unwanted variables* that are likely and the *control techniques* that will be used to minimize their effects and avoid differential influences with respect to independent-variable levels.

11. **Research Design:** Specify whether the design will be within or between subjects/participants; independent groups, correlated groups, or mixed; etc.

12. **Statistical Analysis:** Specify the *descriptive statistics* and the *inferential statistical tests* that will be used. It is important that different statistics are appropriate/valid for different types of measurements and designs and they vary in their assumptions and power/sensitivity.

13. **Results Expected:** *Predict* the data, to whatever extent possible, in more detail than the research hypothesis(es), e.g., the magnitude of effects and error variance, in addition to the hypothesized direction/relation of effects.

14. **Conclusions Possible:** Discuss what *interpretations and implications* would be generated by results that support and do not support the research hypotheses (especially important if testing *competing hypotheses*).

15. **Generalizations Possible:** Discuss *extent* of generalizability of results. This will depend on the breadth of specification of population characteristics (for participants and also independent, dependent, and extraneous variables and values), and on the representativeness/validity of the sampling procedure.

16. **References:** List *all* publications and other sources cited within the proposal, ordering them alphabetically and adhering to the appropriate format.

Relationship of Research Proposals to Research Reports

When writing a proposal, the components listed above are not usually typed in this *outline format* but rather in a *narrative format*. These components of the proposal are often organized under the same section headings that are typically used in a finished research report/paper (see Chapter V): items 3–5 go in the Introduction, items 6–11 in the Methods section, items 12 and 13 in Results, items 14 and 15 in Discussion, and item 16 in References. An *abstract* might be required, especially if the proposal is long.

The *writing style* should be the same as that for a research report, and this style is described in the *Publication Manual of the American Psychological Association*, 4th ed. (1994), which is available in most university bookstores and can also be obtained directly from the APA Order Department, P.O. Box 2710, Hyattsville, MD 20784. Because of its quality and comprehensiveness, it is used in many disciplines.

The methods, results, and discussion sections of a *research proposal* should be written in the *future tense*. After the study is completed, the proposal then may be used in the writing of the *research report*, particularly the Introduction and Methods sections, as well as the References. The Methods section and some other portions of the paper would have to be changed from the future to the *past tense*.

Appendix 4
Ethical Principles

Conduct of Research with Human Participants

The following has been excerpted from *Ethical Principles in the Conduct of Research with Human Participants*. (Washington, D.C., American Psychological Association, 1982.) Italics used for emphasis are my own.

Preamble

> *The decision to undertake research* rests upon a considered judgment by the individual psychologist about how best to contribute to psychological science and human welfare. Having made the decision to conduct research, the psychologist considers alternative directions in which research energies and resources might be invested. On the basis of this consideration, the psychologist carries out the investigation with *respect and concern for the dignity and welfare of the people who participate* and with cognizance of federal and state regulations and professional standards governing the conduct of research with human participants.

Principle A

> *In planning a study,* the investigator has the responsibility to make a careful evaluation of its *ethical acceptability.* To the extent that the weighing of scientific and human values suggests a compromise of any principle, the investigator incurs a correspondingly serious obligation to seek ethical advice and to observe stringent safeguards to protect the rights of human participants.

Principle B

> Considering whether a participant in a planned study will be a *"subject at risk"* or a *"subject at minimal risk,"* according to recognized standards, is of primary ethical concern to the investigator.

Principle C

> The investigator always retains the *responsibility for insuring ethical practice in research.* The investigator is also responsible for the ethical treatment of research participants by collaborators, assistants, students, and employees, all of whom, however, incur similar obligations.

Principle D

> Except in minimal-risk research, the investigator establishes a *clear and fair agreement with research participants,* prior to their participation, that clarifies the obligations and responsibilities of each. The investigator has the obligation to honor all promises and commitments included in that agreement. The investigator informs the participants of all aspects of the research that might reasonably be expected to influence willingness to participate and explains all other aspects of the research about which the participants inquire. Failure to make full disclosure prior to obtaining informed consent requires additional safeguards to protect the welfare and dignity of the research participants. *Research with children or with participants who have impairments that would limit understanding and/or communication* requires special safeguarding procedures.

Principle E

Methodological requirements of a study may make the use of concealment or deception necessary. Before conducting such a study, the investigator has a special responsibility to (1) determine whether the use of such techniques is justified by the study's prospective scientific, educational, or applied value; (2) determine whether alternative procedures are available that do not use concealment or deception; and (3) ensure that the participants are provided with sufficient explanation as soon as possible.

Principle F

The investigator respects the individual's freedom to decline to participate in or to withdraw from the research at any time. The obligation to protect this freedom requires careful thought and consideration when the investigator is in a position of authority or influence over the participant. Such positions of authority include, but are not limited to, situations in which research participation is required as part of employment or in which the participant is a student, client, or employee of the investigator.

Principle G

The investigator protects the participant from physical and mental discomfort, harm, and danger that may arise from research procedures. If risks of such consequences exist, the investigator informs the participant of that fact. Research procedures likely to cause serious or lasting harm to a participant are not used unless the failure to use these procedures might expose the participant to risk of greater harm or unless the research has great potential benefit and fully informed and voluntary consent is obtained from each participant. The participant should be informed of procedures for contacting the investigator within a reasonable time period following participation should stress, potential harm, or related questions or concerns arise.

Principle H

After the data are collected, the investigator provides the participant with information about the nature of the study and attempts to remove any misconceptions that may have arisen. Where scientific or humane values justify delaying or withholding this information, the investigator incurs a special responsibility to monitor the research and to ensure that there are no damaging consequences for the participant.

Principle I

Where research procedures result in undesirable consequences for the individual participant, the investigator has the responsibility to detect and remove or correct these consequences, including long-term effects.

Principle J

Information obtained about the research participant during the course of an investigation is confidential unless otherwise agreed upon in advance. When the possibility exists that others may obtain access to such information, this possibility, together with plans for protecting confidentiality, is explained to the participant as part of the procedure for obtaining informed consent.

Guidelines for Ethical Conduct in the Care and Use of Animals
Developed by the American Psychological Association's Committee on Animal Research and Ethics (CARE)

Psychology encompasses a broad range of areas of research and applied endeavors. Important parts of these endeavors are teaching and research on the behavior of nonhuman animals, which contribute to the understanding of basic principles underlying behavior and to advancing the welfare of both human and nonhuman animals. Clearly, psychologists should conduct their teaching and research in a manner consonant with relevant laws and regulations. In addition, ethical concerns mandate that psychologists should consider the costs and benefits of procedures involving animals before proceeding with the research.

The following guidelines were developed by the American Psychological Association (APA) for use by psychologists working with nonhuman animals. They are based on and are in conformity with Section 6.20 of the *Ethical Principles of Psychologists and Code of Conduct* of APA. In the ordinary course of events, the acquisition, care, housing, use, and disposition of animals should be in compliance with applicable federal, state, local, and institutional laws and regulations and with international conventions to which the United States is a party. APA members working outside the United States are to follow all applicable laws and regulations of the country in which they conduct research.

APA authors are required to comply with APA ethical standards in the treatment of their sample, human or animal, or to describe the details of treatment. A copy of the APA *Ethical Principles of Psychologists and Code of Conduct* may be obtained from:

APA Order Department
P.O. Box 2710
Hyattsville, MD 20784-0710

Questions about these guidelines should be referred to the APA Committee on Animal Research and Ethics (CARE) at science@apa.org, 202-336-6000, or write:

APA Science Directorate
Research Ethics Officer
750 First St., NE
Washington, DC 20002-4242

Violations of the *Ethical Principles of Psychologists and Code of Conduct* by an APA member should be reported to APA's Ethics Committee by calling 202-336-5930, or by writing:

APA Ethics Office
750 First St., NE
Washington, DC 20002-4242

I. Justification of the Research

   A. Research should be undertaken with a clear scientific purpose. There should be a reasonable expectation that the research will (a) increase knowledge of the processes underlying the evolution, development, maintenance, alteration, control, or biological significance of behavior; (b) determine the replicability and generality of prior research; (c) increase understanding of the species under study; or (d) provide results that benefit the health or welfare of humans or other animals.

B. The scientific purpose of the research should be of sufficient potential significance to justify the use of animals. Psychologists should act on the assumption that procedures that would produce pain in humans will also do so in other animals.

C. The species chosen for study should be best suited to answer the question(s) posed. The psychologist should always consider the possibility of using other species, nonanimal alternatives, or procedures that minimize the number of animals in research, and should be familiar with the appropriate literature.

D. Research on animals may not be conducted until the protocol has been reviewed by an appropriate animal care committee, for example, an institutional animal care and use committee (IACUC), to ensure that the procedures are appropriate and humane.

E. The psychologist should monitor the research and the animals' welfare through-out the course of an investigation to ensure continued justification for the research.

II. Personnel

A. Psychologists should ensure that personnel involved in their research with animals be familiar with these guidelines.

B. Animal use procedures must conform with federal regulations regarding personnel, supervision, record keeping, and veterinary care.[1]

C. Behavior is both the focus of study of many experiments as well as a primary source of information about an animal's health and well-being. It is therefore necessary that psychologists and their assistants be informed about the behavioral characteristics of their animal subjects, so as to be aware of normal, species-specific behaviors and unusual behaviors that could forewarn of health problems.

D. Psychologists should ensure that all individuals who use animals under their supervision receive explicit instruction in experimental methods and in the care, maintenance, and handling of the species being studied. Responsibilities and activities of all individuals dealing with animals should be consistent with their respective competencies, training, and experience in either the laboratory or the field setting.

III. Care and Housing of Animals

The concept of psychological well-being of animals is of current concern and debate and is included in Federal Regulations (United States Department of Agriculture [USDA], 1991). As a scientific and professional organization, APA recognizes the complexities of defining psychological well-being. Procedures appropriate for a particular species may be inappropriate for others. Hence, APA does not presently stipulate specific guidelines regarding the maintenance of psychological well-being of research animals. Psychologists familiar with the species should be best qualified professionally to judge measures such as enrichment to maintain or improve psychological well-being of those species.

A. The facilities housing animals should meet or exceed current regulations and guidelines (USDA, 1990, 1991) and are required to be inspected twice a year (USDA, 1989).

B. All procedures carried out on animals are to be reviewed by a local animal care committee to ensure that the procedures are appropriate and humane. The committee should have representation from within the institution and from the local community. In the event that it is not possible to constitute an appropriate local animal care commttee, psychologists are encouraged to seek advice from a corresponding committee of a cooperative institution.

C. Responsibilities for the conditions under which animals are kept, both within and outside of the context of active experimentation or teaching, rests with the psychologist under the supervision of the animal care committee (where required by federal regulations) and with individuals appointed by the institution to oversee animal care. Animals are to be provided with humane care and healthful conditions during their stay in the facility. In addition to the federal requirements to provide for the psychological well-being of nonhuman primates used in research, psychologists are encouraged to consider enriching the environments of their laboratory animals and should keep abreast of literature on well-being and enrichment for the species with which they work.

IV. Acquisition of Animals

A. Animals not bred in the psychologist's facility are to be acquired lawfully. The USDA and local ordinances should be consulted for information regarding regulations and aproved suppliers.

B. Psychologists should make every effort to ensure that those responsible for transporting the animals to the facility provide adequate food, water, ventilation, space, and impose no unnecessary stress on the animals.

C. Animals taken from the wild should be trapped in a humane manner and in accordance with applicable federal, state, and local regulations.

D. Endangered species or taxa should be used only with full attention to required permits and ethical concerns. Information and permit applications can be obtained from:

Fish and Wildlife Service

Office of Management Authority

U.S. Dept. of the Interior

4401 N. Fairfax Dr., Rm. 432

Arlington, VA 22043

703-358-2104

Similar caution should be used in work with threatened species or taxa.

V. Experimental Procedures

Humane consideration for the well-being of the animal should be incorporated into the design and conduct of all procedures involving animals, while keeping in mind the primary goal of experimental procedures — the acquisition of sound, replicable data. The conduct of all procedures is governed by Guideline I.

A. Behavioral studies that involve no aversive stimulation to, or overt sign of distress from, the animal are acceptable. These include observational and other noninvasive forms of data collection.

B. When alternative behavioral procedures are available, those that minimize discomfort to the animal should be used. When using aversive conditions, psychologists should adjust the parameters of stimulation to levels that appear minimal, though compatible with the aims of the research. Psychologists are encouraged to test painful stimuli on themselves, whenever reasonable. Whenever consistent with the goals of the research, consideration should be given to providing the animals with control of the potentially aversive stimulation.

C. Procedures in which the animal is anesthetized and insensitive to pain throughout the procedure and is euthanized before regaining consciousness are generally acceptable.

D. Procedures involving more than momentary or slight aversive stimulation, which is not relieved by medication or other acceptable methods, should be undertaken only when the objectives of the research cannot be achieved by other methods.

E Experimental procedures that require prolonged aversive conditions or produce tissue damage or metabolic disturbances require greater justification and surveillance. These include prolonged exposure to extreme environmental conditions, experimentally induced prey killng, or infliction of physical trauma or tissue damage. An animal observed to be in a state of severe distress or chronic pain that cannot be alleviated and is not essential to the purposes of the research should be euthanized immediately.

F. Procedures that use restraint must conform to federal regulations and guidelines.

G. Procedures involving the use of paralytic agents without reduction in pain sensation require particular prudence and humane concern. Use of muscle relaxants or paralytics alone during surgery, without general anesthesia, is unacceptable and should be avoided.

H. Surgical procedures, because of their invasive nature, require close supervision and attention to humane considerations by the psychologist. Aseptic (methods that minimize risks of infection) techniques must be used on laboratory animals whenever possible.

1. All surgical procedures and anesthetization should be conducted under the direct supervision of a person who is competent in the use of the procedures.

2. If the surgical procedure is likely to cause greater discomfort than that attending anesthetization, and unless there is specific justification for acting otherwise, animals should be maintained under anesthesia until the procedure is ended.

3. Sound postoperative monitoring and care, which may include the use of analgesics and antibiotics, should be provided to minimize discomfort and to prevent infection and other untoward consequences of the procedure.

4. Animals cannot be subjected to successive surgical procedures unless these are required by the nature of the research, the nature of the surgery, or for

the well-being of the animal. Multiple surgeries on the same animal must receive special approval from the animal care committee.

5. When the use of an animal is no longer required by an experimental protocol or procedure, in order to minimize the number of animals used in research, alternative uses of the animals should be considered. Such uses should be compatible with the goals of research and the welfare of the animal. Care should be taken that such an action does not expose the animal to multiple surgeries.

J. The return of wild-caught animals to the field can carry substantial risks, both to the formerly captive animals and to the ecosystem. Animals reared in the laboratory should not be released because, in most cases, they cannot survive or they may survive by disrupting the natural ecology.

K. When euthanasia appears to be the appropriate alternative, either as a requirement of the research or because it constitutes the most humane form of disposition of an animal at the conclusion of the research:

1. Euthanasia shall be accomplished in a humane manner, appropriate for the species, and in such a way as to ensure immediate death, and in accordance with procedures outlined in the latest version of the American Veterinary Medical Association (AVMA) Panel on Euthanasia.[2]

2. Disposal of euthanized animals should be accomplished in a manner that is in accord with all relevant legislation, consistent with health, environmental, and aesthetic concerns, and approved by the animal care committee. No animal shall be discarded until its death is verified.

VI. Field Research

Field research, because of its potential to damage sensitive ecosystems and ethologies, should be subject to animal care committee approval. Field research, if strictly observational, may not require animal care committee approval (USDA, 1989, pg. 36126).

A. Psychologists conducting field research should disturb their populations as little as possible — consistent with the goals of the research. Every effort should be made to minimize potential harmful effects of the study on the population and on other plant and animal species in the area.

B. Research conducted in populated areas should be done with respect for the property and privacy of the inhabitants of the area.

C. Particular justification is required for the study of endangered species. Such research on endangered species should not be conducted unless animal care committee approval has been obtained and all requisite permits are obtained (see IV.D.).

VII. Educational Use of Animals

Laboratory exercises as well as classroom demonstrations involving live animals can be valuable as instructional aids. APA has adopted separate guidelines for the educational use of animals in precollege education, including the use of animals in science fairs and demonstrations. For a copy of APA's *Ethical Guidelines for the Teaching of Psychology in the Secondary Schools,* write to:

APA Education Directorate
High School Teacher Affiliate Program
750 First Street, NE
Washington, DC 20002-4242

A. Psychologists are encouraged to include instruction and discussion of the ethics and values of animal research in all courses that involve or discuss the use of animals.

B. Animals may be used for educational purposes only after review by a committee appropriate to the institution.

C. Some procedures that can be justified for research purposes may not be justified for educational purposes. Consideration should be given to the possibility of using nonanimal alternatives.

Notes:

1. U. S. Department of Agriculture. (1989, August 21). Animal welfare; Final rules. *Federal Register.*

   U. S. Department of Agriculture. (1990, July 16). Animal welfare; Guinea pigs, hamsters, and rabbits. *Federal Register.*

   U. S. Department of Agriculture. (1991, February 15). Animal welfare; Standards; Final rule. *Federal Register.*

2. Write to:

   AVMA
   1931 N. Meacham Road
   Suite100
   Schaumburg, IL 60173
   or call (708) 925-8070

Guidelines for Ethical Conduct in the Care and Use of Animals was developed by APA's Committee on Animal Research and Ethics (CARE). Inquiries about these guidelines should be made to:

APA Science Directorate
Research Ethics Officer
750 First Street, NE
Washington, DC 20002-4242
Phone: 202-336-6000
Fax: 202-336-5953
E-mail: science@apa.org

Example: Human Participant's Informed Consent Statement

Title of Research: ERP Correlates of Cognition for Attention-Deficit/Hyperactivity Disorder

Adult Males With and Without Methlyphenidate

Principle Investigators: Jay E. Gould, Ph.D., Bruce R. Dunn, Ph.D., Frank Andrasik, Ph.D, and Paul Miranne

Department of Psychology

University of West Florida

11000 University Parkway

Pensacola, FL 32514

Phone: (xxx) xxx-xxxx, E-Mail: xxxxxx@xxx.xxx

I. Instructions:

Federal and University regulations require us to obtain a signed consent form for participation in research involving human participants. After reading the statements in section II through IV below, please indicate your consent by signing and dating this form.

II. Statement of Purpose and Procedure:

Thank you for your interest in this research project being conducted by members of The University of West Florida. By this time, one of the investigators should have described the procedures for you in detail. Basically, this stage of the research project involves careful study of the brain-wave activity of adults who have and who do not have Attention-Deficit/Hyperactivity Disorder. The purpose is to better understand this disorder and its treatment with methylphenidate (Ritalin). You will find a summary of the major aspects of the study below, including the risks and benefits of participating. Please feel free to ask questions about the procedure at any time. Carefully read the information provided below. If you wish to participate in this study, sign your name and write the date. Any information you provide to us will be kept in strict confidence. Except for statistical summaries of group data, the information identified with individual participants will only be available to those working on this project — unless prior written consent is given to do otherwise.

I understand that:

(1) I will be asked to disclose certain information about the presence or absence of symptoms of Attention-Deficit/Hyperactivity Disorder (AD/HD), as well as other psychological disorders, by completing an "Adult" and "Childhood" AD/HD Symptoms Scale Self-Report and an additional questionnaire.

(2) I will be asked to disclose certain information regarding my use of drugs, including: prescription drugs (e.g., Ritalin), non-prescription drugs (e.g., cold medicines), and other common drug substances (e.g, nicotine, caffeine, and alcohol), as well as drugs that may not be legal. Again, any information you provide to us will be kept in strict confidence.

(3) It may be necessary to have a brief consultation with certain professionals about the presence or absence of symptoms of AD/HD and other psychological disorders.

(4) If I am participating in the study as an individual with AD/HD, I will be required to come the laboratory on two more occasions: once shortly after taking my medication for AD/HD, and once after not taking this medication and the medication has had more than enough time to wear off. If I am participating as an individual who does not have AD/HD, I too will be required to come to the laboratory on two occasions. The second visit will be about 2 weeks after the first visit.

(5) During each laboratory session, which will last approximately 2 hours, brain-wave activity will be assessed while I perform a variety of cognitive tasks presented on a computer monitor. These tasks will require either memorization or a decision in which the press of a button will be required. An elastic "brain-wave sensor cap" will be affixed on my head, using conductive gel and a strap, to enable recordings associated with the work I am doing during the tasks.

(6) If I have any questions or experience any problems related to participating in this research project, I may call either Dr. Jay E. Gould or Dr. Bruce Dunn at (xxx)-xxxx.

(7) I may discontinue participation in this study at any time without penalty.

III. Potential Risks:

(1) I may experience some mild discomfort when answering personal questions during the assessment.

(2) I may experience some mildly unpleasant or stressful feelings from the cognitive-task phase of the experiment.

(3) I may note some mild tenderness in the scalp area due to placement of the brain-activity sensors. This occurs infrequently and is not long lasting.

(4) Participants who have AD/HD may experience a brief return of their symptoms when, for one session of the study, they temporarily discontinue medication.

IV. Potential Benefits:

(1) Information obtained from my participation in this study may improve the diagnosis of AD/HD, as well as provide a better understanding of the nature of AD/HD and the role of medication in its treatment.

(2) Information obtained from my participation in this study may be useful in offering alternative treatments for AD/HD in the future.

(3) I will receive monetary compensation of $5.00 for attending the first session of the experiment, and $15.00 for attending the second session.

(4) I will be provided with a sample narrative and colorized topographical image, or "brain map," of typical electrophysiological activity under some aspect of the tasks utilized in the study. This is meant to enhance my educational experience of participating in this research.

(5) A summary of the results of this study will be made available to me.

V. Statement of Consent:

I certify that I have read and fully understand the Statement of Procedure given above and agree to participate as a subject in the research described therein. Permission is given voluntarily and without coercion or undue influence. It is understood that I may discontinue participation at any time without penalty or loss of any benefits to which I may otherwise be entitled. I will be provided with a copy of this consent form.

_____   _____
Participant's Name (Please Print)   Participant's Date of Birth

_____   _____
Participant's Signature   Date

_____   _____
Participant's Address (Please Print)   Witness' Signature

_____

_____   IRB Approval Date:_____

Appendix 5
Summary of Extraneous Variables and Controls

Potential Extraneous Variables

Extraneous variables are *extra/unwanted variables* operating in the experimental situation, which like independent variables might affect the dependent variable(s). *Three general sources* of extraneous variables incorporate the 12 types listed below: environmental factors, participant factors, and experimenter factors.

Environment: *Physical and social conditions of the experimental setting*, other than the independent variable(s), that might affect a participant's performance — similar to extraneous variable of history, except not dependent on there being a pretest.

History: *External environmental events*, other than an independent variable, that occur *between any pretest and posttest measurements* of a dependent variable.

Instrumentation: Change that occurs over time in the measurement of a dependent variable due to *variation in mechanical or human observer factors.*

Statistical Regression: Change attributed to the tendency of participants with *extremely high or low scores* on a test to move toward the mean when measured again — due to unreliability of the test instrument and/or the participants.

Maturation: Change due to *conditions internal to an individual* that vary as a function of the *passage of time*, rather than due to environment (thus not learning).

Selection: *Use of differential selection procedures* for participants assigned to the various comparison groups — leads to *individual differences* between groups.

Mortality: *Participant loss* from various comparison groups — possibly differential.

Sequence/Order: *Participation in more than one condition* for comparison, leading to practice or fatigue effects, and/or possible changes in motivation or attention.

Testing: *When tested more than once,* an initial test (a pretest) can affect the participant scores of a subsequent test either directly or indirectly via sensitization or resistance to a treatment (similar to sequence/order effects).

Participant Bias: Subject *expectancy, attitude, or motive* (associated, respectively, with placebo effect, Hawthorne effect, and motive for positive self-presentation).

Participant Sophistication: *Familiarity* of research participants with the field of psychology and/or its experimental procedures (affects participant bias).

Experimenter Features: *Expectancy/bias, performance, appearance, personality* of the researcher(s); also includes undesired variation/effects of procedures, instructions, and apparatus not already noted (e.g., under testing and instrumentation)

Techniques for Making Equivalent Secondary/Extraneous Variance

Systematic/consistent variation of a dependent variable produced by *extraneous variables* is referred to as secondary or extraneous variance.

Reasons for controlling secondary variance are (1) *confounding*, whereby extraneous and independent variables can be *inextricably intertwined, or mixed together*, so that their effects on a dependent variable cannot be separated/distinguished; and (2) *masking*, whereby extraneous variables *hide/obscure* the effects of an independent variable on a dependent variable.

Eleven control techniques are listed below, from the simplest to the most complex. *The goal of all the control techniques* (except conservative arrangement) is to have *equivalence of effects* of all secondary, extraneous variables across all independent-variable conditions — *through elimination of any differential influences* — whenever the extraneous variables cannot themselves be eliminated.

Eliminating Extraneous Variables: *Very desirable, straightforward approach.* Included are the deception, disguised-experiment, single-blind, double-blind, partial-blind, and automation control techniques for participant and experimenter bias.

Holding Extraneous Variables Constant: There are many instances in which secondary, extraneous variables *cannot be eliminated* — participant factors are a particularly obvious case, and environmental factors also sometimes cannot be eliminated. The next best thing might be to hold the extraneous variable/factor constant for all participants and thus for all conditions.

Balancing Extraneous Variables: It is inconvenient or even impossible in some instances to hold constant a secondary, extraneous variable. It is undesirable to hold an extraneous variable constant, even when possible, if *generalizability of results* is desired. With balancing, the different values of the secondary, extraneous variable would simply be made to occur an *equal number of times* and thus to the *same extent* for all treatment and control conditions.

Matching Participants on Extraneous Variables: *Extended and more precise* way of balancing for a secondary, extraneous variable when it is an extraneous participant factor and when there are *several values* of the extraneous variable, as opposed to just two or three. When there are only two independent-variable conditions/levels, participants would be matched into pairs so that those in each pair had the most similar (ideally identical) values on the extraneous variable. One member of each pair would then be assigned to each of the two conditions using randomization.

Yoking Participants on Extraneous Variables: Matching participants for extraneous *experimental and/or environmental factors* (characteristics) rather than matching participants on some extraneous participant characteristic. Control group is usually exposed to the same quantity and perhaps temporal distribution of experimental conditions. Treatment consists of a particular order or relationship of experimental conditions or some additional factor. Yoking participants together to match them can be either physical (like oxen pulling a wagon) or electronic.

Using Participants as their Own Control with Counter-balancing: *Extreme form of matching* for secondary, extraneous participant factors, i.e., individual differences. Each subject participates at all levels of an independent variable and thus serves as his/her own control for extraneous-participant variables. In effect, each participant is *matched with himself/herself. Advantages* are *greater reduction* of secondary and error variance, and thus confounding and masking. This is because the between-conditions variance related to individual participant differences is better controlled by having the same participants in the various conditions, as opposed to different participants even if they're matched. Far more extraneous participant factors are controlled, as well as being better controlled, by within-participants comparisons. Moreover, *fewer participants* are required, because more data are gathered per individual when each participates under more than one condition. *Disadvantages* are the need to control for *order/sequence effects* (e.g., practice and fatigue) using counterbalancing and also the problem of *carryover effects* that might remain due to differential/asymmetrical transfer of practice, fatigue, etc.

Randomizing Extraneous Variables: Randomization is a procedure ensuring that each event in some set of events has an *equal probability/likelihood of occurring*. It is a control technique used to enhance the likelihood that secondary, extraneous variables — both known and unknown, measurable and unmeasurable — which for some reason cannot be eliminated, held constant, or balanced, will be *equal in value and thus effect* for the different independent-variable levels of a study. Participants would be randomly assigned to conditions, and the conditions would be run in a random sequence. Confounding due to systematic biasing by extraneous participant, environmental, and experimenter factors would be controlled by such randomization. Masking, however, would still be possible

Pretesting Participants for Extraneous Variables: Pretesting is a dependent-variable measure taken *prior* to administration of an independent-variable treatment (or a control condition) against which to compare a posttreatment observation (a posttest) on the dependent variable. *Functions* are (1) *participant comparability* on extraneous variables — pretest scores can be used to check on initial equivalence of participants in the different conditions with respect to the dependent-variable measure(s), and then they can be used to adjust posttest scores to take into account any initial differences on the dependent variable. (2) Establishment of a *baseline or trend* — the pretest dependent-variable value provides a *reference level,* i.e., a baseline or, if several pretests are made, a trend against which the posttest score(s) can be compared to determine the effect of a treatment; furthermore, *multiple pretest measures* provide controls for several extraneous variables related to repeated measures that are associated with pretesting itself. (3) *Increased sensitivity/power* to detect effects of an independent variable — accomplished by matching participants based on their pretest scores, or blocking and balancing for the initial value and measuring its effect, and/or adjusting (correcting) for potential ceiling or floor truncation effects.

Control Groups for Extraneous Variables: A control group consists of participants who do not *receive an experimental treatment* but who in all other respects are treated exactly the same as those participants in the experimental/treatment group(s). This group can be thought of as a special treatment group that receives an independent-variable value of zero or that receives a standard value (one that is common/typical/normal). *Functions* are (1) Establishment of a *baseline/reference level* against which to compare the dependent-variable values of the treatment groups and thus to determine the effects, if any, of the independent-variable manipulation; (2) *control* of extraneous variables when experimental manipulations unavoidably involve the introduction of secondary, extraneous variables along with the independent variable; and (3) *evaluation* of effects of extraneous variables when measurement, not just control, is desired.

Treatment Groups and Systematizing Extraneous Variables: Addition of another treatment group is sometimes the best control for confounding by secondary, extraneous variables. *Functions* are (1) *control* of extraneous-variable confounding, (2) *evaluation* of the effects of extraneous variables, and (3) control of extraneous variable masking through *systematization* of the secondary, extraneous-variable. The latter involves building the variable into the research design and treating it as an independent variable, at least for the purpose of statistical analysis. More information is thereby obtained, and as a consequence statistical error variance is reduced, thus enhancing detection of independent-variable effects.

Conservatively Arranging Extraneous Variables: *Rather than controlling* for a secondary, extraneous variable by making it equivalent for all conditions, the conditions instead are arranged so that the effect of an extraneous variable could only *weaken* the measured effect of the independent variable. This is a *conservative approach* because all other techniques for controlling secondary, extraneous variables would result in stronger observed independent-variable effects. This is a *weakness* of this technique. *Strengths and advantages* are that confounding is controlled in a *simpler* manner and one can have even *more confidence* in the research hypothesis if the data support it.

Techniques for Minimizing Error/Random Variance

Inconsistent/random variation of the dependent variable produced by *extraneous variables* is referred to as error or random variance.

Error variance is *self-canceling* around a mean of zero as the number of measures increases; i.e., it *averages toward zero.* It must be minimized, however, because these chance variations, when they haven't averaged to zero, can result in dependent-variable differences between conditions that might be *mistaken* for independent-variable effects. Alternatively, error variance can *mask* the effect(s) of an independent variable when not minimized by the following methods:

Reduction of individual differences among participants;

Elimination, holding constant, or minimization of randomly operating environmental factors;

Maintenance of consistent experimental procedures, i.e., reduction of random variation;

Use of reliable (stable/consistent) and valid (appropriate/correct) measurement and analysis techniques as well as apparatus;

Replication of measurements/observations on participants and use of central tendency measures: mean, median, or mode;

Increase in the number of participants used in the study;

Use of matched or within-participants/-subjects designs;

Use of change/difference/gain scores or analysis of covariance.

Techniques for Maximizing Primary/Experimental Variance

Systematic/consistent variation of the dependent variable produced by manipulation of an *independent variable* is referred to as primary or experimental variance.

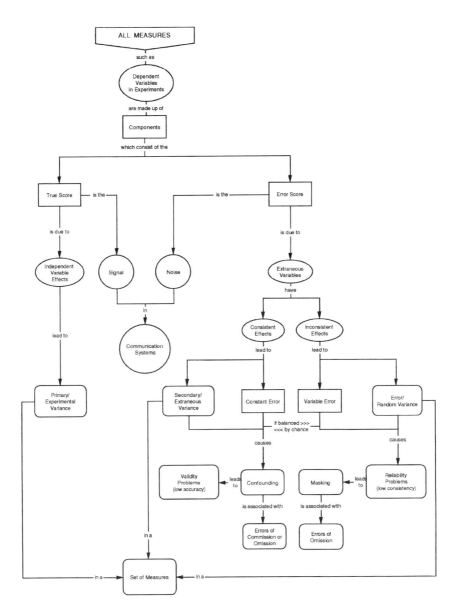

FIGURE A5.1
Components of measures and their relationships to types of variance and errors, reliability and validity, masking and confounding, etc.

Primary variance is *desired*, it is the purpose of an experiment, and it should be maximized because the secondary and error variance of extraneous variables can never be completely eliminated. In essence, the goal is to *maximize the signal-to- noise ratio* though control of extraneous variables, as covered earlier, and through one of the following methods for maximizing independent-variable effects:

Use of *extreme values* of the independent variable;

Use of *optimal values* of the independent variable when extreme values are not optimal;

Use of *several values* of the independent variable when there is no good basis for choosing optimal values. This increases the probability of using optimal values, and thus the power/sensitivity to detect an effect. Also, a more complete and accurate determination of the relationship between the independent and dependent variables is obtained.

TABLE A.5.1

Types of Variance — Matrix Summary

	Variation of Dependent Variable	
Causal Factor	**Consistent**	**Inconsistent**
Independent Variable	Primary or experimental variance	None
Extraneous Variable	Secondary or extraneous variance	Error or random variance

Note: No form of variance falls into the upper right cell of the matrix.

Appendix 6
Experimental and Quasi-Experimental Designs:
Category-Subcategory Relationships

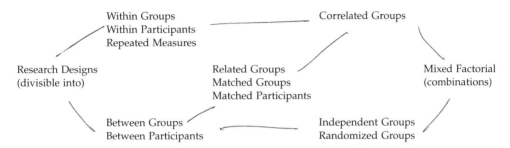

FIGURE A6.1

Research designs can be divided into those that compare the effects of different levels of an independent variable *within* one or more groups of participants/subjects and those that compare the effects *between* different groups of participants. The *between-groups designs* are in turn divided into those consisting of *independent groups* of participants and those consisting of *related*, or *matched, groups of* participants. Both *related-groups designs* and *within-groups designs* are forms of *correlated-groups designs*. Finally, *mixed factorial designs* are a type of multiple treatment design that combines in a single study both *correlated groups* of participants to investigate the effects of one or more independent variables and also *independent groups* of participants for one or more of the other independent variables.

Appendix 7
Schematic Summary of Designs: Experimental, Quasi-Experimental, and Pre-Experimental

In the following list, the most descriptive name for each design is listed first, alternative names are in parentheses, and consistency of naming is strived for across related designs. *All designs are experimental unless otherwise indicated.* Schematic symbols are as follows: X, treatment; O, observation; R, random assignment, BR, block randomization assignment; MR, matching followed by random assignment; Y(R/MR), yoking and then assignment by randomization or matching; X'; general experimental conditions; - - - - -, non-equivalence of design groups.

Pre-experimental Design: Example of What Not to Do

- One-Group Posttest-Only Design (One-Group After-Only Design; One-Shot Case Study)

$$X \qquad O$$

Within-Participants/Groups (Repeated-Measures) Designs

Each participant/subject serves as his/her own control for individual differences, and thus fewer are needed to obtain the same amount of data.

     All design examples are for one factor with two levels:

1. One-Group Pretest-Posttest Design (One-Group Before-After Design)

$$O \qquad X \qquad O$$

   This design group exemplifies quasi-experimental design at best and pre-experimental design otherwise.

2. Time-Series Design (Interrupted Time-Series Design)

$$O \quad O \quad O \ldots \qquad X \qquad O \quad O \quad O \ldots$$

   Time-series design is a quasi-experimental design.

3. Equivalent Time-Samples Design

$$X_1 \, O \qquad X_{0(2)} \, O \qquad X_1 \, O \qquad X_{0(2)} \, O \ldots$$

4. Counterbalanced Designs (Cross-Over Designs)
   - Intraparticipant Counterbalancing/ABBA Design (just one group)

$$X_1 \, O \qquad X_2 \, O \qquad X_2 \, O \qquad X_1 \, O$$

   or alternatively

$$X_2 \, O \qquad X_1 \, O \qquad X_1 \, O \qquad X_2 \, O$$

- Interparticipant Counterbalancing (more than one group)

$$X_1 \: O \qquad X_2 \: O$$

and for a second group of participants

$$X_2 \: O \qquad X_1 \: O$$

- Intra- plus Interparticipant Counterbalancing (combined approach)

$$X_1 \: O \qquad X_2 \: O \qquad X_2 \: O \qquad X_1 \: O$$

and for second group

$$X_2 \: O \qquad X_1 \: O \qquad X_1 \: O \qquad X_2 \: O$$

Between-Participants, Independent (Randomized), Two-Group Designs

All design examples are for one factor with two levels:

1. Nonequivalent Posttest-Only Control-Group Design (Nonequivalent Posttest-Only Design; Static-Group Comparison Design)

$$\begin{array}{cc} X & O \\ \hline & O \end{array}$$

   This design group exemplifies pre-experimental design (naturalistic, ex post facto) in most cases and quasi-experimental design only under certain conditions

2. Nonequivalent Pretest-Posttest Control-Group Design (Nonequivalent Control-Group Design; Pretest-Posttest Static-Group Comparison Design)

$$\begin{array}{ccc} O & X & O \\ \hline O & & O \end{array}$$

   This group exemplifies quasi-experimental design.

3. Multiple Time-Series Design

$$\begin{array}{ccccccc} O & O & O\ldots & X & O & O & O\ldots \\ \hline O & O & O\ldots & & O & O & O\ldots \end{array}$$

   Multiple time-series design is quasiexperimental design.

4. Posttest-Only Control Group Design (After-Only Randomized Two-Group Design)

$$\begin{array}{ccc} R & X & O \\ R & & O \end{array}$$

5. Pretest-Posttest Control Group Design (Before-After Randomized Two-Group Design)

$$R \quad O \quad X \quad O$$
$$R \quad O \qquad\quad O$$

6. Randomized-Blocks Design

$$BR \quad\quad X \quad O$$
$$BR \quad\quad\quad\quad O$$

Between-Participants, Related/Matched/Correlated, Two-Group Designs

All design examples are for one factor with two levels.

1. Match by Correlated-Criterion Design

$$MR \quad\quad X \quad O$$
$$MR \quad\quad\quad\quad O$$

2. Pretest-Match-Posttest Design

$$MR \quad O \quad X \quad O$$
$$MR \quad O \qquad\quad O$$

3. Yoked Control Group Design

$$Y(R/MR) \qquad X(X') \quad O$$
$$Y(R/MR) \qquad (X') \quad O$$

Multiple Treatment Designs

Either there are more than two levels of a single independent variable being studied, *multilevel designs,* and/or there is more than one independent variable (factor) being studied, *multifactor/factorial designs.* In both types of designs, assignment of participants/ subjects to conditions can be by randomization (R) or by matching followed by randomization (MR) for between-participants comparisons or by using repeated measurements for within-participants comparisons. The designs may be posttest-only or pretest-posttest (hence O?), and one of the levels can of course be a control condition (X_0).

Note: The factorial design schematic is of a different form than is used for other designs — it's a matrix — and it uses different symbols that provide different information. The factorial design shown below is for the simplest example, in which there are only two independent-variable factors (A and B) and only two levels for each factor (a_1 and a_2 versus b_1 and b_2). The factor levels/conditions are studied in all four possible combinations (e.g., a_1b_1), with each indicated within a cell of the matrix. "M" represents the respective mean dependent variable score for each factor level.

1. Multilevel Designs

R/MR?	O?	X_1	O
R/MR?	O?	X_2	O
R/MR?	O?	$X_{3(0)}$	O
.	.	.	.
.	.	.	.
.	.	.	.

2. Factorial (Multifactor) Designs

		A		
		a_1	a_2	
B	b_1	a_1b_1	a_2b_1	Mb_1
	b_2	a_1b_2	a_2b_2	Mb_2
		Ma_1	Ma_2	

Appendix 8

Sampling and Generalization: Processes Involved

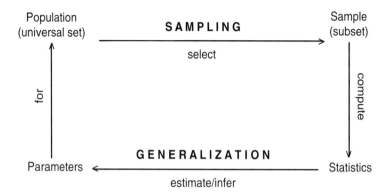

FIGURE A8.1

From a population, a sample can be selected from which statistics can be computed to estimate or infer parameters (e.g., the mean) of the population.

Populations and Samples: Forms of Each

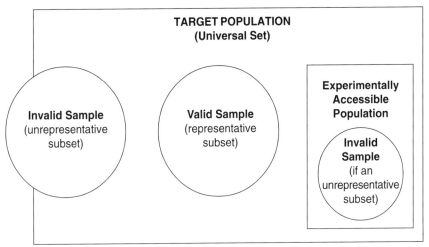

FIGURE A8.2

Populations are shown as rectangles and samples as circles. Valid samples (unbiased) are representative (center circle). Invalid samples (biased) either *include* elements of the target population (left circle) or they *exclude* elements of the target population (right circle), which can occur if samples are taken from an experimentally accessible population that is not representative of the target population.

Appendix A9. 1

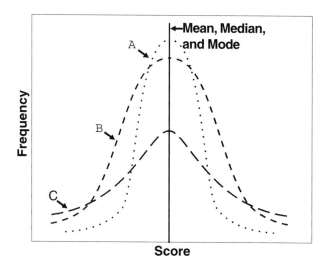

FIGURE A9.1

Three examples of the *normal curve* with distributions that differ in *variance*: "*A*" has the least variance (variability) of score values, and "*C*" has the most. Note that normal curves are *symmetrical* and *bell-shaped*, with the most frequent score values being in the middle range. In addition, the *mean, median, and mode* all have the same score value for any single distribution, although for different distributions this value can vary — unlike the examples in this figure.

Normal Distributions

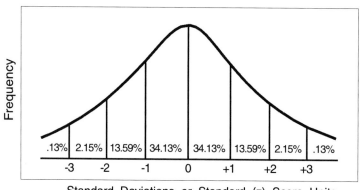

FIGURE A9.2

Regardless of the mean and variance of a set of scores, if they are *normally distributed,* then specific proportions of scores always fall within a certain number of *standard deviations from the mean* of the distribution. This is a very useful characteristic (see text).

Appendix 10

TABLE A10.1

Hypothesis Testing: Two Possible Outcomes of Research

	Results	Null Hypothesis	Research Hypothesis	Potential Errors
1	Statistically significant	Reject (fail to accept)	Fail to reject (accept)	Type I or α (false alarm)
2	Statistically insignificant	Fail to reject (accept)	Reject (fail to accept)	Type II or β (miss)

TABLE A10.2

Decision Matrix Showing Type I and Type II Errors

	True Population Situation	
Decision	Null Hypothesis Is True	Null Hypothesis is False
Reject (fail to accept) the null hypothesis	Type I or α Error $(P = \alpha)$	Correct Decision $(P = 1 - \beta)$
Fail to reject (accept) the null hypothesis	Correct decision $(P = 1 - \alpha)$	Type II or β Error $(P = \beta)$

Appendix 11
Sampling Distributions: Differences Between Pairs of Means

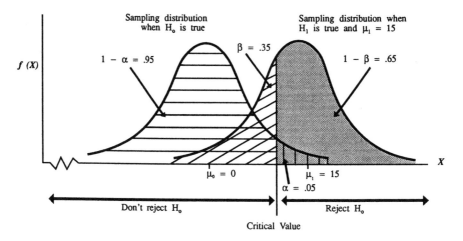

FIGURE A11.1

Theoretical frequencies (*f*) are plotted in each curve for the values of repeatedly measured differences between pairs of sample means (X). The sample means would be those obtained for two independent-variable conditions. The illustrated overlapping distributions of differences are for two hypothetical situations: when the *null hypothesis* (H_0) is true versus when the *research hypothesis* (H_1) is true. Regions/areas under the two curves (indicated by different cross hatching or shading) indicate the probabilities of making a *Type I error* (α) or a *Type II error* (β), as well as the probability of correctly failing to reject the null hypothesis when it is true ($1 - \alpha$) and the probability of correctly rejecting the null hypothesis when it is false ($1 - \beta$). The latter represents the *power* to correctly detect the treatment effects of the independent-variable conditions. As covered further in Appendices 12–14, the various illustrated probabilities would be affected by the chosen *significance-level/confidence-level cutoff(s)*, or *critical value(s)*, for rejecting H_0 ($\alpha = 0.05$ in this example); the use of a *one-tailed* (in this example) versus *two-tailed statistical test*; the *magnitude of the treatment effect* (mean difference $\mu_1 = 15$ in this example) and the *sampling variance*. Note that the magnitudes of the treatment effect and the sampling variance determine the degree of overlap of the two hypothetical sampling distributions. Sampling variance itself is determined by the variability caused by extraneous variables and also by sample size.

Appendix 12
Type I and Type II Error Probabilities: Inverse Relationship and Power

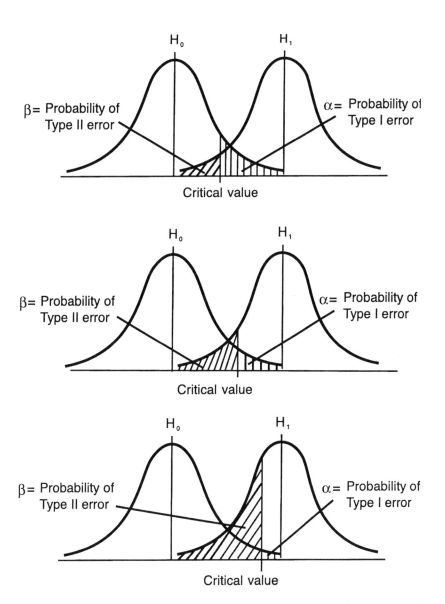

FIGURE A12.1

H_0 and H_1 are *theoretical sampling distributions* for differences between pairs of means when the *null hypothesis* is true versus when the *research hypothesis* is true, respectively. The three sets of distributions illustrate the effects of changing the chosen *significance level* (α) and thus the *critical value/cutoff* for rejecting H_0. From top to bottom, as the significance level is made more conservative by raising it (e.g. from 0.05 to 0.01), the probability of a *Type I error* (α) decreases while the probability of a *Type II error* (β) increases, and thus *power* ($1 - \beta$) decreases.

Appendix 13
Increasing Power by Reducing Overlap of Sampling Distributions

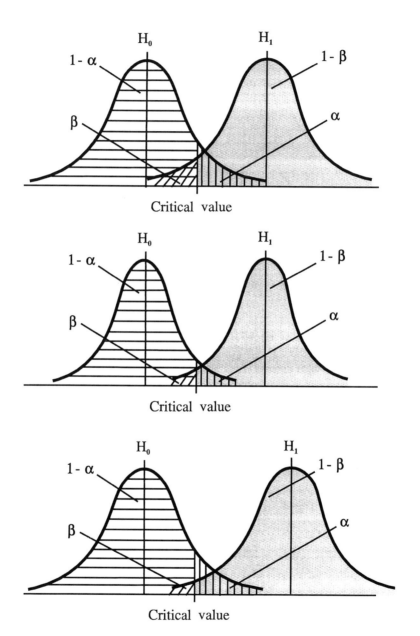

FIGURE A13.1

H_0 and H_1 are *theoretical sampling distributions* for differences between pairs of means when the *null hypothesis* is true versus when the *research hypothesis* is true, respectively. The three sets of these distributions illustrate the increase in *power* $(1 - \beta)$ that results from decreasing the *sampling distribution variance* — which makes the distributions narrower (shown in the middle set), or from increasing the *treatment effect* — which shifts the H_1 distribution further from the H_0 distribution (shown in the bottom set). Both strategies increase the proportion of the H_1 distribution falling beyond the *critical value/cutoff* for rejecting H_0. Importantly, note that the probability of a *Type II error* (β) is reduced without any increase in the probability of a *Type I error* (α).

Appendix 14
Two-Tailed Versus One-Tailed Tests: Effect on Power

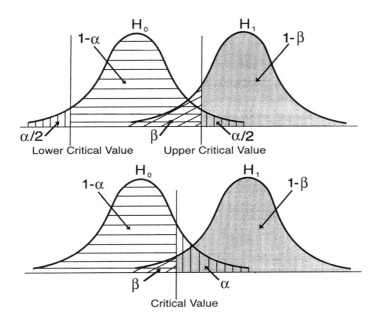

FIGURE A.14

H_0 and H_1 are *theoretical sampling distributions* for differences between pairs of means when the *null hypothesis* is true versus when the *research hypothesis* is true, respectively. The two sets of these distributions illustrate the increase in *power* ($1 - \beta$) that results when instead of a *two-tailed test* (shown in the upper set), a *one-tailed test* (shown in the bottom set) is used — assuming that the research hypothesis *correctly predicts* the direction of the treatment effect (to the right of/greater than zero in this example). Note that this is accomplished by reducing the probability of a *Type II error* (β) without any increase in the probability of a *Type I error* (α).

Appendix 15
Guide for Selecting a Statistical Test as a Function of the Number of Independent Variables and Conditions and the Level of Measurement

Level of Measurement (and data type) of the Dependent Variable	One Independent Variable				Two Independent Variables		
	Two Conditions		More Than Two Conditions		Factorial Designs		
	Independent groups designs	Matched groups or within participants designs	Independent groups designs	Matched groups or within participants designs	Independent groups designs	Matched groups or within participants designs	Independent groups & matched groups or within participants mixed designs
Interval or Ratio (score data)	t-Test or One-way ANOVA for independent measures	t-Test or One-way ANOVA for correlated measures	One-way ANOVA for independent measures	One-way ANOVA for correlated measures	Two-way ANOVA for independent measures	Two-way ANOVA for correlated measures	Two-way ANOVA for mixed measures
Ordinal (rank-order data)	Mann-Whitney U Test or Kolmogorov-Smirnov Test	Wilcoxon Test or Sign Test	Kruskal-Wallis Test	Friedman Test			
Nominal (frequency data)	Chi-Square Test or Fischer Exact Test	McNemar Change Test	Chi-Square Test	Cochran Q Test	Chi-Square Test		

Note: Consult statistics books for explanations of these and other tests. There are also, e.g., Analyses of Variance (ANOVAs) for factorial designs with more than two independent variables and with various combinations of independent- and correlated-measures.

Index

A

ABBA technique, 198
Abstract phenomena, 100
Acceptance, 12
Accuracy
    consistency versus, 97
    of parameter estimation, 331
Admiration, 25
Advertising, 252
Aesthetics, 4
Affection, 25, 33
Alcoholism, 33
Alpha error, 356, 371
Alternate-forms technique, 94, 95, 109
Alternatives, ability to choose between, 21
Altruistic behavior, role-playing studies of, 148
Ambiguous experiences, 25
American Psychological Association (APA), 112, 118, 119, 151, 156
Analyses of variance (ANOVA), 85, 124, 252, 258, 263, 266, 287, 290, 299, 306, 359
    three-way mixed-groups, 307
    trend analysis, 259
Anatomy, 171
Anger, 104
Animal(s)
    ability of to escape capture, 277
    guidelines for ethical conduct in care and use of, 156
    participants, bias affecting, 174
    philosophical justifications for using nonhuman, 160
    reproductive behavior of, 153
    research, facts versus myths about, 152
    rights groups, 152
    subjects, nonhuman, 122
    unnecessary death of, 155
    use of in research, 161
Animal Welfare Act, 156
ANOVA, see Analyses of variance
Antecedent conditions, 18
Anticipation, 17
Anxiety, measurement of, 91
APA, see American Psychological Association
A posteriori tests, 316
Applied research, 60, 61, 72
Artificiality, 47, 250
Association, strength of, 368
Assumed consent, 159
Assumption(s), 213

of analyses, 245
independent reality, 19
key, 98
most parsimonious, 180
of science, 19, 27
Astronomy, 6
Attention
    changes, 192
    selective, 173, 202
Attractiveness, judged, 15
Attribute ratios, 16
Authority, 8, 10, 26
Authorship, 137
Automation, 183

B

Background variance, 178
Bad theory, 68
Balanced squares, 195, 197
Balancing, 210
    based on pretest values, 216
    special type of, 191
Baseline data, 248
Basic research, 60, 61, 72
Bayesian moving average model, 259
Behavior
    altruistic, role-playing studies of, 148
    description of, 33
    determined, 20
    interpretation of, 202
    mob, 33
    of organisms, 17
    reproductive, of animals, 153
    of schizophrenics, 31
    scientist control of, 29
    shoplifting, 138
    unethical, 32
Behavioral medicine, 153
Belief(s)
    correcting erroneous, 27
    superiority of one, 11, 27
Bellwether/area/block sampling, 345–346, 348
Beta error, 357, 371
Between-conditions variation, 165, 237
Between-groups factors, 318
Between-participants designs, 243, 304
Bias(es), 129
    animal participants affected by, 174
    effect, experimenter, 202